X-Ray Astronomy
Selected Reprints

Edited by
Claude R. Canizares

published by
American Association of Physics Teachers

©1986 American Association of Physics Teachers

Published by
American Association of Physics Teachers
5110 Roanoke Place, Suite 101
College Park, MD 20740, U.S.A.

ISBN #0-917853-21-0

Cover Art: From "The Distribution and Morphology of X-Ray-Emitting Gas in the Core of the Perseus Cluster," by A.C. Fabian, E.M. Hu, L.L. Cowie, and J. Grindlay. *The Astrophysical Journal,* **248**, plate 2.
©1981, The American Astronomical Society.   See article beginning on page 131.

Contents

I. General References

Resource Letter XRA-1: X-Ray Astronomy
  Claude R. Canizares 1

Evidence for X Rays from Sources Outside the Solar System
  Riccardo Giacconi, Herbert Gursky, Frank R. Paolini, and Bruno B. Rossi 11

Progress in X-Ray Astronomy
  R. Giacconi 17

The Einstein Observatory: New Perspectives in Astronomy
  R. Giacconi and Harvey Tananbaum 27

II. Stellar Coronae

Results from an Extensive *Einstein* Stellar Survey
  G.S. Vaiana, J.P. Cassinelli, G. Fabbiano, R. Giacconi, L. Golub, P. Gorenstein, B.M. Haisch, F.R. Harnden, Jr., H.M. Johnson, J.L. Linsky, C.W. Maxson, R. Mewe, R. Rosner, F. Seward, K. Topka, and C. Zwaan 39

III. Galactic X-Ray Sources

Masses of Neutron Stars in X-Ray Binary Systems
  R.L. Kelley and S. Rappaport 59

A Status Report on Cygnus X-1
  D.M. Eardley, A.P. Lightman, N.I. Shakura, S.L. Shapiro, and R.A. Sunyaev 63

Transient X-Ray Sources in the Galactic Plane
  L. Cominsky, C. Jones, W. Forman and H. Tananbaum 73

Discovery of Intense X-Ray Bursts from the Globular Cluster NGC 6624
  J. Grindlay, H. Gursky, H. Schnopper, D.R. Parsignault, J. Heise, A.C. Brinkman and J. Schrijver 81

The Discovery of Rapidly Repetitive X-Ray Bursts From A New Source in Scorpius
  W.H.G. Lewin, J. Doty, G.W. Clark, S.A. Rappaport, H.V.D. Bradt, R. Doxsey, D.R. Hearn, J.A. Hoffman, J.G. Jernigan, F.K. Li, W. Mayer, J. McClintock, F. Primini, and J. Richardson 85

High-Resolution X-Ray Observations of the Cassiopeia A Supernova Remnant with the Einstein Observatory
  S.S. Murray, G. Fabbiano, A.C. Fabian, A. Epstein and R. Giacconi 91

X-Ray Spectrum of Cassiopeia A Measured with the *Einstein* SSS
  R.H. Becker, S.S. Holt, B.W. Smith, N.E. White, E.A. Boldt, R.F. Mushotzky and P.J. Serlemitsos 99

Evidence for Elemental Enrichment of Puppis A by a Type II Supernova
  C.R. Canizares and P.F. Winkler 103

IV. Extragalactic X-Ray Sources

Observations of X-Ray Sources in M31
  L. Van Speybroeck, A. Epstein, W. Forman, R. Giacconi, C. Jones, W. Liller and L. Smarr 107

Ariel 5 Observations of the X-Ray Spectrum of the Perseus Cluster
  R.J. Mitchell, J.L. Culhane, P.J.N. Davison and J.C. Ives 117

The Structure and Evolution of X-Ray Clusters
  C. Jones, E. Mandel, J. Schwarz, W. Forman, S.S. Murray and F.R. Harnden, Jr. 123

The Distribution and Morphology of X-Ray-Emitting Gas in the Core of the Perseus Cluster
  A.C. Fabian, E.M. Hu, L.L. Cowie and J. Grindlay 131

Einstein Observations of the X-Ray Structure of Centaurus A: Evidence for the Radio-Lobe Energy Source
  E.J. Schreier, E. Feigelson, J. Delvaille, R. Giacconi, J. Grindlay, D.A. Schwartz, and A.C. Fabian 141

On the Cosmological Evolution of the X-Ray Emission from Quasars
  Y. Avni and H. Tananbaum 147

RESOURCE LETTER

Roger H. Stuewer, *Editor*
School of Physics and Astronomy, 116 Church Street
University of Minnesota, Minneapolis, Minnesota 55455

This is one of a series of Resource Letters on different topics intended to guide college physicists, astronomers, and other scientists to some of the literature and other teaching aids that may help improve course contents in specified fields. No Resource Letter is meant to be exhaustive and complete; in time there may be more than one letter on some of the main subjects of interest. Comments on these materials as well as suggestions for future topics will be welcomed. Please send such communications to Professor Roger H. Stuewer, Editor, AAPT Resource Letters, School of Physics and Astronomy, 116 Church Street SE, University of Minnesota, Minneapolis, MN 55455.

Resource Letter: XRA-1 X-ray astronomy

Claude R. Canizares
Department of Physics and Center for Space Research, Massachusetts Institute of Technology, Cambridge, Massachusetts 02139

(Received 19 September 1983; accepted for publication 19 September 1983)

This Resource Letter provides a guide to the literature on x-ray astronomy. The letter E after an item indicates elementary level or material of general interest to persons becoming informed in the field. The letter I, for intermediate level, indicates material of somewhat more specialized nature; and the letter A indicates rather specialized or advanced material. An asterisk (*) indicates those articles to be included in an accompanying reprint book.

I. INTRODUCTION

X-ray astronomy is one of the youngest branches of astronomical science. Solar x rays were looked for and discovered just after World War II with a detector lifted above the Earth's absorbing atmosphere on a captured V-2 rocket. But the event that marks the true beginning of cosmic x-ray studies is the rocket flight in June 1962 that scanned the night sky with a much more sensitive detector and discovered the intense source named Scorpius X-1.

The intervening two decades have witnessed an exponential growth in instrumentation, observations and understanding of celestial x-ray radiation, at the same time that solar studies continued to develop and mature. In the 1960s a steady progression of rocket and balloon payloads soared aloft to reveal an ever richer x-ray sky, speckled with bright galactic sources, with objects that flared and others that suddenly appeared and then faded from view, and a few tantalizing hints of an even more extensive population of extragalactic x-ray emitters—quasars and radio galaxies. The 1970s opened with the launch of the first small astronomy satellite (SAS-1), renamed *Uhuru*, that for the first time allowed extended x-ray observations over most of the celestial sphere. *Uhuru*'s yield was spectacular: eclipsing x-ray binaries, x-ray pulsars, a candidate black hole, supernova remnants, hot intergalactic gas in clusters of galaxies, x rays from active galactic nuclei, and more. There followed a progression of U.S. and European satellites: OSO-7 and OSO-8 with both solar and celestial instruments, SAS-3 with its capabilities to refine the positions of x-ray sources and to study individual sources for extended periods, Ariel 5 with its all-sky monitor and survey instruments and the first high-energy astronomy observatory (HEAO-1), the largest and most sensitive survey mission ever flown. The culmination came in November 1978 with the launch of HEAO-2, also called the *Einstein* observatory. This was the first, and to date the only, celestial x-ray experiment to use focusing optics that permitted arc-second imaging of the x-ray sky and that achieved sensitivities 2–3 orders of magnitude better than those of the previous instruments. A few years earlier detectors on Skylab had brought the same capabilities to solar studies, followed by the versatile Solar Maximum Mission launched in 1980.

The imaging and spectral capabilities of *Einstein* have brought x-ray astronomy into its maturity. The x-ray pictures of supernova remnants and of galaxy jets have a resolution which approaches that of optical telescopes. With its great sensitivity *Einstein* has shown the pervasiveness of x-ray emission: it occurs in all types of stars, in the interstellar medium, in galactic halos, in intergalactic gas, and in the most distant quasars. X rays now pinpoint the sites of energetic activity in virtually every astronomical object. The spectrometers on *Einstein* have permitted detailed plasma diagnostics of celestial sources that were previously possible only for the solar corona or for laboratory plasmas.

X-ray astronomy was started by physicists totally divorced from traditional astronomical lore. Now the field is an integral part of modern astronomy. Originally taken as the symptom of celestial pathology, now x-ray emission is often used as a probe of fundamental astrophysical processes.

Because the major advances in x-ray astronomy are so recent, it is very difficult to review its literature. Much of the most interesting work is still in progress, found only in the latest numbers of the professional journals. There is a dearth of the synthesizing reviews, articles, and books that normally serve as entry points to the literature of long established fields. My approach has been to include as many reviews as possible and to select those primary sources that have a somewhat larger perspective than usual and so can serve as guides to further reading. Thus, many primary papers of major scientific importance were omitted, but they are referenced in the secondary sources. Those who use this Resource Letter in later years should look for up-to-date reviews in the *Annual Reviews of Astronomy and Astrophysics,* for more recent popular articles in *Scientific American, Physics Today,* and *Sky and Telescope,* and for possible new texts.

My emphasis has been on the x-ray band from ~0.5–10 keV, which occupies most of the attention of x-ray astronomers. References to the very important work at higher or lower energies and to related optical, UV and radio studies can be found in many of the reviews. I have also emphasized celestial as opposed to solar x-ray astronomy, although I believe that there are enough references to reviews of solar studies to mark a path into that area. Finally, the bulk of my references are to observational or phenomenological work. Each reference generally includes citations of more fundamental, theoretical articles.

II. TEXTBOOKS, GENERAL REFERENCES

A. General references

2.1. **"The Past, Present and Future of X-Ray Astronomy: Part I, The Past,"** R. Giacconi, G. Clark, E. Boldt, and H. Tananbaum, J. Wash. Acad. Sci. **71**, 1–69 (1981) (Parts I and II available from Wash. Acad. Sci., 1101 N. Highland St., Arlington, VA 22201). Leaders in the development of x-ray astronomy record the history and development of the field, making this a unique document. Titles are "1962–1972 (up through UHURU)" by Giacconi, "X-ray Astronomy from UHURU to HEAO-1" by Clark, "The High Energy Astronomy Observatory: HEAO-1" by Boldt and "X-Ray Astronomy with the Einstein Observatory" by Tananbaum. The occasion for these reminiscences was a symposium to memorialize the *Uhuru* satellite project. (I)

2.2. **"X-Ray Astronomy,"** B. Rossi, Deadalus (Boston) **106**, 37–58 (1977). One of the prime movers of cosmic x-ray astronomy describes its beginnings and reviews the state of the science in plain language. (E)

*2.3. **"Evidence for X-Rays from Sources Outside the Solar System,"** Riccardo Giacconi, Herbert Gursky, Frank R. Paolini, and Bruno B. Rossi, Phys. Rev. Lett. **9**, 439–443 (1962). This is the "discovery paper" that reports the first observations of extrasolar x rays, made with a rocket-borne detector. The data show both a strong point source (Scorpius X-1) and a diffuse background. (I)

2.4. **Black Holes,** W. Sullivan (Warner, New York, 1979). As background to describing the black-hole candidate Cygnus X-1, Sullivan gives a lively account of the events and personalities involved in the development of cosmic x-ray astronomy. (E)

2.5. **"Extrasolar X-Ray Sources,"** P. Morrison, Ann. Rev. Astron. Astrophys. **5**, 325–350 (1967). For historical perspective it is worth perusing this review written only a few years after the discovery of Scorpius X-1. Most of the physics discussed here is still relevant, but the observations have been superceded several times over. (I)

*2.6. **"Progress in X-Ray Astronomy,"** R. Giacconi, Phys. Today **26** (5), 38–47 (1973). This review for the nonspecialist describes the results of the first x-ray satellite *Uhuru.* These include the discovery of x-ray pulsars, binaries, and clusters of galaxies. (E)

2.7. **"X-Ray Astronomy,"** J. Culhane, Vistas Astron. **19**, 1–68 (1977). One could use this review article as a good introduction to the physics, instrumentation, and pre-*Einstein* results of x-ray astronomy. (I)

2.8. **"X-Ray Astronomy with HEAO-1,"** H. Friedman and K. Wood, Sky and Telescope **56**, 490–494(1978). Here is a nontechnical review of some results from the first high-energy astronomy observatory. Binary pulsars, a second possible black hole (Circinus X-1), and various extragalactic objects are briefly described. (E)

2.9. **"The X-Ray Universe,"** K. Pounds, in *The State of the Universe,* edited by G. Bath (Clarenden, Oxford, 1980), pp. 93–120. This is a brief self-contained introduction to x-ray astronomy which reviews some results prior to *Einstein.* There are few references. (I)

2.10. **"The X-Ray Eyes of Einstein,"** D. Overbye, Sky and Telescope **57**, 527–534 (1979). Here are some of the pretty false-color pictures from EINSTEIN that permit one to "see" the x-ray sky. (E)

2.11. **"The Einstein X-Ray Observatory,"** R. Giacconi, Sci. Am. **242**, 80–102 (1980). (E)

2.12. **"The Einstein Observatory: New Perspectives in Astronomy,"** R. Giacconi and H. Tananbaum. Science **209**, 865–867 (1980). In this reference and the preceding one the principal investigators of the *Einstein* project review the highlights of the first year or so of data from the imaging instruments. This gives a good introduction to the richness of the product. (E).

2.13. **X-Ray Astronomy with the Einstein Satellite,** edited by R. Giacconi (Reidel, Dordrecht, Holland, 1981). Here are the proceedings of a meeting held in January 1980 to present the first detailed results from *Einstein,* galactic to extragalactic. It is a more technical and complete version of the previous reference. (A)

B. Textbooks

Most standard astronomy texts contain very limited material on x-ray astronomy, and this is generally scattered through the work. Such books can be useful in putting x-ray astronomy into a larger context, but the skimpy coverage underrepresents its true role in current astrophysical research. There is also a paucity of texts devoted to high-energy astrophysics in general or x-ray astronomy in particular. Those that do exist are already out of date as they were written too soon to allow inclusion of most of the results from *Einstein.* Nevertheless, they can be very useful as introductions to the field, its techniques, the physics of x-ray emission, the objects that do emit, and many of the major results. The omission of *Einstein* results is generally more severe for studies of extragalactic sources, supernova remnants, and stars than for x-ray binaries.

2.14. **The Physical Universe,** F. H. Shu (University Science, Mill Valley, CA, 1982). This is one of the newest and most intelligent introductory astronomy texts. It includes quite a bit of physics. Both galactic and extragalactic x-ray phenomena are described in the greater astronomical context. (E)

2.15. **Contemporary Astronomy, 2nd ed.,** Jay M. Pasachoff (Saunders, Philadelphia, 1981). This introductory text contains many references to modern x-ray results. These are scattered throughout the book, but they are well indexed. (E)

2.16. **X-Ray Astronomy,** J. Leonard Culhane and P. W. Sanford (Faber and Faber, London, 1981). This complete text includes solar as well as cosmic x-ray astronomy, but contains only a few references to results from *Einstein.* The presentation is somewhat uneven but many topics are presented in detail. (E)

2.17. **Cosmic X-Ray Astronomy,** D. J. Adams (Heyden, Philadelphia, 1981). This is an introductory text covering all aspects of cosmic x-ray astronomy (pre-*Einstein*), including instrumentation and observations. (E)

2.18. **Cambridge Encyclopedia of Astronomy,** edited by S. Mitton (Cape, London, 1977). Although not a text, this is an intelligent compendium of astronomy for the layman with numerous references to x-ray astronomy and rich illustrations. (E)

2.19. **High Energy Astrophysics,** M. S. Longair (Cambridge University, Cambridge, 1981). The main theme of this work is cosmic-ray astrophysics, but the author devotes several pages to the techniques of x-ray astronomy and to some of the physical processes and observational results. (I)

2.20. **The Physics–Astronomy Frontier,** F. Hoyle and J. Narlikar (Freeman, San Francisco, 1980), pp. 164–206. This somewhat unusual astronomy text actually devotes a full chapter to x-ray astronomy. Unfortunately, most of the material is taken from a 1974 reference, so it is quite out of date. Nevertheless, it can serve as an introduction to some aspects of the field. (E)

2.21. **X-Ray Astronomy**, edited by R. Giacconi and H. Gursky (Reidel, Dordrecht, Holland, 1974). Here is a technical exposition of the results from the *Uhuru* project written by many of the participating scientists. Though not quite a text, most of the articles are more readable than those found in the professional journals. (I)

C. Physical processes

Celestial x-ray emission is the sign of energetic phenomena, both thermal and nonthermal. Plasma with temperatures from $\sim 10^6$ to $\sim 10^8$ K produce both continuum and emission lines in solar and stellar coronas, in supernova remnants, and in clusters of galaxies. Synchrotron radiation by relativistic electrons is important in the Crab Nebula and possibly in some radio galaxies, and inverse Compton scattering may be operating in quasers. X-ray absorption by interstellar material is also of importance. These processes are discussed in the references of the Secs. II A and II B as well as those listed here.

2.22. **"The Electromagnetic Spectrum,"** H. Augensen and J. Woodbury, Astronomy **10**, 6–22 (1982). This is a very elementary, brief, but nicely illustrated description of the spectrum of radiation emitted by astronomical objects. (E)

2.23. **Radiation Processes in Astrophysics**, W. H. Tucker (MIT, Cambridge, MA, 1975). This monograph carries an emphasis on high-energy processes and is handy as a reference. (I)

2.24. **Radiative Processes in Astrophysics**, G. B. Rybicki and A. P. Lightman (Wiley, New York, 1979). This is a solid text covering the fundamentals of radiation in astrophysics. X-ray processes are included but not featured (the word "x-ray" does not appear in the index). (A)

2.25. **"Spectra of Cosmic X-Ray Sources,"** S. S. Holt and R. McCray, Ann. Rev. Astron. Astrophys. **20**, 323–366 (1982). This review covers both theory and observation and can serve as a reasonably up-to-date guide to the professional literature. (I)

2.26. **"Interstellar Absorption of Cosmic X-Rays,"** Robert L. Brown and Robert J. Gould, Phys. Rev. D **1**, 2252–2256 (1970).

2.27. **"Interstellar Photoelectric Absorption Cross Sections,"** Robert Morrison and Dan McCammon, Astrophys. J. **270**, 119–122 (1983). Photoelectric absorption by heavy elements in the interstellar medium absorb the low-energy x rays from all cosmic sources. This article and the preceding one give the effective cross sections versus energy. (A)

D. Techniques

In addition to the specific references listed here, the texts and reviews of Secs. II A and II B generally contain descriptions of the instruments of x-ray astronomy.

2.28. **"Observational Techniques,"** H. Gursky and D. Schwartz, in *X-Ray Astronomy*, edited by R. Giacconi and H. Gursky (Reidel, Dordrecht, Holland, 1974), pp. 25–98. This reviews basic techniques, detector properties, signal detection, sensitivity calculations, and some spacecraft design considerations. (I)

2.29. **"Instrumental Techniques in X-Ray Astronomy,"** L. Peterson, Ann. Rev. Astron. Astrophys. **13**, 423–510 (1975). This gives a good review of the physics of x-ray detection and techniques up to and including the HEAO-1 satellite. (I)

2.30. **"The Einstein X-Ray Observatory,"** R. Giacconi *et al.*, Astrophys. J. **230**, 540–550 (1979). This gives a technical description of the telescope, instruments, and their capabilities. (A)

E. Future prospects

The following articles give some idea of the direction of future research and describe upcoming satellite programs.

2.31. **X-Ray Astronomy in the 1980's**, edited by S. S. Holt, NASA Technical Memorandum 83848 (1981). Theorists raise questions about galactic and extragalactic x-ray sources and experiments describe instruments that could give the answers. This was an attempt to define a possible new x-ray mission for NASA. (A)

2.32. **"The Past, Present and Future of X-Ray Astronomy: Part II, The Future"** (see Ref. 2.1). This volume contains papers describing the future plans for U.S., European, and Japanese x-ray missions as they were perceived in December 1980 at the *Uhuru* Memorial Symposium. (I)

2.33. **"High Energy Astrophysics,"** G. W. Clark, Phys. Today **35** (11), 26–33 (1982). The programs for the future as defined by a National Academy of Sciences panel. At the top of the list of priorities is the Advanced X-ray Astrophysics Facility (AXAF). This is a larger, more sensitive successor to *Einstein* which will become a permanent x-ray observatory in space. (E)

2.34. **"AXAF, A Permanent Orbiting X-Ray Observatory: Telescope and Instrumentation Plans,"** M. V. Zombeck, *Proceedings of the Symposium on Advanced Space Instrumentation in Astronomy, XXIV COSPAR Meeting, 1983* in Adv. in Space Res. **2**, 259–270 (1983). Also Opt. Eng. **20**, 297–309 (1981). This is a technical description of a facility that could be launched in the early 1990s. (A)

2.35. **"A Golden Age for Solar Physics,"** A. B. C. Walker, Jr., Phys. Today **35** (11), 60–67 (1982). The future prospects described in this article include an Advanced Solar Observatory with x-ray and UV detectors that could study all aspects of coronal activity. (E)

III. SOLAR AND STELLAR CORONAE

For 15 years, from the late 1940s to 1962, the solar corona was the only object of study in x-ray astronomy. The past decade has seen major advances in the understanding of coronal structures and their role in solar processes. *Einstein* has extended coronal studies to hundreds of normal stars and has forced a complete revision of models of coronal heating mechanisms. A major surprise was the ubiquity of coronal activity. X-ray luminosities range from $\sim 10^{26}$ to 10^{34} erg s^{-1}, but this is always a small fraction of the total power of the star, most of which emerges at optical wavelengths.

A. The solar corona

3.1. **The Sun Our Star**, R. W. Noyes (Harvard University, Cambridge, MA, 1982). All of modern solar research is covered in this book. Its treatment of x-ray phenomena is up-to-date and well placed in context with UV and other measurements of coronal activity. This could serve as an excellent nontechnical introduction to the sun. (E)

3.2. **"A Matter of Degrees,"** R. W. Noyes, in *Revealing the Universe*, edited by J. Cornell and A. P. Lightman (MIT, Cambridge, MA, 1982), pp. 85–100. (E)

3.3. **"Solar Scenarios,"** R. Rosner in Ref. 3.2, pp. 101–114. Observations and theory of the hot solar corona are nicely described in this chapter and the preceding one in this book of popular lectures. (E)

3.4. **"The Solar X-Ray Spectrum,"** J. L. Culhane and L. W. Acton, Ann. Rev. Astron. Astrophys. **12**, 359–81 (1974). One can read this to get an overview of studies of the solar corona. Included are results from SKYLAB, a discussion of plasma diagnostics using x-ray emission lines, and a brief history of coronal studies. (I)

3.5. **"X-Ray Observations of Characteristic Structures and Time Variations from the Solar Corona: Preliminary Results from SKYLAB,"** G. S. Vaiana, J. M. Davis, R. Giacconi, A. S. Krieger, J. K. Silk, A. F. Timothy, and M. Zombeck, Astrophys. J. Lett. **185**, L47–51 (1973). (I)

3.6. **"Recent Advances in Coronal Physics,"** G. S. Vaiana and R. Rosner, Ann. Rev. Astron. Astrophys. **16**, 393–428 (1978). This study of coronal structure (loops, holes, active regions) and the preceding one of temporal variability lead to semiempirical models of coronal physics. (I)

3.7. **"The Solar Maximum Mission,"** E. G. Chipman, Astrophys. J. Lett. **244**, L113–116 (1981). This paper describes the mission and introduces the other papers in this volume that present results from each of the Solar Maximum Mission experiments. The Solar Maximum Mission satellite, timed to permit study of the sun during a maximum in the sunspot activity cycle, contained an array of highly sophisticated x-ray and UV instruments. (A)

3.8. **"X- and UV-Radiation of the Solar Corona,"** G. Elwert, Space Sci. Rev. **33**, 53–82 (1982). Elwert reviews instrumentation, techniques, and selected results. This can serve as a guide to the professional literature. (I)

B. Stellar coronas

3.9. **"Low Luminosity Galactic X-Ray Sources,"** G. S. Vaiana, in *X-ray Astronomy*, edited by R. D. Andresen (Reidel, Dordrecht, Holland, 1981), also published as Space Sci. Rev. **30**, 151–179 (1981). Vaiana reviews the *Einstein* data on coronal activity in all types of stars and lays out the confrontation between those data and the models of coronal x-ray emission. (I)

*3.10. **"Results from an Extensive Einstein Stellar Survey,"** G. S. Vaiana *et al.*, Astrophys. J. **245**, 163–182 (1981). X-ray emission was detected from nearly every type of star in this survey forcing abandonment of previously formed theories of coronal activity. (A)

3.11. **"Stellar Coronae in the Hyades: A Soft X-Ray Survey with the Einstein Observatory,"** R. A. Stern, M. Zolcinski, S. Antiochos, and J. Underwood, Astrophys. J. **249**, 647–661 (1981). About 40 stars in this nearby star cluster, one of the best studied optically, are seen in an x-ray image. (A)

3.12. **"Coronae of Nondegenerate Single and Binary Stars: A Survey of Our Present Understanding and Problems Ripe for Solution,"** J. L. Linsky, in *X-ray Astronomy in the 1980s*, edited by S. Holt, NASA Tech. Memo. 83848 (NASA Goddard Space Flight Center, Maryland, 1981), pp. 13–36. An elegant exposition of what has been learned from *Einstein* and what the new problems are. (A)

3.13. **"Observations of X-Ray Emission from T Tauri Stars,"** Eric D. Feigelson and William M. DeCampli, Astrophys. J. Lett. **243**, L89–94. (A)

3.14. **"Discovery of Three X-Ray Luminous Pre-Main-Sequence Stars,"** Eric D. Feigelson and Gerard A. Kriss, Astrophys. J. Lett. **248**, L35–38 (1981). T Tauri stars described in this article and the preceding one are very young, recently formed stars that have not yet reached their stable "main-sequence" phase. These authors conclude that the x rays may come from extreme coronal activity. (A)

3.15. **"X-Rays from RS Canum Venaticorum Systems: A HEAO-1 Survey and the Development of a Coronal Model,"** F. M. Walter, W. Cash, P. A. Charles, and C. S. Bowyer, Astrophys. J. **236**, 212–218 (1980). (A)

3.16. **"Two Component X-Ray Emission from RS Canum Venaticorum Binaries,"** J. H. Swank, N. E. White, S. S. Holt, and R. H. Becker, Astrophys. J. **246**, 208–214 (1981). Some short-period binary star systems described in this article and the preceding one contain numerous, large coronal loops and have x-ray luminosities several orders of magnitude greater than that of the sun. (A)

3.17. **"Magnetically-Confined Plasmas as a General Astrophysical Phenomenon,"** R. Rosner, L. Golub, and G. S. Vaiana, Astrophys. J. (in press). This is a technical review of the physics of "coronal" plasmas relevant to the sun, other stars, accretion disks, and galaxies. (A)

IV. GALACTIC X-RAY SOURCES

Aside from the coronal activity of normal stars, galactic x-ray emission is generally associated with compact objects or with the renmants of supernova explosions. The compact objects are neutron stars, white dwarfs, or black holes that accrete material from normal companion stars. The energy source for the x-ray emission is the deep gravitational potential of the collapsed object into which the material falls, releasing 10^{34}–10^{39} erg s^{-1}. The x-ray signals show rich temporal structure: binary eclipses, flares, periodic pulsations of 1-1000 s, millisecond flickering, and giant bursts.

The supernova remnants are large, extended regions of hot plasma, stellar ejecta, and interstellar matter, shock heated to $\sim 10^6$–10^8 K by the $\sim 10^{51}$-erg explosion that terminates the evolution of moderately massive stars. The bulk of their emission is in x rays, which carry information about the stellar explosion and about the interstellar medium around it.

A. General references and reviews

The literature on galactic x-ray sources is extensive, but the field has matured sufficiently to have spawned numerous good review articles.

4.1. **Accretion Driven Stellar X-Ray Sources,** edited by W. H. G. Lewin and E. P. J. van den Heuvel (Cambridge University, Cambridge, 1983). This is the most complete and up-to-date collection of papers on all aspects of compact galactic x-ray sources. Both observations and theory are reviewed, and there are extensive references to the literature. (A)

4.2. **Galactic X-Ray Sources,** edited by Peter W. Sanford, Paul Laskarides, and Jane Salton (Wiley, Chichester, 1982). Proceedings of a 1979 workshop dealing primarily with x-ray binaries, containing several good review articles together with shorter papers reporting specialized topics of interest at that time. It is more uneven and less comprehensive than the previous reference but gives a snapshot of the state of the field at the close of the 1970s. (A)

4.3. **"The Astronomy of Accretion and the Physics of X-Ray Sources,"** J. P. Ostriker, Ann. N.Y. Acad. Sci. **302**, 229–243 (1977). The properties of compact galactic sources are reviewed and an attempt is made to classify them phenomenologically. (A)

4.4. **"Accreting Neutron Stars, Black Holes, and Degenerate Dwarf Stars,"** D. Pines, Science **207**, 597–606 (1980). Pines shows how x-ray binaries can be used to probe the physical properties of the collapsed objects they contain. (I)

4.5. **"The Optical Counterparts of Compact Galactic X-Ray Sources,"** H. V. D. Bradt and J. E. McClintock, Ann. Rev. Astron. Astrophys. **21**, 13–66 (1983). This up-to-date review contains extensive references, a catalog, and a description of the various types of compact galactic sources. (I)

4.6. **"Compact X-Ray Sources,"** G. R. Blumenthal and W. H. Tucker, Ann. Rev. Astron. Astrophys. **12**, 23–46 (1974). (A)

4.7. **"Recent Progress in the Theory of Compact X-Ray Sources,"** F. K. Lamb, Ann. N.Y. Acad. Sci. **262**, 331–360 (1975). This guide and the preceding one to the early literature also outline the theoretical framework of accreting binary x-ray sources. (A)

B. Compact sources: Binaries, pulsars, bursters, etc.

4.8. **"X-Ray Sources and their Optical Counterparts,"** C. Jones, W. Forman, and W. Liller, Sky and Telescope **48**, 289–291 (1974); **48**, 372–375 (1974). This ia a good nontechnical introduction to the early work on binary x-ray pulsars. (E)

4.9. **"Accretion Torques in X-Ray Pulsars,"** S. Rappaport and P. C. Joss, Nature **266**, 683–685 (1977). (A)

4.10. **"Accretion by Rotating Magnetic Neutron Stars III. Accretion Torques and Period Changes in Pulsating X-Ray Sources,"** P. Ghosh and F. K. Lamb, Astrophys. J. **234**, 296–316 (1979). This paper and the preceding one show how pulse-period changes are related to the torques applied to the compact star by the accreting material. The observational evidence very clearly supports the view that x-ray pulsars are accreting neutron stars. (A)

4.11. **"A Model for Compact X-Ray Sources: Accretion by Rotating Magnetic Stars,"** F. K. Lamb, C. J. Pethick, and D. Pines, Astrophys. J. **184**, 271–289 (1973). This early reference covers many of the fundamental theoretical concepts of the accretion process. (A)

*4.12. **"Masses of Neutron Stars in X-Ray Binary Systems,"** R. L. Kelley and S. Rappaport, in *Pulsars*, edited by W. Sieber and R. Wielebinski (Reidel, Dordrecht, Holland, 1981), pp. 353–356. (A)

4.13. **"X-Ray Pulsars in Massive Binary Systems,"** S. A. Rappaport, in *Accretion Driven Stellar X-Ray Sources*, edited by W. H. G. Lewin and E. P. J. van den Heuvel (Cambridge University, Cambridge, 1983). (A)

4.14. **"Neutron Stars in Interacting Binary Systems,"** P. Joss and S. A. Rappaport, Ann. Rev. Astron. Astrophys. (in press). This reference and the two preceding ones are up-to-date reviews of observations of x-ray binary pulsars and of the measurement of orbital parameters and neutron star masses. (A)

4.15. "**Accretion Powered X-Ray Pulsars,**" N. E. White, J. H. Swank, and S. S. Holt, Astrophys. J. **270**, 711–734 (1983). Emphasis here is on the changes in the x-ray spectrum with pulse phase. These are used to deduce the geometry of the beamed emission. (A)

4.16. "**Formation and Evolution of X-Ray Binaries,**" E. P. J. van den Huevel, in Ref. 4.1. Models of both low- and high-mass binary systems are reviewed, as are the evolutionary scenarios that might lead to x-ray activity. (A)

4.17. "**Highly Compact Binary X-Ray Sources,**" P. C. Joss and S. A. Rappaport, Astron. Astrophys. **71**, 217–220 (1979). In contrast to binaries with massive stellar members, systems with low-mass stars and neutron stars are proposed to explain optically faint compact x-ray sources. (A)

4.18. "**Evidence for the Binary Nature of Centaurus X-3 from UHURU X-Ray Observations,**" E. Schreier, R. Levinson, H. Gursky, E. Kellogg, H. Tananbaum, and R. Giacconi, Astrophys. J. Lett. **172**, L79–89 (1972). (I)

4.19. "**An Extended Observation of Cen X-3 with the Ariel-5 Sky Survey,**" K. A. Pounds, B. A. Cooke, M. J. Ricketts, M. J. Turner, and M. Elvis, Mon. Not. R. Astron. Soc. **172**, 473–481 (1975). (A)

4.20. "**Further Studies of the Pulsation Period and Orbital Elements of Centaurus X-3,**" G. Fabbiano and E. J. Schreier, Astrophys. J. **214**, 235–244 (1977). This study and the two preceding ones are from the voluminous literature on Centaurus X-3, the first x-ray source to reveal its binary nature. The rotating neutron star accretes matter from the stellar wind of its optically luminous companion. The varying transparency of the wind modulates the x-ray flux dramatically. (A)

4.21. "**Her X-1 and Cen X-3 Revisited,**" R. Giacconi, Ann. N. Y. Acad. Sci. **262**, 312–330 (1975). (I)

4.22. "**X-Ray Spectra of Hercules X-1: II. The Pulse,**" S. H. Pravdo, E. A. Boldt, S. S. Holt, and P. J. Serlemitsos, Astrophys. J. **216**, L23–26 (1977). (A)

4.23. "**Pulse Timing Observations of Hercules X-1,**" J. E. Deeter, P. E. Boynton, and S. H. Pravdo, Astrophys. J. **247**, 1003–1012 (1981). (A)

4.24. "**Einstein Observatory Pulse-Phase Spectroscopy of Hercules X-1,**" R. A. McCray, J. M. Shull, P. E. Boynton, J. E. Deeter, S. S. Holt, and N. E. White, Astrophys. J. **262**, 301–307 (1982). This paper and the preceding three present x-ray spectra and timing data on Her X-1, one of the best studied binary x-ray pulsars. The last paper explores models of the x-ray-matter interactions near the pulsar. (A)

4.25. "**Evidence of Strong Cyclotron Line Emission in the Hard X-Ray Spectrum of Hercules X-1,**" J. Trumper, W. Pietsch, C. Reppin, W. Voges, R. Staubert, and E. Kendziorra, Astrophys. J. Lett. **219**, L105–110 (1978). (A)

4.26. "**Cyclotron Lines in the Hard X-Ray Spectrum of Hercules X-1,**" W. Voges, W. Pietsch, C. Reppin, J. Trumper, E. Kendziorra, and R. Staubert, Astrophys. J. **263**, 803–813 (1982). (A)

4.27. "**An Absorption Feature in the Pulsed Hard X-Ray Flux from 4U0115 + 63,**" Wheaton et al., Nature **282**, 240–243 (1979). This article and the two preceding ones concern cyclotron absorption and emission which probably occurs in the very intense ($\sim 10^{12}$ g) magnetic fields near the surfaces of accreting neutron stars in x-ray pulsars. (A)

4.28. "**Orbital Elements and Masses for the SMC X-1/Sanduleak 160 Binary System,**" F. Primini, S. Rappaport, P. C. Joss, G. W. Clark, W. Lewin, F. Li, W. Mayer, and J. McClintock, Astrophys. J. Lett. **210**, L71–74 (1976). (A)

4.29. "**Pulse Profile and Refined Orbital Elements for SMC X-1,**" F. Primini, S. A. Rappaport, and P. C. Joss, Astrophys. J. **217**, 543–548 (1977). (A)

4.30. "**Detection of Soft X-Ray Emission from SMC X-1,**" A. N. Bunner and W. T. Sanders, Astrophys. J. Lett. **228**, L19–22 (1979). SMC X-1, discussed in this article and the two preceding ones, is a highly luminous x-ray binary with a spinning neutron star orbiting a massive normal star. It is located in the Small Magellanic Cloud (SMC), a small neighbor galaxy visible in the southern sky. There are other, similar systems in our galaxy. (A)

4.31. "**The X-Ray Spectrum of AM Hercules from 0.1 to 150 keV,**" R. E. Rothschild et al., Astrophys. J. **250**, 723–732 (1981). (A)

4.32. "**X and UV Radiation from Accreting Magnetic Degenerate Dwarfs,**" D. Q. Lamb and A. R. Masters, Astrophys. J. Lett. **234**, L117–122 (1979). (A)

4.33. "**The 805 Second Pulsar H2252-035,**" N. E. White and F. E. Marshall, Astrophys. J. Lett. **249**, L25–28 (1981). (A)

4.34. "**Rapid Oscillations in Cataclysmic Variables V. H2252-035, A Single-Sideband X-Ray and Optical Pulsar,**" J. Patterson and C. M. Price, Astrophys. J. Lett. **243**, L83–87 (1981). The AM Her systems discussed in this article and the three preceding ones form a class of close binaries with accreting magnetic white dwarfs. They are both x-ray and UV emitters and display a rich variety of temporal variability. (A)

4.35. "**4U1626-67: A Prograde Spinning X-Ray Pulsar in a 2500s Binary System,**" J. Midleditch, K. O. Mason, J. E. Nelson, and N. E. White, Astrophys. J. **244**, 1001–1021 (1981). (A)

4.36. "**4U1626-67 and the Character of Highly Compact Binary X-Ray Sources,**" F. K. Li, P. C. Joss, J. E. McClintock, S. Rappaport, and E. L. Wright, Astrophys. J. **240**, 628–635 (1980). The object of study in this paper and the preceding one is a remarkable low-mass binary system with a spinning neutron star of 7s period in a 2500s orbit with a low-mass star. (A)

4.37. "**A Status Report on Cygnus X-1,**" D. M. Eardley et al., Comments on Astrophys. **7**, 151–160 (1978). (I)

4.38. "**Cygnus X-1: A Candidate of the Black Hole,**" M. Oda, Space Sci. Rev. **20**, 757–814 (1977). (A)

4.39. "**Submillisecond Measurements of the Low State of Cygnus X-1,**" R. E. Rothschild, E. A. Boldt, S. S. Holt, and P. J. Serelemitsos, Astrophys. J. **213**, 818–826 (1977). (A)

4.40. "**Hard X-Ray Spectrum of Cyg X-1,**" R. A. Sunyaev and J. Trumper, Nature **279**, 506–508 (1979). Cyg X-1, discussed in this article and the three preceding ones, remains the best candidate black hole. The facts and interpretations presented in the earlier papers are largely unchanged today. See also the more popular treatment given in Sullivan's book (Ref. 2.4), in the Cambridge Encyclopedia (Ref. 2.18) and in "Resource Letter BH-1: Black Holes" by Steven Detweiler, Am. J. Phys. **49**, 394–400 (1981). (A)

4.41. "**Simultaneous X-Ray and Optical Observations of Rapid Variability in Scorpius X-1,**" S. A. Illovaisky, C. Chevalier, N. White, K. O. Mason, P. W. Sanford, J. Delvaille, and H. W. Schnopper, Mon. Not. R. Acad. Sci. **191**, 81–94 (1980). (A)

4.42. "**Rapid X-Ray and Optical Flares from Scorpius X-1,**" L. D. Petro, H. V. Bradt, R. L. Kelley, K. Horne, and R. Gomer, Astrophys. J. Lett. **251**, L7–11 (1981). Sco X-1, discussed in this article and the preceding one, was the first nonsolar x-ray source to be detected. It is probably a neutron star in a 0.8-day binary system, but much of its rich temporal behavior is not well understood. (A)

4.43. "**Some Comments on Transient X-Ray Sources,**" K. Pounds, Comments on Astrophysics **6**, 145–156 (1976). (I)

4.44. "**Transient X-Ray Sources in the Galactic Plane,**" L. Cominsky, C. Jones, W. Forman, and H. Tananbaum, Astrophys. J. **224**, 46–52 (1978). There are various kinds of x-ray "novae" whose outbursts last for weeks to months. Some of these, at their brightest, outshine nearly all other sources in the galaxy. This paper and the preceding one are reviews. (I)

4.45. "**X-Ray Stars in Globular Clusters,**" G. W. Clark, Sci. Am. **237**, 43–55 (1977). (E)

4.46. "**An X-Ray Survey of Globular Clusters and their X-Ray Luminosity Functions,**" P. Hertz and J. E. Grindlay, Astrophys. J. **275** (1983). In the globular star clusters of our galaxy discussed in this article and the preceding one binaries can form by stellar capture making x-ray sources, some of which are also x-ray bursters. (I)

*4.47. "**Discovery of Intense X-Ray Bursts from the Globular Cluster NGC6624,**" J. Grindlay, H. Gursky, H. Schnopper, D. R. Parsignault, J. Heise, A. C. Brinkman, and J. Schrijver, Astrophys. J. Lett. **205**, 1127–130 (1976). (I)

4.48. "**The Discovery of Bursting X-Ray Sources,**" J. E. Grindlay, Comments on Astrophys. **6**, 165–175 (1976). (I)

4.49. "**The Sources of Celestial X-Ray Bursts,**" W. H. G. Lewin, Sci. Am. **244**, 72–82 (1981). (E)

4.50. "**X-Ray Bursters and the X-Ray Sources of the Galactic Bulge,**" W. H. G. Lewin and P. C. Joss, Space Sci. Rev. **28**, 3–88 (1981). (I)

4.51. "**The Mystery of the X-Ray Burst Sources,**" W. H. G. Lewin, in *Revealing the Universe*, edited by J. Cornell and A. P. Lightman (MIT,

Cambridge, 1982), pp. 149–172. (E)

4.52. "**Helium Burning Flashes on an Accreting Neutron Star: A Model for X-Ray Burst Sources,**" P. C. Joss, Astrophys. J. Lett. **225**, L123-127 (1978). (A)

*4.53. "**The Discovery of Rapidly Repetitive X-Ray Bursts from a New Source in Scorpius,**" W. H. G. Lewin et al., Astrophys. J. Lett. **207**, L105–108 (1976). (A)

X-ray bursts which are discussed in Refs. 4.47–5.3 are remarkable, recurrent, ~ 10-s-long events that release $\sim 10^{40}$ ergs. They were a total mystery in 1975 but now their principles are understood. All but one are almost surely caused by thermonuclear detonations on the surfaces of neutron stars. The exception is the remarkable "rapid burster" which may be caused by an accretion instability.

4.54. "**Satellite Observations of Cataclysmic Variables,**" K. O. Mason and F. R. Cordova, Sky and Telescope **63**, 25–29 (1982). (E)

4.55. "**Accreting Degenerate Dwarfs in Close Binary Systems,**" F. A. Cordova and K. O. Mason, in Ref. 4.1. The "cataclysmic variables" of optical astronomy are erruptive binary systems with white dwarf members that exhibit a variety of rich phenomena in the x-ray band. (I)

C. Supernova remnants

4.56. "**X-Rays from SNRs,**" J. L. Culhane, in *Supernovae*, edited by D. N. Schramm (Reidel, Dordrecht, Holland, 1977), pp. 29–51. This is a good review of the pre-*EINSTEIN* data together with a brief exposition of the theory of x-ray emission from supernova remnants. (I)

4.57. "**X-Ray Emission from an Inward Propagating Shock in Young Supernova Remnants,**" C. F. McKee, Astrophys. J. **188**, 335–340 (1974). The basic dynamics of young supernovae are described and two regions of x-ray emission are identified. One region is associated with the primary shock wave that propagates into the interestellar medium, the other with a shock wave that runs "inward" through the stellar ejecta. (A)

4.58. "**Einstein Observations of Supernova Remnants,**" F. D. Seward, in *Supernovae: A Survey of Current Research*, edited by M. J. Rees and R. J. Stoneham (Reidel, Dordrecht, Holland, 1982), pp. 519–528. Seward gives a brief overview of the wonderful x-ray images of supernova remnants obtained with the Einstein Observatory. (I)

4.59. **Supernova Remnants and Their X-Ray Emission**, edited by J. Danziger and P. Gorenstein (Reidel, Dordrecht, Holland, 1983). These proceedings of an international meeting contain the most up-to-date research. There are many excellent reviews of the observations and theoretical aspects of young and old remnants, pulsars, and relevant physical processes. (I) (A)

*4.60. "**High Resolution X-Ray Observations of the Cassiopeia A Supernova Remnant with the Einstein Observatory,**" S. S. Murray, G. Fabbiano, A. C. Fabian, A. Epstein, and R. Giacconi, Astrophys. J. Lett. **243**, L69–72 (1979). (A)

4.61. "**The Mass of Tycho's Supernova Remnant As Determined from a High Resolution X-Ray Map,**" F. Seward, G. Gorenstein, and W. Tucker, Astrophys. J. Lett. **266**, 287–297 (1983). (A)

4.62. "**A High Resolution X-Ray Image of Puppis A. Inhomogeneities in the Inter-stellar Medium,**" R. Petre, C. R. Canizares, G. A. Kriss, and P. F. Winkler, Astrophys. J. **258**, 22–30 (1982). The images displayed in this reference and the preceding two reveal many aspects of supernova remnant evolution and of their interactions with the interstellar medium. (A)

*4.63. "**X-Ray Spectrum of Cassiopeia A Measured with the Einstein Observatory SSS,**" R. H. Becker, S. S. Holt, B. W. Smith, N. E. White, E. A. Boldt, R. F. Mushotzky, and P. J. Serlemitsos, Astrophys. J. Lett. **234**, L73–76 (1979). (A)

4.64. "**Iron Line Emission from a High-Temperature Plasma in Cassiopeia A,**" S. H. Pravdo, R. H. Becker, E. A. Boldt, S. S. Holt, R. E. Rothschild, P. J. Serlemitsos, and J. H. Swank, Astrophys. J. Lett. **206**, L41–44 (1976). (A)

4.65. "**A Survey of X-Ray Line Emission from the Supernova Remnant Puppis A,**" P. F. Winkler, C. R. Canizares, G. W. Clark, T. H. Markert, K. Kalata, and H. W. Schnopper, Astrophys. J. Lett. **246**, L27–31 (1981). (A)

*4.66. "**Evidence for Elemental Enrichment of Puppis A by a Type II Supernova,**" C. R. Canizares and P. F. Winkler, Astrophys. J. Lett. **246**, L33–36 (1981). (A)

The x-ray spectra of supernova remnants such as those discussed in Refs. 4.63–4.66 show numerous emission lines that can be used as diagnostics of the physical conditions in the hot plasma. Measurements of elemental abundances reveal the enrichment of the interstellar medium by the supernova ejecta.

4.67. "**Observation of X-Rays from the Crab Pulsar,**" A Toor and F. D. Seward, Astrophys. J. **216**, 560–564 (1977). (A)

4.68. "**X-Ray Spectra of the Crab Pulsar and Nebula,**" S. H. Pravdo and P. J. Serlemitsos, Astrophys. J. **246**, 484–488 (1981). (A)

4.69. "**X-Ray Study of Two Crablike Supernova Remnants: 3C58 and CTB80,**" R. H. Becker, D. J. Helfand, and A. E. Szymkowiak, Astrophys. J. **255**, 557–563 (1982). (A)

4.70. "**A New, Fast X-Ray Pulsar in the Supernova Remnant MSH 15-52,**" F. D. Seward and F. R. Harnden, Astrophys. J. Lett. **256**, L45–47 (1982). (A)

The Crab Nebula is the remnant of a supernova observed in 1054. Both the synchrotron nebula and the pulsar that powers it are strong x-ray sources. Recently several other similar objects have been found as well.

4.71. "**The X-Ray Superbubble in Cygnus,**" W. Cash, P. Charles, S. Bowyer, F. Walter, G. Garmire, and G. Riegler, Astrophys. J. Lett. **238**, L71–76 (1980). The superbubble is a 1500-light-year-diam region filled with 2-million-degree gas representing nearly 10^{52} ergs. It was probably formed by 30–100 supernova explosions. (I)

4.72. "**Detection of X-Rays During the Outburst of Supernova 1980K,**" C. R. Canizares, G. A. Kriss, and E. D. Feigelson, Astrophys. J. Lett. **253**, L17–22 (1982). X-rays from an exploding supernova in another galaxy could arise from several mechanisms. (A)

4.73. "**Thermal X-Ray Emission from Neutron Stars,**" D. J. Helfand, G. A. Chanan, and R. Novick, Nature **283**, 337–343 (1980). (A)

4.74. "**Unpulsed X-Ray from Pulsars,**" D. J. Helfand, in *Pulsars*, edited by W. Sieber and R. Wielebinski (Reidel, Dordrecht, Holland, 1981), pp. 343–350. Neutron stars with temperature above half a million degrees which are discussed in this reference and the preceding one, emit x rays as blackbody radiation. Measurements from *Einstein* are important in understanding the internal composition of the neutron star. (A)

D. Soft x-ray background

The soft x-ray background is a patchy, diffuse emission that is mostly of galactic origin but may also contain a component from outside the galactic disk. The galactic part is emitted by diffuse million-degree plasma that fills much of interstellar space and that was heated by supernovae.

4.75. "**The Diffuse Soft X-Ray Sky,**" Y. Tanaka and J. A. M. Bleeker, Space Sci. Rev. **20**, 815–888 (1977). This is a detailed introduction and review of the relevant physical processes and of the data up to 1977. (A)

4.76. "**The Soft X-Ray Diffuse Background and the Structure of the Local Interstellar Medium,**" P. M. Fried, J. A. Nousek, W. T. Sanders, and W. L. Krasuhaar, Astrophys. J. **242**, 987–1004 (1980). (A)

4.77. "**The Soft Diffuse X-Ray Background,**" D. McCammon, D. N. Burrows, W. T. Sanders, and W. L. Kraushaar, Astrophys. J. **269**, 107–135 (1983). Maps of x-ray brightness displayed in this article and the preceding one cover nearly the entire celestial sphere. (A)

V. EXTRAGALACTIC X-RAY SOURCES

Normal galaxies are sources of x rays simply because they contain the same kinds of objects that emit x rays in the Milky Way (Sec. IV). In the nearest galaxies these can be resolved by *Einstein* and form important samples of "galactic" objects. Active galactic nuclei have x-ray luminosities up to 10^{47} erg s^{-1} compared to a total of $\sim 10^{39}$ ergs s^{-1} for all the sources in a normal galaxy. What powers these nuclei in "Seyfert galaxies" and quasars is not known, but popular models invoke accretion onto black holes with masses $\sim 10^8$ times that of the sun. X-ray emission probably occurs very close to the center of the nucleus and so may carry more information about its nature and physical processes than do optical or radio emission. The short timescale variability (as short as several hundred seconds) of several x-ray active nuclei supports this view.

Hot intergalactic gas in clusters of galaxies is another source of extragalactic x rays. Rich clusters contain 1000 or more galaxies, while poor ones might have only a few dozen. Evidence suggests that many clusters contain as much mass in the low-density (10^{-2} to 10^{-3} particles cm^{-3}) 10^7–10^8 K plasma as in all the luminous stars of the member galaxies. Both stars and gas are gravitationally bound by the still more plentiful "dark matter" whose nature is not known. The x-ray studies contribute to the understanding of cluster structure, dynamics, and evolution.

Finally, the diffuse x-ray background is apparently isotropic radiation that, like the microwave blackbody radiation, is clearly of cosmological origin. Whether it arises from a superposition of a great many discrete sources, like quasars or young galaxies at high redshift, or from a pervasive intergalactic plasma, it is sure to carry information about the large-scale properties of the universe.

A. General reference and reviews

5.1. **"Extragalactic X-Ray Sources,"** H. Gursky and D. A. Schwartz, Ann. Rev. Astron. Astrophys. **15**, 541–568 (1977). The authors review all aspects of extragalactic x-ray astronomy and give numerous references to the pre-*Einstein* literature. (I)

5.2. **"X-Ray Astronomy in the Einstein Era,"** R. Giacconi, in *X-Ray Astronomy*, edited by R. D. Andresen (Reidel, Dordrecht, Holland, 1981), pp. 3–32. (I)

5.3 **"Extragalactic X-Ray Astronomy with the Einstein Observatory,"** R. Giacconi, in *Tenth Texas Symposium on Relativistic Astrophysics*, edited by R. Ramaty and F. C. Jones [Ann. N.Y. Acad. Sci. **375**, 210–234 (1981)]. This article and the preceding one also review the explosion of new results from *Einstein* on galaxies, clusters of galaxies, and quasars. (I)

B. Normal and nearly normal galaxies

5.4 **"X-Ray Emission from Normal Galaxies,"** K. S. Long and L. P. van Speybroeck in Ref. 53. *Einstein* has resolved individual sources in nearby galaxies, extending the study of "galactic" sources beyond the Milky Way. (I)

*5.5. **"Observations of X-Ray Sources in M31,"** L. Van Speybroeck, A. Epstein, W. Forman, R. Giacconi, C. Jones, W. Liller, and L. Smarr, Astrophys. J. Lett. **234**, L45–50 (1979). The x-ray images reveal many dozens of x-ray sources in our neighbor galaxy, the Andromeda nebula. (A)

5.6. **"A Soft X-Ray Study of the Large Magellanic Cloud,"** K. S. Long, D. J. Helfand, and D. A. Grabelsky, Astrophys. J. **248**, 925–944 (1981). Many of these sources in the nearest galaxy to the Milky Way are supernova remnants, most were previously unknown. (A)

5.7. **"X-Ray Observations of Peculiar Galaxies with Einstein,"** G. Fabbiano, E. Feigelson, and G. Zamorani, Astrophys. J. **256**, 397–409 (1981). These galaxies appear to be undergoing bursts of star formation. The many young supernova remnants and newly formed binary x-ray sources enhance the x-ray luminosities by one or two orders of magnitude over those of "normal" galaxies. (A)

C. Cluster of galaxies

5.8. **"Rich Clusters of Galaxies,"** P. Gorenstein and W. Tucker, Sci. Am. **239**, 110–128 (1978). Here is a nontechnical introduction to the subject and a description of the pre-*Einstein* data. (E)

*5.9. **"Ariel 5 Observations of the X-Ray Spectrum of the Perseus Cluster,"** B. J. Mitchell, J. L. Culhane, P. J. N. Davison, and J. C. Ives, Mon. Not. R. Astron. Soc. **175**, 29P–34P (1976). (A)

5.10. **"OSO-8 Spectra of Clusters of Galaxies I: Observations of 20 Clusters, Physical Correlations,"** R. F. Mushotzky, P. J. Serlemitsos, B. W. Smith, E. A. Boldt, and S. S. Holt, Astrophys. J. **225**, 21–39 (1978). (A)

5.11. **"Parameters and Predictions for the X-Ray Emitting Gas of Coma, Perseus and Virgo,"** J. N. Bahcall and C. L. Sarazin, Astrophys. J. Lett. **213**, L99–103 (1977). The detection of emission lines from ionized iron showed that the clusters described in this article and the preceding two contain hot gas enriched with heavy elements. The x-ray spectra give temperatures for the hot intracluster gas. (A)

*5.12. **"The Structure and Evolution of X-Ray Clusters,"** C. Jones, E. Mandel, J. Schwartz, W. Forman, S. S. Murray, and F. R. Harnden, Jr., Astrophys. J. Lett. **234**, L21–25 (1979). (I)

5.13. **"X-Ray Images of Clusters of Galaxies,"** W. Forman and C. Jones, Ann. Rev. Astron. Astrophys. **20**, 547–585 (1982). This reference is a guide to the *Einstein* literature with a discussion of the various types of x-ray clusters and the possible connection to cluster evolution. The preceding one gives some of the results on a variety of cluster types. (I)

5.14. **"Hot Gas in Clusters of Galaxies,"** S. F. Gull and K. J. E. Northover, Mon. Not. R. Astron. Soc. **173**, 585–603 (1975). (A)

5.15. **"On the Equilibrium Distribution of Gas in Clusters of Galaxies,"** Susan M. Lea, Astrophys. Lett. **16**, 141–144 (1975). (A)

5.16. **"X-Ray Line Spectroscopy for Clusters of Galaxies—I,"** John N. Bahcall and Craig L. Sarazin, Astrophys. J. **219**, 781–794 (1978). This paper and the preceding two detail the fundamental theory of static distributions of hot plasma gravitationally bound to the cluster. (A)

5.17. **"X-Ray Spectra of Clusters of Galaxies,"** C. R. Canizares, Ref. 2.13, pp. 215–226. (A)

5.18. **"Observation of the Core of the Perseus Cluster with the Einstein Solid State Spectrometer: Cooling Gas and Elemental Abundances,"** R. F. Mushotzky, S. S. Holt, B. W. Smith, E. A. Boldt, and P. J. Serlemitsos, Astrophys. J. Lett. **244**, L47 (1981). (A)

5.19. **"X-Ray Spectroscopy of the Galaxy M87: Radiative Accretion of the Hot Plasma Halo,"** C. R. Canizares, G. W. Clark, J. G. Jernigan, and T. H. Markert, Astrophys. J. **262**, 33–41 (1982). Spectroscopy as discussed in this article and the preceding two has revealed that the gas in some clusters is cooling and accreting onto a central galaxy at rates of up to several hundred solar masses per year. (A)

5.20. **"Subsonic Accretion of Cooling Gas in Clusters of Galaxies,"** A. C. Fabian and P. E. J. Nulsen, Mon. Not. R. Astron. Soc. **180**, 479–484 (1977). (A)

5.21. **"Radiative Regulation of Gas Flow Within Clusters of Galaxies: A Model for Cluster X-Ray Sources,"** L. L. Cowie and J. Binney, Astrophys. J. **215**, 723–732 (1977). (A)

5.22. **"Radiative Accretion Flow onto Giant Galaxies in Clusters,"** W. G. Mathews and J. N. Bregman, Astrophys. J. **224**, 308–319 (1978). (A)

5.23. **"Star Formation in a Cooling Flow,"** A. C. Fabian, P. E. J. Nulsen, and C. R. Canizares, Mon. Not. R. Astron. Soc. **201**, 933–938 (1982). (A)

Radiative accretion of cluster gas was suggested on theoretical grounds, as described in Refs. 5.20–5.22. Reference 5.23 describes the likely end point of most of the flow, namely, formation of low mass stars.

5.24. **"X-Ray Measurements of the Mass of M87,"** D. Fabricant, M. Lecar, and P. Gorenstein, Astrophys. J. **241**, 552–560 (1980). The x-ray surface brightness around this elliptical galaxy is interpreted and points to an enormous mass (several times 10^{13} solar masses) for the galaxy and its halo. (A)

*5.25. **"The Distribution and Morphology of X-Ray Emitting Gas in the Core of the Perseus Cluster,"** A. C. Fabian, E. M. Hu, L. L. Cowie, and J. Grindlay, Astrophys. J. **248**, 47–54 (1981). Detailed studies of the x-ray surface brightness distribution reveal the temperature and density distributions of the emitting gas and show evidence for accretion of the gas onto the central galaxy. (A)

5.26. "**Evolution of the Cluster X-Ray Luminosity Function Slope,**" J. P. Henry, A. Zoltan, and U. Briel, Astrophys. J. **262**, 1–8 (1982). Observations of distant clusters can constrain models of cluster formation and evolution. (A)

D. Active galactic nuclei, quasars

5.27. "**Seyfert Galaxies as X-Ray Sources,**" M. Elvis, T. Maccacaro, A. S. Wilson, M. J. Ward, M. V. Penston, R. A. E. Fosbury, and G. C. Perola, Mon. Not. R. Astron. Soc. **183**, 129–157 (1978). (A)

5.28. "**X-Ray Observations of Seyfert Galaxies with the Einstein Observatory,**" G. A. Kriss, C. R. Canizares, and G. R. Ricker, Astrophys. J. **242**, 492–501 (1980). The detection of x rays from numerous galaxies with active nuclei is reported in this article and the preceding one. Correlations of x-ray power with other physical parameters are explored to look for clues about the underlying emission mechanism. (A)

5.29. "**In Search of X-Ray Quasars,**" H. Bradt and B. Margon, Sky and Telescope **56**, 499–503 (1978). Prior to *Einstein* only three quasars were known to emit x rays. This is a nontechnical review of their discovery and properties. (E)

5.30. "**The X-Ray Jets of Centaurus A and M87,**" E. D. Feigelson and E. J. Schreier, Sky and Telescope **65**, 6–10 (1983). (E)

*5.31. "**Einstein Observations of the X-Ray Structure of Centaurus A: Evidence for the Radio Lobe Energy Source,**" E. J. Schreier, E. Feigelson, J. Delvaille, R. Giacconi, J. Grindlay, D. A. Schwartz, and A. C. Fabian, Astrophys. J. Lett. **234**, L39–44. (A)

5.32. "**The X-Ray Structure of Centaurus A,**" E. D. Feigelson, E. J. Schreier, J. P. Delvaille, R. Giacconi, J. E. Grindaly, and A. P. Lightman, Astrophys. J. **251**, 31 (1981). Long narrow jets of relativistic particles as described in this article and the preceding two, are seen emerging from many active galactic nuclei. Previously detected only at radio wavelengths, they now appear on x-ray images from *Einstein*. (A)

5.33. "**The Farthest and the Brightest,**" H. Tananbaum, in *Revealing the Universe*, edited by J. Cornell and A. P. Lightman (MIT, Cambridge, MA, 1982), pp. 117–130. (E)

5.34. "**Too Much from Too Little?,**" A. P. Lightman, in Ref. 5.33, pp. 131–145. The two chapters in this reference and the preceding one are nontechnical descriptions of quasars, their emission, and the possible explanations for their prodigious luminosities. (E)

5.35. "**X-Ray Studies of Quasars with the Einstein Observatory,**" G. Zamorani *et al.*, Astrophys. J. **245**, 357–374 (1981). (A)

5.36. "**X-Ray Properties of Quasars,**" W. H.-M. Ku, D. J. Helfand, and L. B. Lucy, Nature **288**, 323–328 (1980). *Einstein* detected x rays from a great many quasars. Much of the data on quasars are presented in this paper and the preceding one, together with some discussion of the relationship of x ray to optical and radio emission. (A)

*5.37. "**On the Cosmological Evolution of the X-Ray Emission From Quasars,**" Y. Avni and H. Tananbaum, Astrophys. J. Lett. **262**, L17–21 (1982). (A)

5.38. "**The Cosmological Evolution and Luminosity Function of X-Ray Selected AGNs,**" T. Maccacaro *et al.*, Astrophys. J. Lett. **266**, L73–78 (1983). X-ray data discussed in this paper and the preceding one support the conclusion (based on optical data) that quasars were more luminous at earlier epochs. (A)

5.39. "**Rapid X-Ray Variability in the Seyfert Galaxy NGC6814,**" A. F. Tenant, R. F. Mushotzky, E. A. Boldt, and J. H. Swank, Astrophys. J. **251**, 15–25 (1981). (A)

5.40. "**Evidence for 200 Second Variability in the X-Ray Flux of the Quasar 1525 + 227,**" T. Matilsky, C. Shrader, and H. Tananbaum, Astrophys. J. Lett. **258**, L1–5 (1982). Rapid variability as discussed in this paper and the preceding one sets very strong constraints on the size of the emitting regions in quasars and active galaxy nuclei. These are some extreme examples of short timescale phenomena in extragalactic sources. (A)

5.41. "**The Luminosity of Serendipitous X-Ray Quasars,**" B. Margon, G. Chanan, and R. A. Downes, Astrophys. J. Lett. **253**, L7–11 (1982). (A)

5.42. "**Optical and X-Ray Properties of X-Ray Selected Active Galactic Nuclei,**" G. A. Kriss and C. R. Canizares, Astrophys. J. **261**, 51–63 (1982). Many of the previously unknown x-ray sources seen by *Einstein* (serendipitously), as discussed in this article and the preceding one,

have turned out to be quasars. X-ray detection is a very effective way to find quasars. (A)

5.43. "**X-Ray Emitting BL Lacertae Objects Located by the Scanning Modulation Collimator Experiment of HEAO-1,**" D. A. Schwartz, R. E. Doxsey, R. E. Griffiths, M. D. Johnson, and J. Schwarz, Astrophys. J. Lett. **229**, L53–57 (1979). (A)

5.44. "**Quasi-Simultaneous Observations of BL Lac Object MRK501 in X-Ray, UV, Visible, IR and Radio Frequencies,**" Y. Kondo *et al.*, Astrophys. J. **243**, 690–699 (1981). (A)

5.45. "**Studies of BL Lac Objects with the Einstein X-Ray Observatory: The Absolute Volume Density,**" D. A. Schwartz and W. H.-M. Ku, Astrophys. J. **266**, 459–465 (1983). So called "BL Lac Objects" are active galactic nuclei that show peculiar properties including very rapid and extreme variability. This technical paper and the two preceding ones explore several hypotheses concerning BL Lacs and point to other literature. (A)

5.46. "**Spectroscopy of Compact Extragalactic X-Ray Sources,**" S. Holt, in Ref. 2.13, pp. 173–187. (I)

5.47. "**X-Ray Spectra and Time Variability of Active Galaxies,**" R. Mushotzky, in *Eleventh Texas Symposium on Relativistic Astrophysics* [Ann. N.Y. Acad. Sci. (in press)]. The x-ray spectra of active galactic nuclei, as described in this reference and the preceding one, reveal a remarkable uniformity. This may be a clue to the emission mechanism itself. (I)

5.48. "**Physical Processes for X-Ray Emission in Galactic Nuclei,**" M. J. Rees, in *X-Ray Astronomy*, edited by R. D. Andresen (Reidel, Dordrecht, Holland, 1981), pp. 87–100. [Also published as Space Sci. Rev. **30**, 1–4 (1981)]. (A)

5.49. "**X-Rays from Quasars and Active Galaxies,**" A. P. Lightman, Space Sci. Rev. **33**, 335–357 (1982). This technical paper and the preceding one explore some of the theoretical considerations that are relevant to the x-ray emission from quasars and that may point to greater understanding of the emission mechanism. (A)

E. The cosmic x-ray background

5.50. "**The Cosmic X-Ray Background,**" E. Boldt, Comments on Astrophys. **9**, 97–116 (1981). A general review with evidence for a possible diffuse hot gas in the universe. (I)

5.51. "**The Origin of the Cosmic X-Ray Background,**" B. Margon, Sci. Am. **248**, 104–119 (1982). The possible origins of the background radiation are explored with particular attention to the contribution of x-ray quasars. (E)

5.52. "**The X- and Gamma-Ray Backgrounds,**" A. C. Fabian, in *Tenth Texas Symposium on Relativistic Astrophysics*, edited by R. Ramaty and F. C. Jones [Ann. N.Y. Acad. Sci. **375**, 235–253 (1981)]. Fabian reviews the observations and explores the various possible contributions to the background radiation. (A)

5.53. "**A Medium Sensitivity X-Ray Survey Using the Einstein Observatory,**" T. Maccacaro *et al.*, Astrophys. J. **253**, 504–511 (1982). This study of the nature of moderately faint x-ray sources addresses the question of the total population of extragalactic x-ray sources visible above a given threshold flux. (A)

VI. HANDBOOKS, CATALOGS, FILMS, ETC.

A. Handbooks

6.1. **Astrophysical Quantities, 3rd ed.**, C. W. Allen (Althone, London, 1973). This is the most widely used and valuable reference in astronomy, but its coverage of x-ray related information is poor. (I)

6.2. **Astrophysical Formulae**, K. R. Lang (Springer-Verlag, New York, 1974). This impressive compendium covers all of astrophysics and includes some useful formulas relevant to x-ray astronomy. (I)

6.3. **Glossary of Astronomy and Astrophysics**, J. Hopkins (University of Chicago, Chicago, 1976). This slim volume might ease the way through the jargon of astrophysical literature. (E)

6.4. **High Energy Astrophysics Handbook**, M. Zombeck (Cambridge University, Cambridge, 1982). A wealth of formulas, tables, and graphs relevant to all high-energy astrophysics, with a lot of coverage of x rays. (I)

B. X-ray source catalogs

6.5. "The Fourth Uhuru Catalog of X-Ray Sources (4U)," W. Forman, C. Jones, L. Cominsky, P. Julien, S. Murray, G. Peters, H. Tananbaum, and R. Gaicconi, Astrophys. J. Suppl. **38**, 357–412 (1978).

6.6. "Positions and Identifications of Galactic X-Ray Sources," H. V. Bradt, R. E. Doxsey, and J. G. Jernigan, in *X-Ray Astronomy*, edited by W. A. Baity and L. E. Peterson (Pergamon, Oxford, 1979). See also Ref. 4.5.

6.7. "The MIT/OSO-7 Catalog of X-Ray Sources: Intensities, Spectra and Long Term Variability," T. H. Markert *et al.*, Astrophys. J. Suppl. **39**, 573–632 (1979).

6.8. "The Ariel V (3A) Catalogue of X-Ray Sources—I. Sources at Low Galactic Latitude. ($[b] < 10°$)," R. S. Warwick *et al.*, Mon. Not. R. Astron. Soc. **197**, 865–892 (1981).

6.9. "The Ariel V (3A) Catalog of X-Ray Sources—II. Sources at High Galactic Latitude ($[b] > 10$)," I. M. McHardy, A. Lawrence, J. P. Pye, and K. A. Pounds, Mon. Not. R. Astron. Soc. **197**, 893–920 (1981). References 6.5–6.10 are some of the major catalogs listing source name, position, x-ray flux, optical identification (if any), etc.

6.10. "HEAO A-2 Soft X-Ray Source Catalog," J. J. Nugent *et al.*, Astrophys. J. Suppl. **51**, 1–28 (1983).

C. Pictures, slides, film

6.11. **A Meeting with the Universe NASA EP-177,** edited by B. M. French and S. P. Maran (NASA, Washington, DC, 1981). Some nice x-ray related illustrations placed in the broader astronomical context are included in this large-format book. (E)

6.12. **Vision of "Einstein,"** slide collection, Photographic Services, Smithsonian Institution, Washington, DC 20560 (1981). Here are 57 slides covering all x-ray astronomy and including many of the exciting false-color images from *Einstein*. A booklet gives a fairly detailed explanation of each slide. (E)

6.13. **"Cosmic Fire,"** distributed by King Features (235 E. 45 St., New York, NY 10017). This WGBH "Nova" film is a rework of an earlier BBC production. It touches on many of the major problems in x-ray astronomy. Some of the active practitioners in the field explain what they do, how they do it, and why. (E)

ACKNOWLEDGMENTS

I thank Tom Markert for his suggestions and Nancy Ferrari and Lisa Magnano for their expert preparation of the manuscript. I also thank Haldan Cohn for his helpful comments.

PHYSICAL REVIEW LETTERS

VOLUME 9   DECEMBER 1, 1962   NUMBER 11

EVIDENCE FOR X RAYS FROM SOURCES OUTSIDE THE SOLAR SYSTEM*

Riccardo Giacconi, Herbert Gursky, and Frank R. Paolini
American Science and Engineering, Inc., Cambridge, Massachusetts

and

Bruno B. Rossi
Massachusetts Institute of Technology, Cambridge, Massachusetts
(Received October 12, 1962)

Data from an Aerobee rocket carrying a payload consisting of three large area Geiger counters have revealed a considerable flux of radiation in the night sky that has been identified as consisting of soft x rays.

The entrance aperture of each Geiger counter consisted of seven individual mica windows comprising 20 cm² of area placed into one face of the counter. Two of the counters had windows of about 0.2-mil mica, and one counter had windows of 1.0-mil mica. The sensitivity of these detectors for x rays was between 2 and 8 Å, falling sharply at the extremes due to the transmission of the filling gas and the opacity of the windows, respectively. The mica was coated with lampblack to prevent ultraviolet light transmission. The three detectors were disposed symmetrically around the longitudinal axis of the rocket, the normal to each detector making an angle of 55° to that axis. Thus, during flight, the normal to the detectors swept through the sky, at a rate determined by the rotation of the rocket, forming a cone of 55° with respect to the longitudinal axis. No mechanical collimation was used to limit the field of view of the detectors. Also included in the payload was an optical aspect system similar to one developed by Kupperian and Kreplin.[1] The axes of the optical sensors were normal to the longitudinal axis of the rocket. Each Geiger counter was placed in a well formed by an anticoincidence scintillation counter designed to reduce the cosmic-ray background. The experiment was intended to study fluorescence x rays produced on the lunar surface by x rays from the sun and to explore the night sky for other possible sources. On the basis of the known flux of solar x rays, we had estimated a flux from the moon of about 0.1 to 1 photon cm^{-2} sec^{-1} in the region of sensitivity of the counter.

The rocket launching took place at the White Sands Missile Range, New Mexico, at 2359 MST on June 18, 1962. The moon was one day past full and was in the sky about 20° east of south and 35° above the horizon. The rocket reached a maximum altitude of 225 km and was above 80 km for a total of 350 seconds. The vehicle traveled almost due north for a distance of 120 km. Two of the Geiger counters functioned properly during the flight; the third counter apparently arced sporadically and was disregarded in the analysis. The optical aspect system functioned correctly. The rocket was spinning at 2.0 rps around the longitudinal axis. From the optical sensor data it is known that the spin axis of the rocket did not deviate from the vertical by more than 3°; for purposes of analysis, the spin axis is taken as pointing to zenith. The angle of rotation of the rocket corresponds with the azimuth

439

Φ and is measured from north as zero and increasing to the east. The data were reduced by using the optical aspect information to determine the azimuth as a function of time. Each complete rotation of the rocket was divided into sixty equal intervals, and the number of counts in each of these intervals was recorded separately.

The total data accumulated in this manner during the entire flight are shown in Fig. 1 for the operating Geiger counters. The observed region of the sky is shown in Fig. 2. The counting rates show an altitude dependence on both the ascending and descending portions of the flight. These are shown in Fig. 3, the numbers representing three-second sums. The rocket had begun tumbling during descent. The data in that portion of the flight are difficult to interpret and have not been included in the analysis.

The residual cosmic-ray background could not be determined directly. However, the strong angular dependence of the counting rate and the large difference between the counting rates of the counters provided with windows of different thickness clearly show that most of the recorded counts are due to a strongly anisotropic and very soft radiation. Thus, the possible existence of a small cosmic-ray effect is not an essential element in the discussion of the results.

The large peak that appears at about 195° in both counters shows that part of the recorded radiation is in the form of a well collimated beam. The fact that the counting rate does not go to zero on either side of the peak shows that this beam is superimposed on a diffuse background radiation. The background radiation itself is not isotropic, but appears to have a higher intensity in directions to the east of the peak than in the direction to the west of the peak, suggesting a secondary maximum centered around 60°. The statistical significance of this conclusion may be evaluated by comparing the total number of counts recorded by counter No. 2 in an angular interval east of the maximum (from Φ = 102° to Φ = 18°) with that recorded in an equivalent angular interval west of the maximum (Φ = 282° to

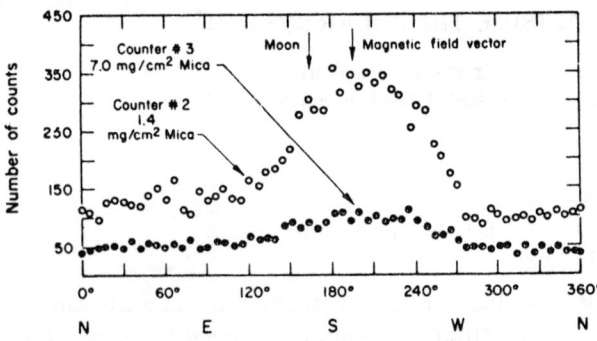

FIG. 1. Number of counts versus azimuth angle. The numbers represent counts accumulated in 350 seconds in each 6° angular interval.

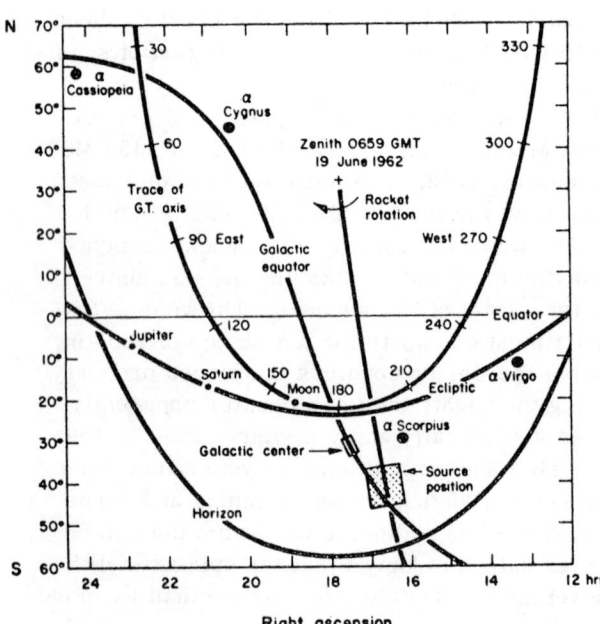

FIG. 2. Chart showing the portion of sky explored by the counters.

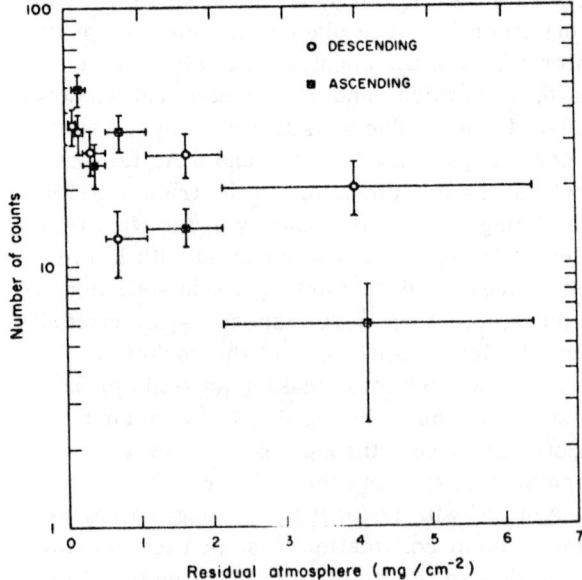

FIG. 3. Counting rate versus residual atmosphere. The numbers represent counts accumulated in counter No. 2 in three-second intervals for azimuth angles from 102° to 282°.

$\Phi = 6°$). The two numbers are 2005 and 1582, respectively, yielding a difference of 423 ± 60 counts. A similar excess, although statistically less significant, appears in counter No. 3 and is 90 ± 40 counts.

At the location where the measurements were obtained, the magnetic field has an inclination of 63° and a declination of 13° east of north. Thus, the field lines are at an azimuth of 193°, which is about the same as the azimuth of the observed radiation peak. This coincidence makes one wonder whether the radiation might not consist of charged particles spiraling along the field lines. On the basis of the minimum energy necessary for the penetration of the thin- and thick-window counters, the radiation would have to consist of electrons with energies of the order of several tens of keV, or protons with energies of the order of one MeV. On the other hand, it would be unlikely that protons form the main component of the observed radiation, considering that they must possess a much higher energy than electrons in order to penetrate the windows. In any event, both protons and electrons are strongly deflected by the earth's magnetic field and must exhibit axial symmetry with respect to the magnetic field; i.e., at a given pitch angle the flux of particles must be independent of azimuth around the field. The magnetic field makes an angle of 27° with the spin axis of the rocket and the detector axis makes an angle of 55° with the spin axis. Hence, particles moving with pitch angles between 28° and 82° would be normal to the detector axis at two different roll angles and would tend to give a double peaked distribution of counts. Particles with a 90° pitch angle would be detected with maximum efficiency from the north. Thus, the sharpness and azimuth of the observed peak requires that the pitch angles of these particles be very small. It is hard to find a resonable source for particles with the required small pitch angles at the location of our measurements. In particular, particles spilling out of the inner radiation belt ought to have a broad pitch-angle distribution. One may add that it is not easy to account for the sharpness of the observed peak even under the extreme assumption that all particles had a zero pitch angle, i.e., came as a parallel beam in the direction of the field lines. The shape of the peak depends on the absorption curve of this radiation. The curve in Fig. 4 represents the counting rate as a function of roll angle for a beam of particles with zero pitch angle computed under the assumption of an exponential absorption, with an absorption coef-

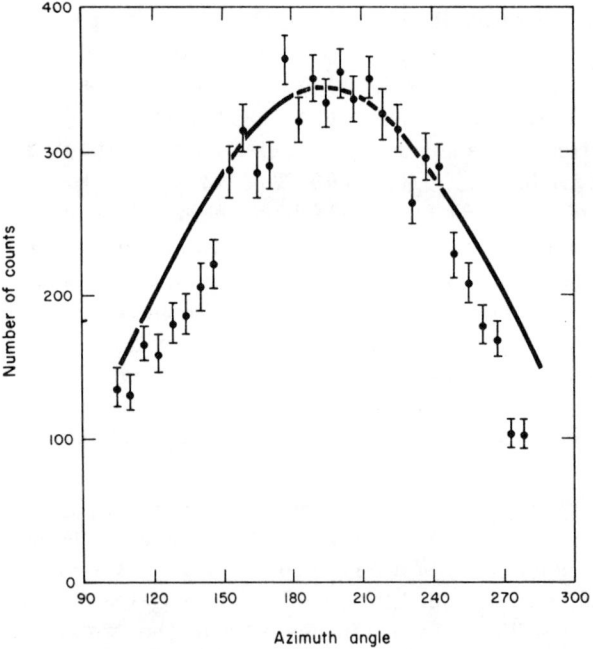

FIG. 4. Comparison of experimental results with the computed angular dependence for a unidirectional beam of electrons exhibiting exponential absorption in the counter window.

ficient consistent with the relative responses of the thick-window and thin-window counters. The observed angular dependence is not consistent with the calculated curve, and it is difficult to ascribe the discrepancy to the arbitrary assumption of an exponential absorption. Moreover, it is clear that the presence of particles with finite pitch angles would broaden the predicted distribution.

It is also clear that the radiation responsible for the asymmetry of the background cannot consist of charged particles. Thus, we conclude that the bulk of the observed radiation is not corpuscular, but electromagnetic in nature.

The counters were so constructed as to be insensitive to visible or ultraviolet light. The data themselves provide a definite test on this point since a strong visible light source, the moon, and two comparatively strong ultraviolet light sources, Virgo and presumably the moon, went through the field of view of the counters, and yet were not detected. Thus, if the radiation is electromagnetic, it must consist of soft x rays.

Consider first the radiation responsible for the peak and assume that it originates from a point source, and that it is monochromatic with a wavelength λ. The difference in counting rates observed in the two counters with the two different mica windows depends on the absorption coefficient $\mu_{mica}(\lambda)$ for the given radiation and on the minimum angle θ formed during the rotation of the rocket by the normal to each counter with the direction of the source. The experimental data give the product $\mu_{mica}(\lambda)\sec\theta$. The variation of the counting rate with altitude depends on the absorption coefficient $\mu_{air}(\lambda)$ for the given radiation and on the angle ψ between the horizon and the source (ψ and θ are related by a numerical constant), and yields the product $\mu_{air}(\lambda)\sec\psi$. By comparing these two pieces of information it is possible to determine values of λ and ψ for which both relations are satisfied. One obtains a λ of about 3 Å and a ψ of about 10°.

The shape of the peak also depends on $\mu_{mica}(\lambda)$ and ψ. Use of the previously determined λ, under the assumption of a point source, allows a semi-independent determination of ψ which yields a value of 20°. The difference between the two values of ψ can be accounted for by a source width of approximately 10°. Thus, the peak appears to be due to a source emitting x rays of a 3Å wavelength whose origin is about 10° above the horizon. The location of this source is shown as "source position" on the sky map in Fig. 2. The measured flux from this source is 5.0 photons cm^{-2} sec^{-1}.

The diffuse character of the observed background radiation does not permit a positive determination of its nature and origin. However, the apparent absorption coefficient in mica and the altitude dependence is consistent with radiation of about the same wavelength as that responsible for the peak. Assuming the source lies close to the axis of the detectors, one obtains the intensity of the x-ray background as 1.7 photons cm^{-2} sec^{-1} sr^{-1} and of the secondary maximum (between 102° and 18°) as 0.6 photon cm^{-2} sec^{-1}. In addition, there seems to be a hard component to the background of about 0.5 cm^{-2} sec^{-1} sr^{-1} which does not show an altitude dependence and which is not eliminated by the anticoincidence.

The question arises whether the source of the observed x radiation could be associated with the earth's atmosphere and ascribed to some form of auroral activity. The rarity of occurrence of auroras of the magnitude required to account for the observed intensities at the latitude of the measurement makes this possibility very unlikely.

In addition, the following comments are apropos. The bulk of the measurements were obtained between altitudes of 100 to 225 km and over a range of distances from the firing point of the rocket of 15 to 100 km. The variation of the measured intensity within these limits is consistent with a source at infinity. A lower limit on the position of the source places it at a distance greater than 1000 km. This number, combined with the measured elevation angle of 10°, places a lower limit on the altitude of the source which is about 400 km above the earth's surface. Auroral electrons of energy sufficient to produce the observed radiation would, on the other hand, penetrate down to an altitude of about 100 km and would expend the bulk of their energy at altitudes lower than 200 km. We conclude that the hypothesis of an auroral source for the observed radiation is not consistent with the data.

From Fig. 2, showing the locations of the source as well as of the moon and planets, it is clear that the observed source does not coincide with any obvious scattering body belonging to our solar system. Further, the intensity of solar x radiation at the observed wavelength is much too low in this period of the solar cycle to account for the observed intensities of the peak or of the background on the basis of back-scattered solar radiation. It would thus appear that the radiation does not originate in our solar system.

From Fig. 2 we see that the main apparent source is in the vicinity of the galactic center at a G.T. azimuthal angle of about 195°. We also see that the trace of the G.T. axis lies close to the galactic equator for a value of the azimuthal angle near 40°, which is the region where the background radiation is recorded with greater intensity. This apparent maximum of the background radiation is the general region of the sky where two peculiar objects—Cassiopeia A and Cygnus A—are located. It is perhaps significant that both the center of the galaxy where the main apparent source of x rays lies, and the region of Cassiopeia A and Cygnus A where there appears to be a secondary x-ray source, are also regions of strong radio emission. Clark[2] has pointed out that the probable mechanism for the production of the nonthermal component of the radio noise, namely, synchrotron radiation from cosmic electrons in the galactic magnetic fields, can also give rise to the x rays we observe.

In the cosmic-ray air shower experiment presently being carried out in Bolivia,[3] tentative evidence has been obtained for the existence of cosmic γ rays in the energy region of 10^{14} eV at a rate of 10^{-3}-10^{-4} of the charged cosmic-ray flux at the same energy with an indication of enhanced emission in the galactic plane. Clark has shown that cosmic electrons must be produced along with γ-rays by the decay of mesons that arise in the interactions of cosmic rays with interstellar matter. Since electrons at these energies lose their energy predominantly via synchrotron radiation in the galactic magnetic field, one should observe roughly the same total energy in synchrotron radiation at the earth as in γ-ray energy. For electrons of 2×10^{14} eV in a field of 3×10^{-6} gauss, the peak of the synchrotron emission is at 3 Å; in a stronger field this will happen at lower electron energies. It has been shown[4] that x rays in this wavelength region are not appreciably absorbed over interstellar distances.

With this one experiment it is impossible to completely define the nature and origin of the radiation we have observed. Even though the statistical precision of the measurement is high, the numerical values for the derived quantities and angles are subject to large variation depending on the choice of assumptions. However, we believe that the data can best be explained by identifying the bulk of the radiation as soft x rays from sources outside the solar system. Synchrotron radiation by cosmic electrons is a possible mechanism for the production of these x rays.

Ordinary stellar sources could also contribute a considerable fraction of the observed radiation.

We gratefully acknowledge the continued encouragement and technical assistance for this program by Dr. John W. Salisbury, Chief, Lunar and Planetary Exploration Branch of Air Force Cambridge Research Laboratories (AFCRL), and by personnel from both AFCRL and the White Sands Missile Range.

---

*The research reported in this paper was sponsored by the Air Force Cambridge Research Laboratories, Office of Aerospace Research, under Contract AF 19 (604)-8026.

[1] J. E. Kupperian and R. W. Kreplin, Rev. Sci. Instr. 28, 19 (1957).

[2] G. W. Clark (to be published).

[3] Joint Program, MIT, University of Tokyo, University of Michigan, and University at La Paz. The MIT Cosmic Ray Group has graciously made available to us the preliminary results of the air shower experiment. We have also profited considerably from discussions with Professor M. Oda of the University of Tokyo and with Professor G. W. Clark of MIT.

[4] S. E. Strom and K. M. Strom, Publs. Astron. Soc. Pacific 73, 43 (1961).

physics today

MAY 1973 VOL 26 NO 5

Reprinted with permission from *Physics Today*, 37 (5), pp. 38-47, "Progress in X-ray astronomy," R. Giacconi. © 1973 American Institute of Physics.

Progress in x-ray astronomy

Recent satellite studies of celestial x-ray sources have enabled some binary-system components to be identified as white dwarfs, neutron stars, and in one case a black hole.

Riccardo Giacconi

The most significant progress in x-ray astronomy in the past few years has been brought about by the advent of satellite observatories and by the great number of new radio and optical identifications of cosmic x-ray sources. Since the discovery, with rocket-borne instruments, of extrasolar sources of x radiation ten years ago, it has been clear to most experimenters that a very considerable advance in our knowledge could be obtained with satellite instrumentation. The launch on 12 December 1970 of the first small astronomy satellite, UHURU, entirely devoted to x-ray observations, was expected to lead to the detection of fainter sources with finer angular resolution and positional accuracy, thereby expanding our catalog of celestial x-ray emitters. What was unexpected was the qualitative change in our understanding of the nature of cosmic x-ray emitting objects that several new discoveries by UHURU have brought about.

The discovery of pulsating x-ray sources in binary systems has offered us the first clear evidence of the role of accretion as the energy source for the generation of the very large x-ray fluxes observed from stellar objects. The presence in these systems of white dwarfs, neutron stars or black holes gives us a most fortunate astrophysical laboratory in which to study their properties. Evidence is becoming

Riccardo Giacconi is executive vice-president of American Science and Engineering, Inc., of Cambridge, Massachusetts.

more convincing, almost on a weekly basis, in favor of the interpretation of the Cygnus X-1 in terms of a black hole orbiting a massive companion.

In extragalactic x-ray research the increase in the number of observed sources has been so great that for the first time an attempt at classification of different types of extragalactic emitters is possible, as is an evaluation of their contribution to the isotropic x-ray background. The detection of stellar sources in external galaxies of the local group has established an unequivocal value for the intrinsic x-ray luminosities of stellar objects. Detection of diffused emission from clusters of galaxies furnishes important clues on the dynamics of the clusters and on the existence of intracluster gas, and it may provide a very powerful tool with which to investigate cosmological questions.

As the scope of x-ray astronomy expands and the observations yield a greater and greater wealth of experimental data, it becomes impossible in a short review to give a comprehensive description of the experiments and of the results. I will base most of this article on results obtained from UHURU, with a sprinkling of data from the two satellite experiments by MIT and La Jolla on the OSO-7 satellite and some from rocket flights.

I will not attempt to describe results obtained on x-rays of lower energies than observable from UHURU, about 1-20 keV, except as they pertain to the discussion of particular objects. This is not due to any lack of progress in the field, but rather to the fact that the sky appears quite different below and above 1 keV. At lower energies the diffused background appears to be dominated by galactic contributions, in net contrast to what we believe to be the case at higher energies. Also the individual sources observed, such as extended supernova remnants, appear in general to be characterized by lower temperatures than are observed for compact stellar sources and are, therefore, unobservable at higher energies. The production mechanisms giving rise to the observed radiation also appear to be different in the two regions of the spectrum. Finally, with the exception of old supernova remnants, there is no evidence for the existence of sources below 1 keV, not observed at higher x-ray energies.

The night sky in x rays

Before discussing individual properties of x-ray sources, let us consider some of their general properties—such as number-count, ranges of intrinsic luminosity and spatial distribution in b and l (galactic longitude and latitude). UHURU has given us the most complete catalog of x-ray sources in the 2–20 keV range. The 2U Catalog lists 125 sources detected during the first 70 days of operation of the satellite, with the coverage of approximately half of the sky. The intensity of the observed sources ranges from 2×10^4 counts per second for Sco X-1 to 2 cts/sec for M-31, as observed by the UHURU in-

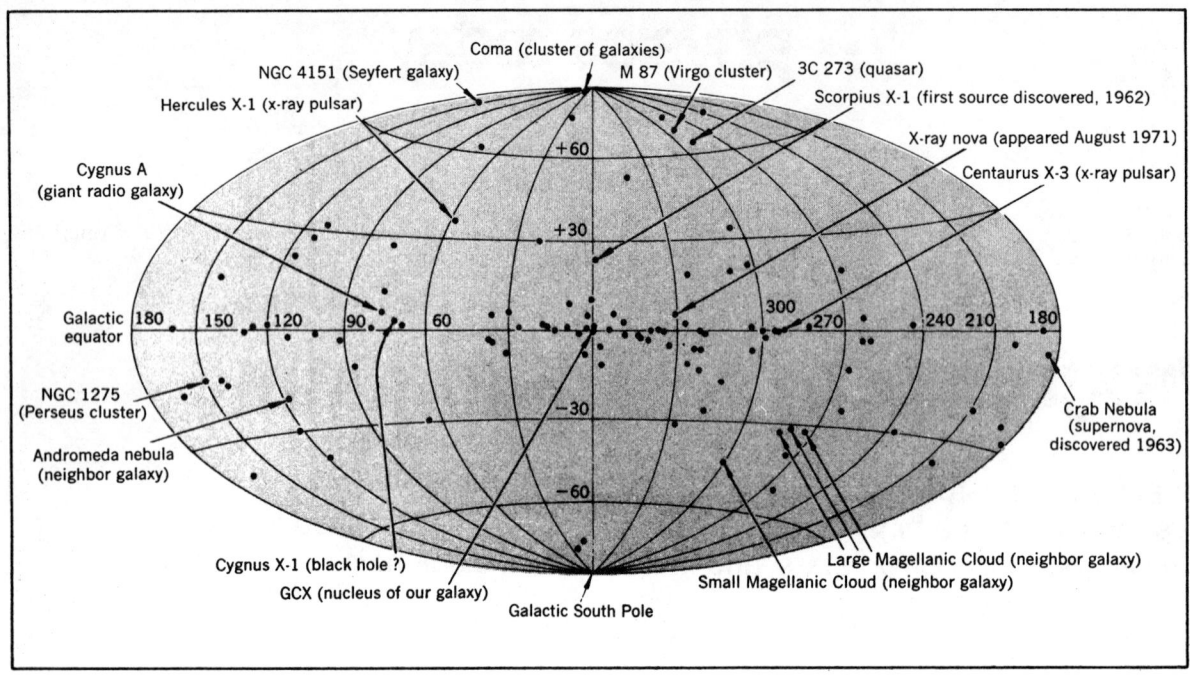

The x-ray sky as seen by the UHURU satellite. The dots indicate x-ray sources in the 2U catalog. This is an equal-area projection in galactic coordinates. Figure 1

struments in the 2 to 6 keV range. A map of the sky in galactic coordinates containing all the 2U sources is shown in figure 1.

The sources divide naturally into two groups: those that are clustered along the galactic plane ($|b| < 20$ deg) and those at high latitude ($|b| > 20$ deg). Sources along the galactic plane are on the average much more intense than those at high latitude. This is just the opposite to what one would expect for a single galactic population. If the latitude of a source was taken as indicative of distance, high-latitude sources should be nearby and brighter than those in the plane, as is the case for common stars. In fact, we shall see that almost all the low-latitude sources are galactic, whereas almost all the high-latitude sources are extragalactic. (Of 44 sources with $|b| > 20$ deg, 14 have been identified with extragalactic objects and two with strong nearby galactic objects: Sco X-1 and Her X-1.)

Using the 2U Catalog and taking into account in detail the fraction of the sky surveyed at each sensitivity, we can construct a log N–log S distribution for the two classes of sources, as shown in figure 2. The qualitative difference we observe between the distribution of intensity in sources at low and high latitude is striking. We have investigated the possibility that it might be due to experimental biases, and we find that the possible corrections one could estimate would not be sufficient to alter this effect significantly.

Examining in detail the distribution in galactic coordinates (as well as the number count) of the sources clustered along the galactic equator, we find that we observe two kinds of galactic sources. These are a small number of sources in the central region of our galaxy with very large intrinsic luminosity (10^{38} erg sec^{-1}) and a large number of less intense sources (10^{36} erg sec^{-1}) disposed along the spiral arms of our galaxy. The bulk of the emission from our galaxy comes from the integrated emission of only a dozen very powerful sources. One hundred or so more sources contribute about 10% of the total. About two thirds of all existing sources in the galaxy with intrinsic luminosity above 10^{36} erg sec^{-1} have already been observed.

The 2U Catalog lists 44 sources at high latitude $b > 20$ deg. Of these, two have high intensity (2×10^4 cts/sec for Sco X-1, and 10^2 cts/sec for Her X-1) and are identified with galactic objects. Five are identified with individual stars in the Large and Small Magellanic Clouds at about 20 cts/sec. Ten with intensity ranging from 20 cts/sec (Virgo Cluster M-87) to 2 cts/sec (M-31) are identified with individual galaxies or clusters, and the remainder is unidentified.

Because all of the weak sources that have been identified have been found to be extragalactic, we are led to question whether the unidentified sources with $b > 20$ deg might also be extragalactic. Examining their l and b distributions, we note that they are consistent with isotropy. This result could be reconciled with galactic origin only if the sources were members of a halo population, or were very nearby objects. The first possibility is excluded by the fact that we observe no excess in the direction of the galactic center; the second, by the fact that we do not observe an intense galactic ridge. Unless we invent an *ad hoc* galactic population with appropriate intrinsic luminosity and distribution, we are led to conclude that all high-latitude sources are extragalactic.

This view is confirmed by the number-count distribution for these sources, which is consistent with what we expect for isotropically distributed extragalactic sources in a static Euclidian Universe.

Galaxies of the local group

In normal galaxies, including our own, the x-ray emission appears to be dominated by the integrated output of individual stellar sources. The x-ray luminosity (10^{39} erg/sec) appears to be a small fraction of the total light emitted at all wavelengths $L_x/L_{rad} = 10^{-5}$. In all other identified extragalactic sources the total energy emitted in x rays is much larger in absolute value, ranging from 10^{42} to 10^{46} erg/sec in the 1 to 10 keV range, and is also larger relative to the total emission at all wavelengths.

In figure 3 a summary of the observed classes of x-ray emitting objects is given. On the left are distances in megaparsecs. Plotted in each class are

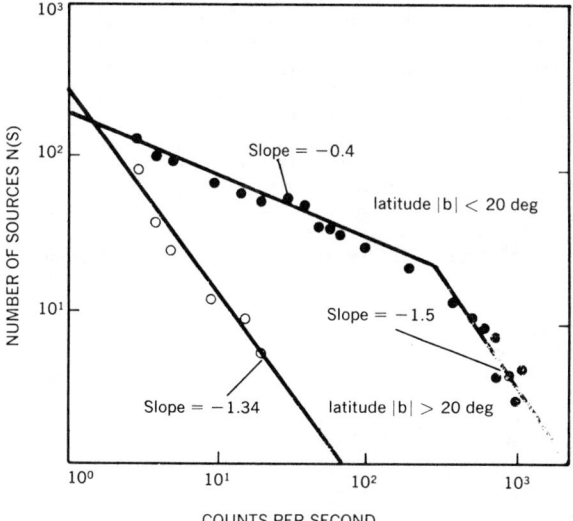

Number counts of x-ray sources in the 2–6 keV energy range plotted as a function on intensity. The low-latitude sources, b <20 deg. are shown as dots; high-latitude sources, b >20 deg, are shown as open circles. Also shown are the best-fit straight lines. All data are corrected for non-uniform sky coverage. The scale on the right-hand ordinate, N(S) per steradian, applies only to the high-latitude sources.   Figure 2

Identified extragalactic x-ray sources. The top of the colored region for each class indicates the distance of the nearest known member of that class. The Seyfert galaxies are the only group for which the nearest member, NGC 4051, is not detected, but a more distant member is detected. The table (inset) shows the ratio of x-ray luminosity to optical luminosity, L_x/L_v, for nine representative identified sources.   Figure 3

the observed objects at their distance; the color tint is extended to the distance of the nearest known object in each class. We note that we observe close to the nearest or brightest member in each class. The identified extragalactic sources include Seyfert galaxies, QSO's, giant radio galaxies and clusters.

The x-ray data available on each source are the location, the energy spectrum and the angular size. We find an important distinction in the emission characteristics of individual extragalactic objects and clusters of galaxies. For sources identified with individual galaxies we measure a definite low-energy cutoff, whereas only upper limits (consistent with no cutoff) are observed in cluster sources. In Table 1 we summarize these results, and also we list the measured angular sizes for the different sources. We note that for all sources identified with individual galaxies we observe spectra that are cut off at low energy, and these sources have small angular size. For cluster sources we observe no cutoff and extended size.

In the case of 4151 the stated limit of 15 arc minutes corresponds to about 50-kiloparsecs diameter at the distance of 4151. In fact, the emitting region may be much more compact. The very large cutoff, which corresponds to a column density of absorbing matter in excess of 10^{23} atoms cm^{-2}, could be taken to indicate that the x rays are coming from a small region at the nucleus of NGC 4151. In fact, if we assume that the source is as large as the computed upper limit, then the total mass one can derive for the absorbing gas is prohibitively large (3.5 × $10^{11} M_\odot$). Similar considerations apply in the case of NGC 5128 and 3C 273. Figure 4 shows the observed location of the source in NGC 5128. It is clear that the emission is concentrated in a region not much larger than the optical galaxy.

We conclude that the x rays we observe from these individual galaxies are emitted in small regions near the nucleus of the galaxy. Presumably they are not due to the collective emission of a number of individual stars because the x-ray output is 10^3 times larger than in normal galaxies and the L_x/L_v ratio is also much larger. At present we do not have sufficient data to distinguish between thermal bremsstrahlung emission and synchrotron or inverse Compton emission processes for the emission occurring in individual galaxies.

The detection of x-ray emission from extended regions in clusters of galaxies is potentially one of the most important new results from UHURU. We have identified about ten sources associated with clusters. In figure 5 the positional information of the x-ray source region for the clusters in Virgo, Centaurus, Perseus and Coma is shown. We find that the center of emission is centered on active galaxies when present, as in Virgo and Perseus. They have great physical size ($\approx$1 Mpc). The x-ray emission appears to be strongly dependent upon cluster richness, and for richness class 2 we observe a narrow range of luminosities

Table 1. Sizes and X-Ray Luminosities of Extragalactic X-Ray Sources

Source	Size Angular	Size Kiloparsec	Luminosity (erg/sec)	Low-energy cut-off (keV)
Abell 2256	35′ ± 15′	2800	5 × 10^{44}	≤1.7
Perseus—NGC 1275	35′ ± 3′	740	3 × 10^{44}	≤1.1
Coma	36′ ± 4′	1050	2 × 10^{44}	≤1.0
Cen—NGC 4696	37′ ± 8′	500	2 × 10^{43}	≤1.9
Virgo—M 87	50′ ± 5′	200	7 × 10^{42}	≤1.0
NGC 4151	≤15′	≤60	1 × 10^{42}	5.7 ± 1.1
NGC 5128	≤10′	≤20	6 × 10^{41}	3.0 ± 0.4
GCX*	≈2°	≈0.3	≈10^{37}	2.8 ± 0.2
3C 273	?	?	2.7 × 10^{45}	2.6 ± 0.7

* Extended source at the center of our own galaxy, included for comparison

between 10^{44} and 10^{45} erg/sec. They are, therefore, much more luminous than individual galaxies.

The absence of cutoffs in their spectra seems to require that they are truly diffused rather than a collection of individual sources. The spectral information available from UHURU alone is insufficient to distinguish between possible emission mechanisms. In particular, one cannot distinguish between a power-law spectrum and an exponential spectrum. However, rocket observations of Coma and Virgo at long wavelength (10–20 Å) appear to favor a turnover at low energy, as would be expected for an exponential thermal bremsstrahlung spectrum.[1,2] These results also appear to indicate the presence of a more complex structure in the emission region of Coma and Virgo than is indicated by our simple fit to a uniformly illuminated extended source. In particular, there appears to be evidence for a region of small angular size containing about one third of the total flux centered on the active galaxy.

If the mechanism for emission is thermal bremsstrahlung from a hot intercluster gas, we find that the gas would have a total mass comparable to the mass contained in the observed galaxies and a temperature of some 70 × 10^6 K. This gas could be heated by energy output from active galaxies. It might be that the x-ray emission is not simply related to cluster richness, but to specific parameters describing the dynamics of the clusters. A. Solinger and W. Tucker[3] have suggested, for example, that the x-ray emission be related to the fourth-power velocity dispersion of the galaxies of the cluster and although not proven, this suggestion appears consistent with all available data.

It is to be expected that further study of the powerful extended sources in clusters will tell us much about the evolution of clusters and perhaps about cosmological phenomena. The rich clusters, being very luminous and of great angular size, will be identifiable out to distances where redshift effects dominate. We might be able, with the advent of large orbiting x-ray observatories, to detect clusters to $Z = 3$ and learn about their evolution in the early stages of the formation of the Universe.

Galactic sources

The major new results in the study of galactic x-ray sources have been produced by the refinement in our ability to measure positions in the sky, which has led to several new identifications with radio and optical objects, and by the opportunity afforded by satellite observatories to monitor with high time resolution (0.1 sec) the x-ray flux of specific sources for periods of time spanning intervals of up to two years.

Of the approximately 65 galactic sources with intensity greater than 8 cts/sec (4 × 10^{-4} Sco) in the energy range 2–8 keV, approximately 60% have positional errors in the 2U Catalog of less than 0.02 deg^2 and some 20% have positional errors less than 0.003 deg^2. While this is not better than the best that could be done for the strongest individual sources from a modulation collimator rocket experiment,[4] the extension of these precise measurements to many more and weaker sources has led to significant new results.

With respect to time variation, we had considerable indications prior to the UHURU flight for the existence of significant time variability in several of the galactic sources.[5] However, the short glimpses obtained with different payloads always raised the question of instrumental effects and did not give the characteristic time scale for the variation.

Of the 65 galactic sources, seven can be identified with supernova remnants, and their flux averaged over a second or longer is not observed to vary, although, of course, NP0532 is known to exhibit x-ray pulsations at 33 m/sec. Of the remainder, more than 60% have been observed to vary on the time scales of 0.1 sec or longer. In only five cases, of the 21 where no variation was observed, was the intensity greater than 10^{-3} Sco X-1; thus, for several it might simply be that we have not yet achieved sufficient precision to determine their variability. The precision of the available locations and this general prevalence of short-time variability excludes the possibility that most of the galactic x-ray sources may be connected with extended supernova remnants. At most, 20% of them may ultimately fall into this category. The remainder must be associated with compact objects.

Considerable progress has been made in the past few years in the study of x-ray emission connected with supernova remnants, particularly with rocket measurements at long wavelengths. The energy source has been determined in the case of Crab to reside, probably, in the rotational energy of the pulsar, which is observed to slow down at a rate sufficient to provide the required 10^{38} erg/sec of energy. A particularly important experiment by Robert Novick[6] has established the existence of polarization of the x-ray radiation, thereby strongly supporting the view that x rays are emitted *via* synchrotron radiation from energetic electrons. The x-ray emission from the diffused nebulae surrounding the older supernovas can apparently be explained as thermal bremsstrahlung emission from gas heated by the transformation of kinetic energy of the shell ejected by the supernova explosion. Satellite experi-

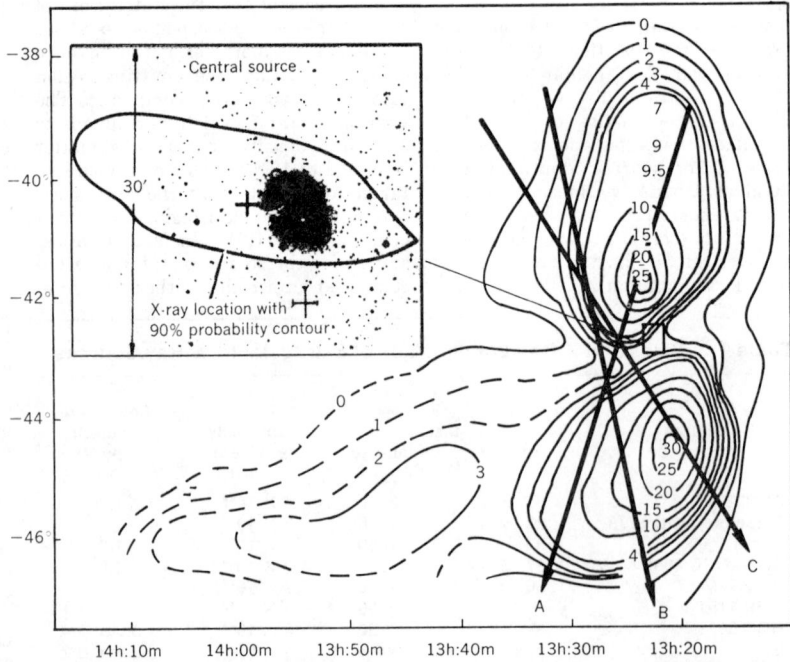

Centaurus A. The optical galaxy NGC 5128 is located within the x-ray source error box. The size of the source is less than 10 arc min. This still allows the x-ray emission to be coming from the inner radio lobes shown in the inset, or from the optical galaxy. The nucleus of the optical galaxy is the most likely location of the source, because it is so strongly cut off at low x-ray energies.

Figure 4

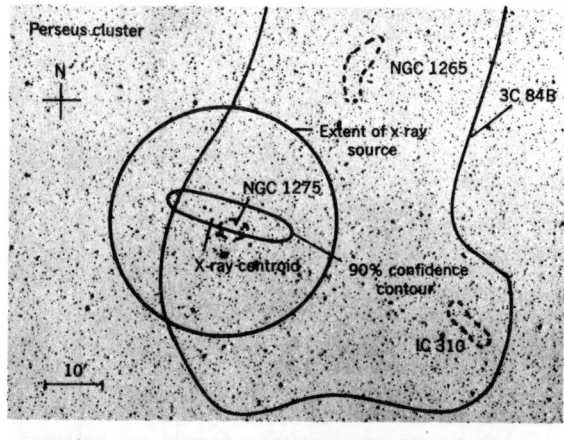
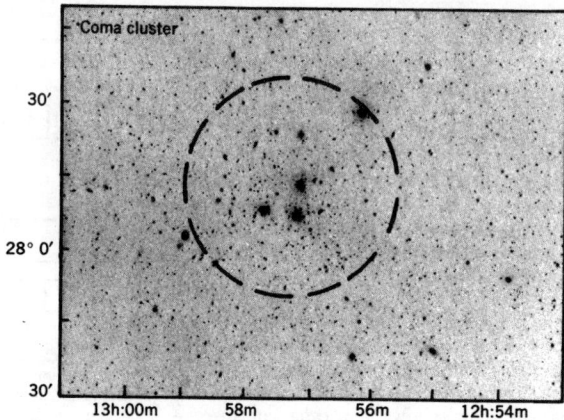
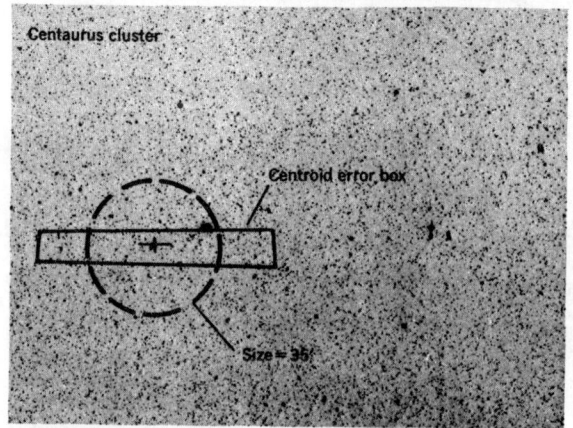
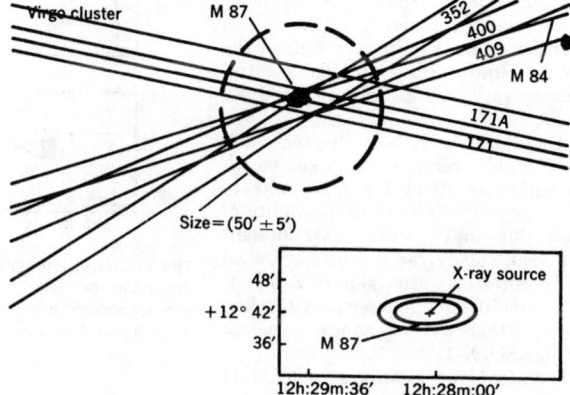

The four strongest x-ray sources in clusters of galaxies. In each case, the extent of the source is indicated by the simplest figure, a circle, since we have no information on the shape. The Virgo picture also shows the one-directional locations for the centroid of the extended source obtained from five scans, labelled by the orbit number pertaining to each scan. The inset is the two-dimensional error box obtained by combining the five one-directional locations. The inner contour corresponds to 68% confidence, and the outer contour is for 90% confidence. The Centaurus source, 2U 1247-41, contains NGC 4696 within its centroid error box. The Perseus source contains NGC 1275, the exploding Seyfert galaxy, within the centroid error box. In Coma, both the giant elliptical NGC 4874 and the kinematic center of the cluster lie within the error box. Figure 5

ments have contributed very slightly to the development in the field; therefore, I will concentrate on the new results obtained on compact x-ray sources.

As outlined above, the majority of the sources in our galaxy are characterized by the following properties:
▶ Their luminosity is the range 10^{36}–10^{38} ergs/sec.
▶ The number in the galaxy is exceedingly small; perhaps no more than 100, of which about half have been observed.
▶ The sources in the outer regions of the galaxy are found in the spiral arms.
▶ The sources exhibit large (>2) variability in the 2–20 keV range with time scales ranging from 0.1 seconds to months.
▶ The excitation process by which x rays are emitted favors the production of 2–10 keV photons, or most of the sources are embedded in an absorbing envelope with column density of order of 10^{22} atoms/cm² of hydrogen. These last statements stem from the fact that no x-ray source is observed only at energies greater than 20 keV. The flux at $E > 20$ keV is much smaller than in the 1–20 keV region. At most one or two sources are observed only at energies less than 1 keV, a result that cannot be explained on the basis of interstellar absorption.

The preceding comments indicate that the compact x-ray sources are among the most luminous objects in our galaxy and that they are exceedingly rare or represent short-lived x-ray emitting phases in stellar evolution. Thus, one can conceive of very exotic circumstances that could furnish the energy source and the production mechanism for the observed x-ray fluxes.

In several of the sources, however, we are as yet unable to determine the nature of the energy input and of the acceleration or heating mechanism. This includes not only weak unidentified sources for which data may be insufficient, but also some of the strongest and best identified such as Sco X-1, Cyg X-2, GX 17 + 2, and so on.

A notable exception exists, however, in this puzzling situation. One of the most important recent results from UHURU is the discovery of x-ray sources in the binary system. Although we may not yet be able to understand completely all the details of these systems, we have an energy source, accretion, and well defined ideas as to the dimensions and nature of the systems themselves.

Sco X-1-like sources

Sco X-1 was the first x-ray source observed. It was correctly identified in 1966 from positional data only. The optical Sco X-1 has been subject to intensive study since its discovery, as has been reviewed by W. H. Hiltner.[7] As part of the UHURU observing program, more than one week was devoted to correlated x-ray optical studies of this source.

An important characteristic has been

revealed in the infrared region[8]; namely, that the flux density falls off as expected from free-free absorption in the emitting region, from which the size of the emitting region can be inferred as between 10^4–10^5 km, depending on the distance to the object. The observed absence of x-ray line emission is consistent with an object of the size; the lines become broadened and weakened by electron scattering.[9]

Optically, Sco X-1 varies between 12–13 magnitude. The intensity varies by a few percent on a time scale of minutes (flickering) and flares by a factor of two on a time scale of hours. The spectrum shows a strong continuum plus a number of high excitation emission lines. There is no evidence for a periodic component in the optical intensity or in the Doppler shift of the spectral lines, although the spectral lines do vary in both intensity and position.

In radio, Sco X-1 is the seat of a faint, highly variable radio source radiating from about 0.2 flux units to below present limits of detectability of 0.005 flux units. There are present two weak steady radio sources, several arc minutes on either side of Sco X-1. To a precision of arc seconds the line joining these two sources passes through Sco X-1.[10]

In spite of an enormous amount of observational material, the only information available is on the size of the x-ray emitting region; there is no information on the mass or dimensions of the system.

Cyg X-2 is believed to be similar to Sco X-1, but there is not nearly so much information available concerning it. Its intensity is only about 1/40 of Sco X-1. The optical identification was proposed in 1967,[11] based on the discovery of a star with characteristics similar to that of Sco X-1 within the area of uncertainty of the x-ray source. The distance is estimated to be 500 to 700 pc. Assuming this distance, we estimate the x-ray luminosity to be of order 10^{36} erg/sec; the optical luminosity is lower by a factor of 100 and is dominated by the G-type star. The optical emission from the x-ray region could be, as in Sco X-1, about 10^{-3} of x-ray emission.

Several other x-ray sources have properties, in terms of the spectrum and variabilities of the x-ray intensity, similar to Sco X-1 and Cyg X-2 (for instance, GX17 + 2); however, there has been no other identification of a similar optical object. If, however, the ratio $L_x/L_{opt} = 10^3$ then this is not so surprising. Sco X-1 and Cyg X-2 are not only bright in x rays, but they are located at high galactic latitude out of regions of high optical obscuration. The visible emission from a source 10^{-2} of Sco X-1 located in the galactic

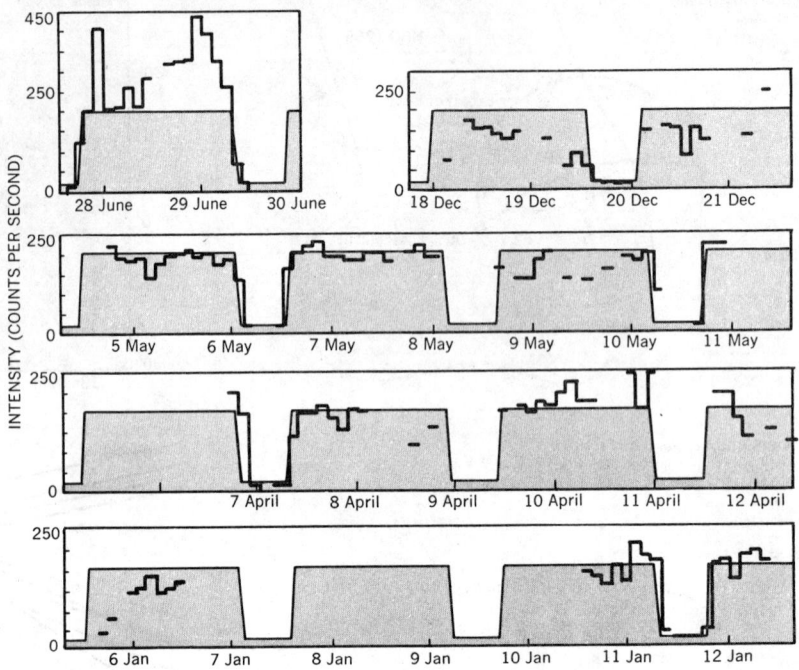

The observed intensity from Cen X-3 averaged over intervals of one tenth of a day and corrected for elevation in the field of view, plotted as a function of time. Statistical errors are not shown; they are typically about 1%. Also shown, by the color tint, is the average intensity predicted by the light curve.   Figure 6

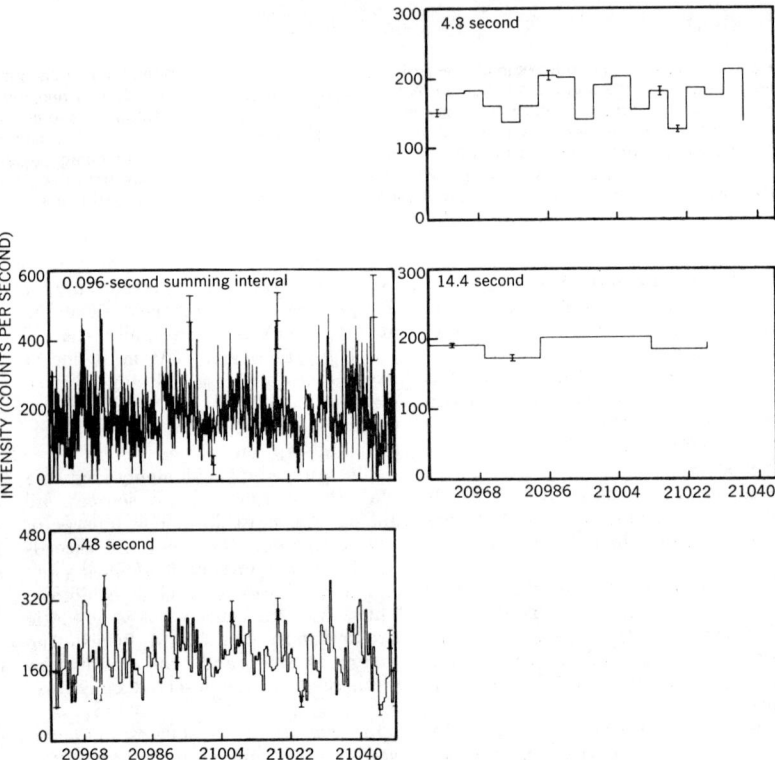

Observations of Cygnus X-1 on 10 July 1971, corrected for triangular collimator response. The four versions show data summed over 0.096 sec, 0.48 sec, 4.8 sec and 14.4 sec intervals. Typical one-standard-deviation error bars have been added.   Figure 7

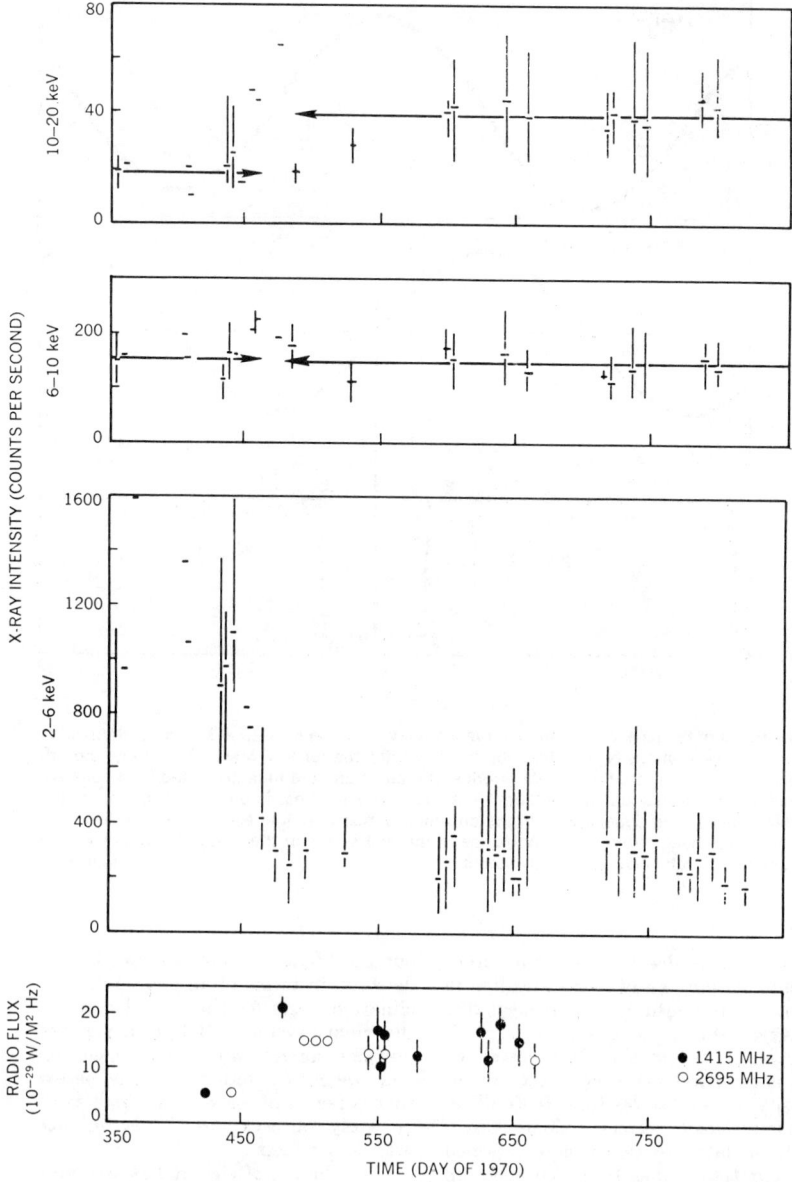

Cygnus X-1 observations over a 16-month period. The x-ray data are shown for three energy bands, 2–6 keV, 6–10 keV and 10–20 keV. The transition discussed in the text occurred around day 450. The radio data are shown at the bottom. Figure 8

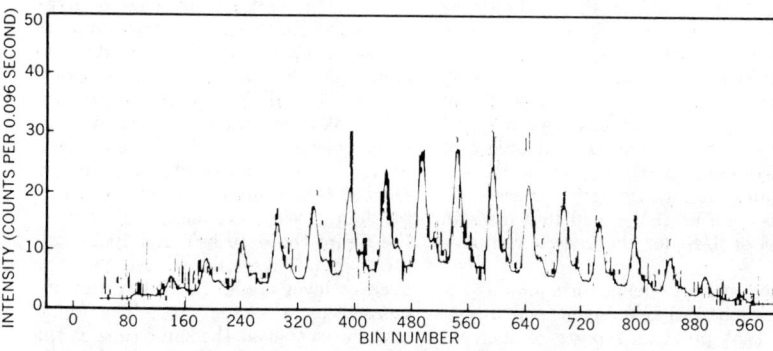

X-ray variability of Cen X-3. This curve shows the 4.8-sec pulsations as they appear in the raw data (each bin is 0.096 sec). The fitted curve is the result of a Fourier analysis and contains a sine wave and harmonic components. Figure 9

plane would appear to be of 17 m if its optical emission were 10^{-3} of its x-ray emission. With several magnitudes of obscuration, such a star would be hopelessly difficult to find.

X-ray binaries

Some of the observed properties of the known or suspected x-ray binaries are listed below to indicate the degree of certainty or uncertainty surrounding their assignment to this category. We have a total of seven sources for which we have evidence of binary nature.

Cyg X-1—a randomly pulsating x-ray object with extremely rapid intensity variation on the time scale of 50 milliseconds or less. It has been identified with a variable radio source, which in turn leads to an optical identification with a spectroscopic binary of 5.6 days period (HDE226868).

Cen X-3—an eclipsing (2.1 days) periodically pulsating (4.8 seconds) x-ray source. Doppler shifts in the period of pulsation in phase with the 2.1-day occultation period give us, from the x-rays alone, conclusive evidence for binary nature. No radio source or optical counterpart is yet known.

Her X-1—an eclipsing (1.7 days), periodically pulsating x-ray source (1.24 sec). Doppler shifts in the period of pulsation, in phase with occultation period, demonstrated the binary nature. There is recent identification with a 1.7-day-period eclipsing optical binary (NZ Her), but there is no radio identification yet. Her X-1 is the best-known object after the Crab pulsar.

2U(1700-37)—a 3.4-day x-ray eclipsing binary. The suggested candidate star is HD153919 07f[6m], on the basis of positional coincidence (similar to Cyg X-1).

2U(0115-73)—a 3.9-day eclipsing x-ray binary. The source is in the Small Magellanic Cloud. Its importance is that we have some idea of its distance, and we can derive a value for the intrinsic luminosity: $L_x \approx 10^{38}$ erg sec^{-1} cm^{-2}.

2U(0900-40)—a variable x-ray source, identified on the basis of positional coincidence with a B-type supergiant that has a companion with period of 9 days (SAO 220767). UHURU data and OSO-7 data confirm periodic behavior with a 9-day period.[12]

Cyg X-3—a 4.8-hour eclipsing x-ray source or periodically pulsating x-ray source. It has been identified with the source of a giant radio burst on 2 September 1972 on the basis of positional coincidence only.

All of the above sources, with the exception of Cygnus X-1, share a common feature in that they show definite intensity variations periodically recurring over times of the order of 0.2 to 9 days; Cen X-3 was the first source ob-

44

served to show eclipses in the x-rays. The light curve, derived from observations that have now extended over two years, is shown in figure 6. All sources we have considered as binaries show deep and broad eclipses except for Cygnus X-1. Most of these sources show extreme variability in their x-ray fluxes. Another common feature is their spectral shape and the existence of large and variable low-energy cut-off, except again for Cygnus X-1.

If we consider the seven sources we believe to be binaries, the only one whose spectrum is not cut off is Cyg X-1. This is also the only one that does not occult. If we consider this to be more than a chance occurrence, we may offer a plausible explanation. To view occultation, the observer must be close to the plane of the binary orbit. Furthermore, most of any matter being transferred between the members of the system (as with an accretion disk), as well as any matter leaving the system, will be concentrated in the orbital plane. Thus, low-energy absorption at the source would be most pronounced in the orbital plane and would thus be greatest for the occulting sources. I might also add that C. T. Bolton[13] has recently stated that the Cyg X-1 optical binary candidate does have a rather low inclination, in agreement with the lack of x-ray occultations and thus of low-energy absorption.

To summarize the foregoing: There are a number of x-ray sources that we believe to be binaries either because of their x-ray behavior or because the candidate object has been shown to be a member of a binary system. All of the sources that exhibit eclipses in the x-rays also show very flat spectra and large and variable low-energy cut offs. All sources appear to exhibit large fluctuations in intensity. Periodic pulsations of seconds occur in Cen X-3 and Her X-1, aperiodic fluctuations in times of milliseconds in Cyg X-3, intensity changes in times of seconds or less on 2U1700-37 and in 2U0900-40. This is a very strong indication that x rays originate from compact regions of the order of less than 10^{10} or 10^9 cm.

When we attempt a more detailed examination of the characteristic of the x-ray emitting binaries, we find that they appear to fall into two different types: irregular pulsating objects, like Cyg X-1, and X-ray pulsars, like Cen X-3 and Her X-1.

Cygnus X-1

Perhaps the most significant of the UHURU results for the galactic x-ray sources has been the discovery of pulsations from Cygnus X-1, which led to further study of this object and to the present belief that we are dealing with a black hole. The discovery of the pulsations also led to a search for other rapidly varying x-ray sources leading to many of the results I have been discussing. Figure 7 contains data already reported in the literature showing substantial variations in x-ray intensity on time scales from 100 milliseconds to tens of seconds. Some 80 seconds of data are shown here summed on four time scales from 100 msec up to 14 sec. With the relatively good x-ray location determined by an MIT rocket flight and by UHURU, a radio source was discovered by L. Braes and G. K. Miley[14] and by R. M. Hjellming and C. M. Wade.[15] It is this precise radio location that led to the optical identification, by L. Webster and P. Murdin[16] and by C. T. Bolton,[17] of Cygnus X-1 as a 5.6-day spectroscopic binary system. The central object of this system is a 9th-magnitude B0 supergiant, and conservative mass estimates such as $12 M_\odot$ lead to a mass in excess of $3 M_\odot$ for the unseen companion.

There is no evidence for a 5.6-day eclipse in the 2–6-keV region, and we believe that previous reports of such an effect at higher energies was caused by the large-scale time variability and not by a 5.6-day effect. This does not rule out the identification, and can be understood in terms of an appropriate inclination angle for the orbital plane of the binary system. Bolton in a recent preprint agrees with this conclusion and presents a refined 5.6-day period that is not in phase with the high-energy x-ray observations, in accordance with our findings.

With the use of UHURU as an observatory, we have now accumulated over 16 months of data on Cygnus X-1, which are shown in figure 8. We have plotted the 2–6 keV intensity against time. The vertical lines for a given day show the range of variability observed on that day. For some days we have only the average intensity shown by a dash available in our analyzed results. We see that a remarkable transition occurred in March and April 1971, with the source changing its average 2–6 keV intensity level by a factor four. We have also indicated in the figure the 6–10 keV and 10–20 keV x-ray intensities, and we see that the average level of the 10–20 keV flux increased by a factor two. The figure also shows that at the same time as the x-ray intensity changed, a weak radio source appeared at the Cyg X-1 location and was detected by the Wester-

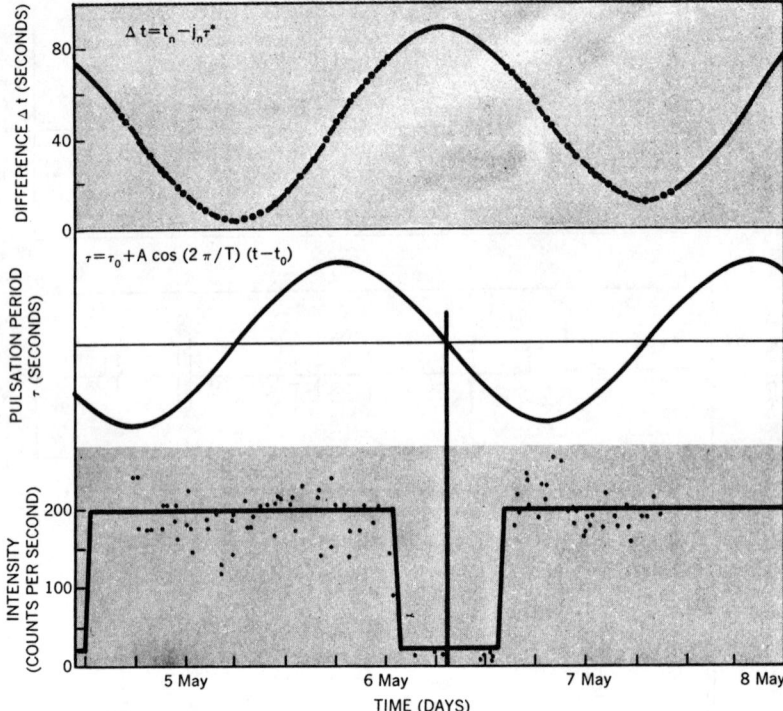

Comparison of the period function and the intensity variation for Cen X-3. The bottom curve shows observed intensity and the light-curve predictions for 5–7 May. At the top, the difference Δt between the time of occurrence of a pulse and the time predicted for a constant period is plotted as a function of time. The sinusoid is the best-fit analytic function. In the center is shown the dependence of the pulsation period t on time as derived from the best-fit phase function. Note the coincidence of the null points of the period function with the centers of the high and low intensity states.   Figure 10

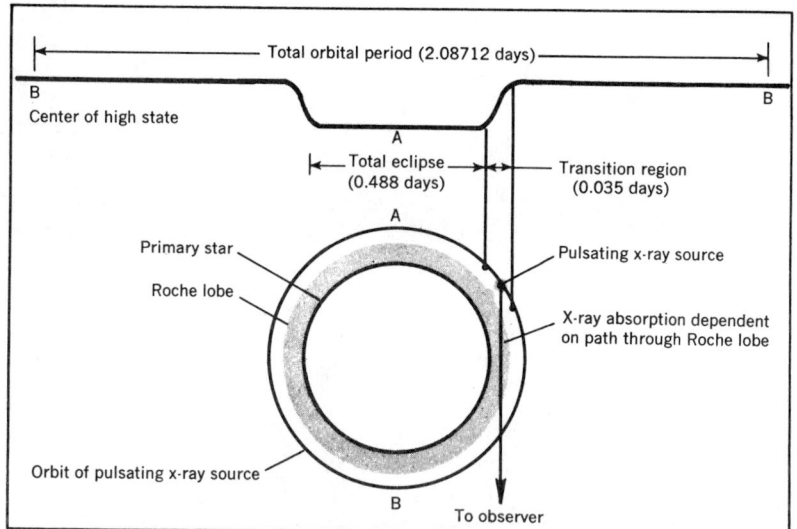

An occulting binary x-ray system, shown schematically. It consists of a compact x-ray object orbiting a central star. Numerical data apply to the Cen X-3 system. Figure 11

bork and NRAO groups. Reviewing the arguments then: Cyg X-1 undergoes large intensity changes in times as short as 100 millisecs, so requiring the x-ray emitting region to be compact. The very good x-ray position, plus the correlated x-ray–radio variation, demonstrate the x-ray–radio identification. The optical–radio identification is based on position agreement better than 1 arc second. Then the optical data taken conservatively require at least three solar masses in the unseen companion, which is the compact x-ray source. Thus we have the strong possibility that Cygnus X-1 is a black hole—it is compact and its mass exceeds $3M_\odot$.

Cen X-3 and Her X-1

We come now to two x-ray sources, Cen X-3 and Her X-1, that can be identified as binaries solely from their x-ray properties. The first source is Cen X-3, shown in figure 9. Here we see a regular pulsing source with a period of 4.8 seconds. The histogram shows the actual counts observed and the heavier curve is a sine-wave-plus-harmonics fit to the data. The x-ray emission is at least 90% pulsed with the 4.8-sec period. Figure 10 shows some of the data that demonstrate that Cen X-3 is an occulting binary system. The bottom portion shows three days of intensity data accumulated in May 1971 with a clear-cut downward transition followed by an upward transition about a half-day later. Many such transitions have been observed, and all are fit to a 2.08712 ± 0.00004 day period. By measuring the arrival time of individual 4.8-second pulses we have also determined that the pulsation frequency is Doppler shifted in phase with the 2.097-day occultation cycle, as is shown in the top portion of the figure where the pulse arrival time delays are fit by a sine wave. Under the model of an occulting binary system, we have made very precise determinations of the projected orbital velocity, 415.1 ± 0.4 km/sec, the projected orbital radius (1.191 ± 0.001) × 10^{12} cm, and the mass function of the system (3.074 ± 0.008) × 10^{34} gm.

Figure 11 is a schematic representation of this system with a compact x-ray object orbiting a central star. We find from the Doppler velocity that the mass of the central star must be at least 15 solar masses, and calculations, such as those of R. E. Wilson[18] for close binaries, lead to masses of the order of a few tenths of a solar mass for the x-ray emitting object. These calculations assume that the sharpness of the occultation means that the radius of the Roche lobe is greater than, or equal to, the size of the occulting region. At present there has not been an optical identification of Cen X-3. Similar, but more precise, information is available on Her X-1, where we also have a positive identification.

Following the suggestion of W. Liller that HZ Her could be a strong candidate object for Her X-1 due to the strong ultraviolet excess and variability, many observations[19,20,21] have definitively established that HZ Her exhibits eclipses in the visible with period and phase in exact agreement with the x-ray observations.

There appears therefore to be absolutely no doubt, on the basis of the 1.7-day eclipse and of a very recent and finer position determination by the MIT group, that the identification is correct. Recently, D. Crampton and J. B. Hutchings succeeded in obtaining measurements of the radial velocities of several lines in the optical spectrum from 3500 to 5000 Å for the visible companion. The radial velocity of the x-ray star is measured from the Doppler effect on the 1.24-second pulsations. Under this condition we can compute the mass of each of the stars in the system. Crampton and Hutchings deduce masses of approximately 2.5 $M_\odot$ for the occulting object and about 1.3 $M_\odot$ for the x-ray emitting object.

W. Forman, C. Jones and W. Liller estimate distances of 2.5 to 4.5 kpc for the source, depending on the spectral type. This corresponds to an x-ray luminosity of about 10^{36} ergs/second, although this could decrease to 10^{35} ergs/sec by accounting for the limited solid angle that the pulsed source fills. We have analyzed the x-ray period for December 1971, January 1972, March 1972 and April 1972, and can set an upper limit of 2 microsec change (slow-down) in period in four months. For a one-solar-mass neutron star of 10-km radius, the rotational kinetic energy is 2.5 × 10^{44} ergs, and with our upper limit on the period slowing down we find that only 2.5 × 10^{32} ergs/sec could be converted into x-rays by a 100% efficient mechanism if the source were slowing down at the determined upper limit rate. This, then, demonstrates that rotation cannot be the energy source, and one turns to accretion as the most likely alternative energy source, with rotation still presumably providing the clock mechanism.

Some conclusions

Several significant results have come about from the UHURU mission that have greatly expanded the scope of galactic and extragalactic x-ray astronomy.

In extragalactic research we may have the beginning of an explanation for the diffused x-ray background. We may have detected the presence of high-temperature, intercluster gases. The relationship of the mass and temperature of this gas to other cluster parameters, as determined by visible-light observations of the galaxies in the cluster, may furnish us a tool with which to study cluster dynamics and to investigate evolutionary effects of cluster matter at redshifts of cosmological significance.

In galactic x-ray astronomy, the discovery of x-ray sources in binary systems gives us some understanding of the mechanisms (accretion) leading to the production of copious x-ray fluxes in stellar objects. Because the systems contain massive luminous stars, optical identifications are relatively straightforward and, thus, we can determine distances and absolute luminosities.

When such identifications occur, we have available a radial velocity for the x-ray star, directly from the x-ray data, and for the massive companion from the Doppler shift of the visible-light emission or absorption lines. We also have additional information from the length of the occultation. We can, therefore, determine the masses of both objects and thus we can distinguish between different possible compact objects, such as white dwarfs, neutron stars and black holes. We are indeed very fortunate that exotic objects of such great astrophysical significance may be located in binary systems where we can conveniently study them.

* * *

This article is adapted from a talk given at the Sixth Texas Symposium on Relativistic Astrophysics, New York, last December.

References

1. P. Gorenstein, B. Harris, H. Gursky, Astrophys. J. **172**, L41 (1972).
2. G. Garmire, private communication (1972).
3. A. Solinger, W. Tucker, Astrophys. J. **175**, L107 (1972).
4. S. Rappaport, R. Doxsey, W. Zaumen, Astrophys. J. **168**, L43 (1971).
5. H. Friedman, E. T. Byram, T. A. Chubb, Science **156**, 374 (1967).
6. H. Kestenbaum, J. R. P. Angel, R. Novick, Astrophys. J. **164**, L87 (1971).
7. W. A. Hiltner, D. E. Mook, Ann. Rev. Astron. and Astrophys. **8**, 139 (1970).
8. G. Neugebauer, J. Oke, E. Becklin, G. Garmire, Astrophys. J. **155**, 1 (1969).
9. R. Novick, M. C. Weisskopf, R. Berthelsdorf, R. Linke, R. S. Wolff, Astrophys. J. **174**, L1 (1972).
10. R. M. Hjellming, C. M. Wade, Astrophys. J. **164**, L1 (1971).
11. R. Giacconi, P. Gorenstein, H. Gursky, P. D. Usher, J. R. Waters, A. Sandage, P. Osmer, J. V. Peach, Astrophys. J. **148**, L119 (1967).
12. M. P. Ulmer, W. A. Baity, W. A. Wheaton, L. E. Peterson, Astrophys. J. **178**, L121 (1972).
13. C. T. Bolton, Nature Physical Science **240**, 174 (1972).
14. L. Braes, G. K. Miley, Nature **232**, 246 (1971).
15. R. M. Hjellming, C. M. Wade, Astrophys. J. **168**, L21 (1971).
16. L. Webster, P. Murdin, Nature **235**, 37 (1972).
17. C. T. Bolton, Nature **235**, 271 (1972).
18. R. E. Wilson, Astrophys. J. **174**, L27 (1972).
19. W. Forman, C. Jones, W. Liller, Astrophys. J. **177**, L103 (1972).
20. J. N. Bahcall, N. A. Bahcall, Astrophys. J. **178**, L1 (1972).
21. D. Crampton, J. B. Hutchings, Astrophys. J. **178**, L65 (1972).

The Einstein Observatory: New Perspectives in Astronomy

Riccardo Giacconi and Harvey Tananbaum

X-ray astronomy has become a major branch of observational astronomy since its beginning less than 20 years ago. The pace of discovery in this field can be compared to the great surge in radio astronomy over the past quarter-century.

The first exploratory phase occurred with rocket flights in the period 1962 to 1970 and with crude satellite instruments for exploratory surveys in the period 1970 to 1978. The astonishingly rich returns from even these relatively insensitive instruments were due in large part to the discovery of extremely luminous galactic and extragalactic x-ray sources whose existence was previously unsuspected. Perhaps most important among these results were the discovery of x-ray emission from mass-exchange binary systems containing a collapsed star (white dwarfs, neutron stars, and possibly black holes), which has provided us with a unique laboratory for the study of relativistic astrophysics, and the discovery of the intergalactic high-temperature gas associated with clusters of galaxies, a previously unobserved component of the universe containing as much luminous mass as all other known objects. The rich phenomenology of the high-energy sky revealed by these early missions had only little overlap with the interests of the mainstream of traditional astronomy, since only exceptionally luminous galactic sources and only the nearest and most powerful extragalactic sources could be observed.

The situation changed radically with the launch on 13 November 1978 of the Einstein (HEAO-2) Observatory, which introduced the use of focusing high-resolution optics to x-ray astronomy. The improvement in sensitivity and resolution brought about by this mission has resulted in a qualitative change in the scope of the discipline. At one stroke, x-ray observations have been extended to all known classes of stars in our galaxy and to the most distant discrete objects known in the universe. Although the analysis of the data is still in preliminary form, it is already clear that x-ray observations are again revealing new and unsuspected aspects of astronomical objects.

X-ray observations have emerged so rapidly as a powerful tool for astronomy because of the prevalence of high-energy phenomena in the universe and the important and at times determinant role that they play in its dynamics and evolution. This view of a universe in constant turmoil, created in an enormous initial explosion, with components being torn apart again and again by impulsive events, is one that has developed over the last few decades through observations at all wavelengths from the microwave band to the gamma-ray region of the spectrum—the highest energy radiation accessible for study. Modern astronomy has adopted an all-wavelength approach to the study of the heavens and to the extraction of crucial information about the processes that determine the evolution of stars and galaxies.

In particular, many of the violent processes of formation and collapse of stars and galaxies give rise to high-temperature gases or high-energy particles. When this occurs, copious x-ray fluxes are produced, either through thermal emission from gases with temperatures of millions of degrees or by interaction of the high-energy particles with electromagnetic fields or photons. The x-rays thus produced can readily traverse the vast interstellar and intergalactic distances separating us from their point of origin, although they cannot penetrate the earth's atmosphere. The development of space observatories has made it possible to use this rich new channel of information.

Summary. High-sensitivity x-ray measurements with the recently launched Einstein Observatory are having a major impact on wide areas of astronomical research. The x-ray luminosity of young O, B, and A stars and late K and M stars is found to be several orders of magnitude greater than predicted by current theories of coronal heating. Detailed x-ray images and spectra of supernova remnants are providing new information on the temperature, composition, and distribution of material ejected in supernova explosions as well as of the material comprising the interstellar medium. Observations of galaxies are yielding insights on the formation and evolution of stellar systems and galaxies over a wide range of variables. X-ray time variations are being used to probe the underlying energy source in quasars and active galactic nuclei. The distribution of mass in clusters of galaxies is being traced through detailed x-ray images, and the data are being used to classify clusters and trace their formation and evolution. Substantial progress is being made in several areas of cosmological research, particularly in the study of the diffuse x-ray background.

Riccardo Giacconi is a professor of astronomy at Harvard University and an associate director of the High Energy Astrophysics Division, Harvard/Smithsonian Center for Astrophysics, Cambridge, Massachusetts 02138. Harvey Tananbaum is a research scientist at the Harvard/Smithsonian Center for Astrophysics.

The Einstein Observatory

The first surveys and discoveries in x-ray astronomy were made primarily with mechanically collimated detectors. However, the sensitivity of these early observations fell short of that required to realize the full potential of x-ray astronomy for the study of low-luminosity objects in our galaxy, for the study of extragalactic objects, and for serious cosmological research. This limitation in sensitivity was essentially due to the background-limited nature of the instrumentation; sensitivity could increase only in proportion to the square root of the product of the detector area and the observing time. Further progress in the field required instruments built on entirely different principles.

At visible light and radio wavelengths, astronomers are well aware of the increase in sensitivity that results from the use of focusing optics. By concentrating the signal into a relatively small detector element, through focusing, the observer achieves a great reduction in background and increases the angular resolution as well. However, the application of this concept to x-ray astronomy is quite difficult, since under normal conditions lenses and mirrors tend to absorb rather than focus x-rays. It had been discovered in the 1930's that one could achieve high efficiency of reflection if x-rays were made to reflect from a surface at very small angles of incidence (grazing incidence). This effect is possible because the index of refraction of some materials is slightly smaller at x-ray wavelengths than the index in vacuum. Radiation incident at angles less than the critical angle (grazing angle) can therefore undergo total external reflection. By 1952 Wolter (1) had generated optical designs for an x-ray microscope based on this principle, but no practical implementation was possible because of the inability to polish the very small surfaces involved to the very high accuracies required. By 1960 Giacconi and Rossi (2) had recognized the great advantage that focusing optics could bring to x-ray astronomy, as well as the feasibility of constructing the required optical surfaces on the large physical scale appropriate for telescopes.

During the ensuing decade the required fabrication techniques and technology were developed and first applied to studies of the sun, culminating with the many thousands of highly resolved pictures of the solar corona which the astronauts brought back from Skylab (3). As a result of the developments in x-ray astronomy as a whole and in the use of focusing optics for solar studies, NASA accepted a 1970 proposal for an x-ray telescope mission to study extrasolar sources. This proposal was submitted by a consortium including Columbia University, Massachusetts Institute of Technology (MIT), Goddard Space Flight Center (GSFC), and our scientific group, then at American Science and Engineering and now at the Harvard-Smithsonian Center for Astrophysics (CFA). The mission was finally realized with the launch on 13 November 1978 of the Einstein Observatory.

The scientific portion of the satellite is diagrammed in Fig. 1, which shows the 0.6-meter-aperture x-ray telescope, the optical bench, and the focal plane transport assembly, a lazy-Susan arrangement that permits one of several imaging instruments or spectrometers to be brought to the focus (4). Three star sensors are flown for on-the-ground aspect determination and as part of the on-orbit attitude control system, which also utilizes gyroscopes, reaction wheels, and reaction control gas to unload momentum from the reaction wheels.

The grazing incidence telescope consists of four nested paraboloids of revolution followed by four nested hyperboloids. The nearly cylindrical surfaces are nested inside one another to increase the effective collecting area. The mirrors are made of fused quartz, which is figured and polished to very high precision. A very thin nickel coating provides the desired index of refraction for focusing the x-rays. Two reflections are required for each x-ray (one by a paraboloid and one by the following hyperboloid) to achieve a reasonable resolution over the 1° field of view of the telescope. At low energies (0.25 kiloelectron volts) the total effective area of the mirror is ~ 400 square centimeters, and this decreases to 30 square centimeters at the high-energy cutoff (4 kiloelectron volts). Since the grazing angles are $\gtrsim 1°$, approximately 60×400 cm^2 of total surface area was actually figured and polished for the set of four paraboloids and four hyperboloids. The total area of each of these sets corresponds to slightly more than the area of the Hale Observatory–Mount Palomar 200-inch optical reflector.

Our group at CFA developed two types of complementary imaging detectors to employ at the focus of the telescope. A high-resolution imager (HRI) utilizes two multichannel plates operating in cascade to convert individual x-rays to an electron cloud, whose centroid is determined by using a crossed-grid charge detector and associated electronics. The position and time of each event (up to 100 per second) are inserted as digital data into the telemetry stream and can be combined with the aspect data on the ground to reconstruct the x-ray image of the sky. This detector has the highest spatial resolution available, 1 arc second, to be used in conjunction with the 3.5 arc second angular resolution (full width at half-maximum) of the telescope. It covers the central 25 arc minutes of the telescope field of view, has a detection quantum efficiency that varies from 30 to 6 percent over our energy band, and has no inherent spectral

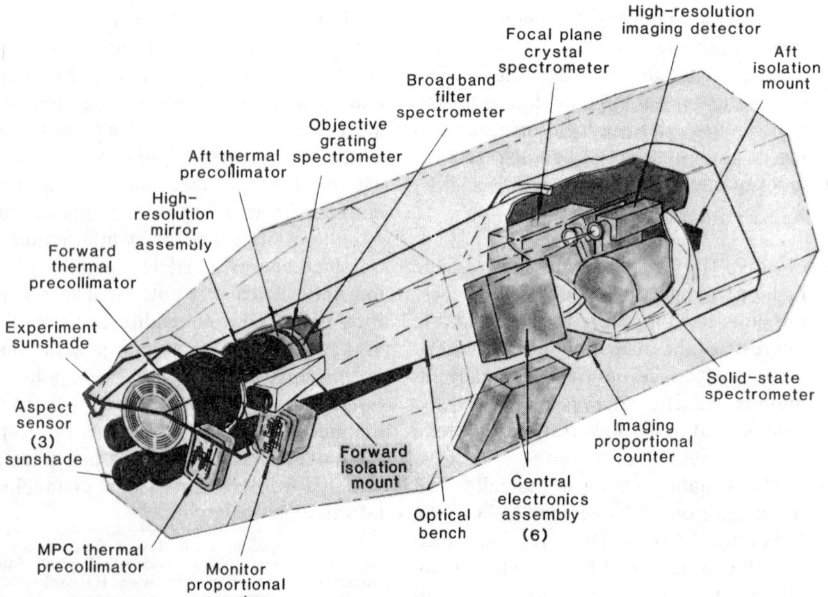

Fig. 1. Schematic diagram showing the major scientific elements comprising the Einstein Observatory. [This figure originally appeared in *Astrophysical Journal*]

resolution, although it can be used in conjunction with interchangeable broadband filters and transmission gratings that can be placed into the x-ray beam.

The second type of imaging detector is an imaging proportional counter (IPC), which has 1 arc minute spatial resolution, a 1° field of view to cover the full telescope aperture, a high-quantum efficiency, and moderate spectral resolution. Imaging is accomplished by use of two planes of cathode wires with an anode plane in between. An incident x-ray is converted to an electron, which creates an avalanche at the anode while still preserving spatial information. Induced signals are sensed on the cathodes and used to generate the position information.

The observatory also carries two sensitive spectrometers for use at the telescope focus. One is a cooled, solid-state silicon spectrometer developed by GSFC, which combines high efficiency with good resolution (~ 150 electron volts). The other is a curved crystal Bragg spectrometer developed by MIT, which features very high resolution (in some cases of order 1 eV) accompanied by relatively low efficiency.

Although Einstein was originally proposed and approved as a principal investigator experiment, we have long recognized the desirability of making its great capabilities available to all interested astronomers. Therefore, the observatory has been operated as a national facility since its launch, and more than 330 guest observers were approved during the first 1½ years. A series of guest and consortium observing targets, now numbering more than 7000, are entered and stored in our computer. Approximately ten different targets are selected each day so that more than 3000 have been observed in the first year. Commands are generated to configure and point the observatory for sequences approximately 12 hours long. Once per orbit the on-board tape recorder transmits ~ 100 minutes of data to a NASA ground station. These data are then relayed to GSFC and forwarded to the CFA. At the CFA a specialized data handling system is used to manipulate the data stream, determine the instantaneous pointing direction of the telescope as a function of time, and produce x-ray images of the sky. These images can then be searched for sources before further detailed scientific analyses are carried out. The data can also be displayed and manipulated on one of several television screens provided as part of the computer system.

The satellite has operated without serious problems since its launch and, as discussed below, has fulfilled our expectations with regard to increased sensitivity and angular resolution. The only serious limitation of the observatory is its finite lifetime. At the time of launch, the nominal lifetime was 1 year. Through judicious target scheduling designed to minimize net perturbing torques, we have been able to extend the life expectancy of the finite amount of gas available to remove momentum from the reaction wheels. At present we expect the gas to last through 1981—2 years beyond the nominal 1 year. It is also probable that the continued high level of solar activity will cause the satellite orbit to decay by 1982, leaving us without an ongoing high-sensitivity x-ray astronomy facility possibly until the launch of the Advanced X-ray Astrophysics Facility (AXAF) around 1987 or 1988.

Fig. 2. High-resolution x-ray photograph of nearby binary star system, Alpha Centauri. The brighter x-ray source corresponds to the K star and the other x-ray source to the G star, contrary to theoretical expectations for the relative x-ray emission from these two different classes of stars.

X-rays from Stellar Systems

Considerable progress has been made in the last 50 years in understanding the formation and evolution of stars, particularly during the long period of time they spend in the quiet burning of their nuclear fuels after they reach the main sequence. However, the answers to many important problems remain incomplete. These include the details of the stellar formation process, of the energy transport processes in main sequence stars, of the collapse of massive stars, and of the behavior of matter in superdense collapsed stars.

The full range of these problems is being attacked by astronomers, using all the observational tools at their disposal. For example, infrared observations have given us the opportunity to study many aspects of the early phases of stellar evolution (5). Spectroscopic studies in the visible range of wavelengths and more recently in the ultraviolet from the Copernicus satellite have provided fundamental clues to the behavior of stars throughout their evolutionary track (6).

In its first decade, x-ray astronomy appeared to be mainly capable of giving information about the later stages of stellar evolution. The discovery by UHURU of binary x-ray sources (7), which contain a collapsed star accreting matter from its normal companion, provided startling new information regarding the endpoints of stellar evolution. It first demonstrated that the formation of a neutron star (presumably through a supernova explosion) need not disrupt the binary system in which the star is contained. It provided an astrophysical laboratory in which the properties of the collapsed object itself could be studied as they manifested themselves in response to varying torques and forces due to changes in the accretion flow. The internal structure of neutron stars could be studied and their angular momentum, crust-to-core coupling, and other characteristics measured.

The binary x-ray systems provided the first independent measurements of neutron star masses. For one system, Cygnus X-1, combined x-ray and optical observations revealed the existence of a compact object less than 10 kilometers in radius with a mass of more than 6 solar masses and incompatible with any equilibrium solution predicted by general relativity except for a black hole. Cygnus X-1 still furnishes the most promising and convincing evidence we have for the existence of black holes of stellar mass (8).

The discovery of x-ray bursts (9) with the ANS satellite led to an intensive study of these objects by SAS-3, HEAO-1, and most recently Hakucho, the Japanese x-ray satellite. In these systems the general characteristics of the luminous x-ray sources described above, namely their very high absolute luminosity (~ 10^{37} to 10^{38} ergs per second) and very rapid time variability, appear to be carried to an extreme. Extremely luminous (10^{38} to 10^{40} erg/sec) flashes of radiation appear to originate from otherwise inconspicuous objects located sometimes in the center of globular clusters but often outside as well. Although we still cannot fully decide between models involving massive accreting black holes and binary systems containing a neutron star, the present consensus seems to prefer the latter interpretation; in that case these objects would be of the same general class as the binaries previously described. With the HEAO-1 spacecraft other types of lower luminosity binaries, such as RS CVn systems containing a K star and a G star, were discovered to be x-ray emitters. A few lower luminosity binaries containing white dwarf secondaries had also been detected as x-ray sources by rocket and satellite experiments, as had a few isolated stars (10).

With its qualitative improvement in sensitivity, the Einstein Observatory has brought about a revolutionary change in the prospects for stellar studies through x-ray observations. All classes of known stars have now been detected in their x-ray light (11), with x-ray luminosities ranging from 10^{26} to 10^{34} erg/sec. The predictions for main sequence stars based on coronal heating mechanisms through the acoustic noise of convective motions have been found to be widely at odds with observations. From young stars of the O, B, and A spectral types, as well as from the late K and M stars, the observed x-ray fluxes are 1000 to 1 million times greater than expected. Figure 2 shows, for example, an IPC image of Alpha Centauri. The x-ray emission from the K star is found to be more intense than that from the G star, contrary to theoretical expectation. In Fig. 3 the observed ratio of the x-ray to visible light flux is shown as a function of spectral type (as well as blue color excess in the visible) for main sequence stars. Previous theoretical calculations based primarily on our knowledge of one star, the sun, predicted substantial x-ray fluxes only for late A, F, and G stars, whereas the observations show substantial x-ray emission for all spectral types. These findings will form the foundation for new theoretical modeling of the energy transport mechanisms from the core and heating of the outer atmosphere of stars. We suspect that any new model will have to take fully into account the roles of the magnetic field and of stellar rotation.

For early stars and stars possibly still embedded in their parent nebulas the Einstein observations also seem to open a new avenue of study. The discovery of x-ray emission from young O and B stars in associations (12), from the Orion variables and the stars in the Orion Trapezium (13), from the Eta Carinae nebula (14), and other peculiar objects show that high-energy photons are produced at the seat of stellar formation and can reach us through the gaseous envelope that prevents us from seeing the objects in visible light. If stellar winds furnish the means of shedding excess material and if bubbling up and annihilation of magnetic fields furnish the means of removing magnetic energy and possibly angular momentum, we can expect x-ray observations to provide crucial clues to the formation process of stars in the pre-main sequence period.

Important new information has also been provided by the Einstein Observatory on the remnants of supernova explosions. In some of them the burnt cinder of the explosion, a pulsar, can easily be detected. The cover, for example, shows an x-ray picture of the Crab Nebula pulsar and the surrounding nebulosity. The entire x-ray emission in this case

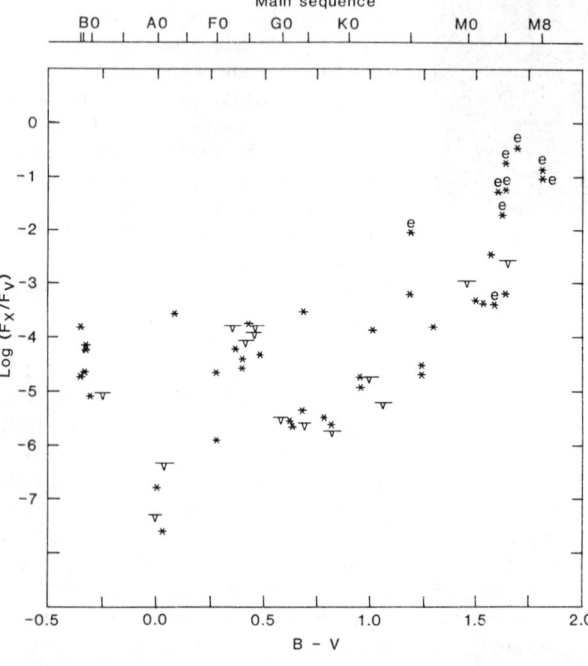

Fig. 3. Ratio of x-ray to visible light flux plotted logarithmically against (visible light) spectral type to summarize the Einstein x-ray observations for various main sequence stars. Spectral type ranges from O and B stars, which are the hottest, bluest, and youngest, to M stars, which are the coolest, reddest, and oldest. The $B-V$ scale is a visible light color difference index. For several stars not detected in the x-ray observations, upper limits are indicated.

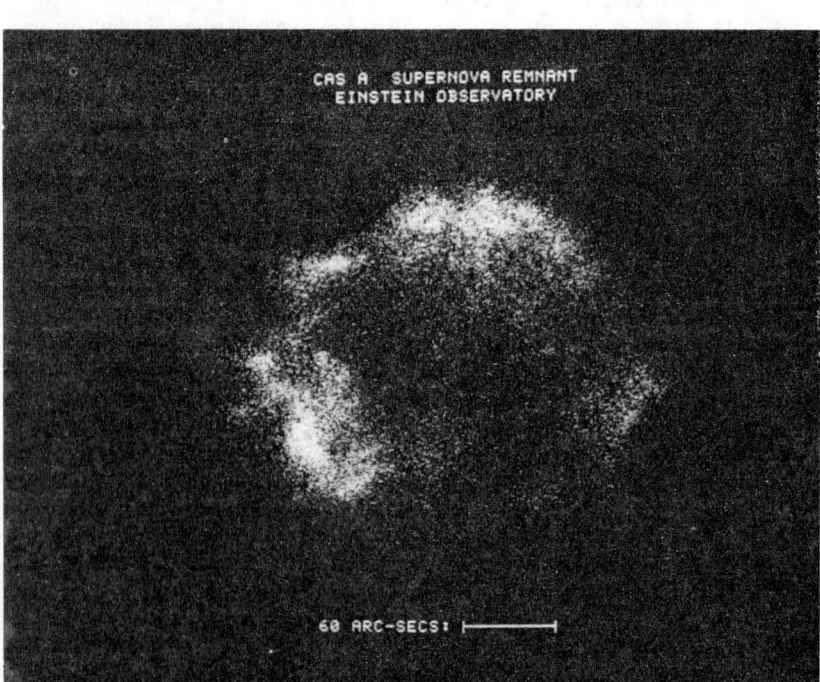

Fig. 4. High-resolution x-ray image of the shell-like supernova remnant Cas A. Various features seen in this image are discussed in the text. The data are digitized into 2 arc second by 2 arc second pixels (picture elements), and the overall exposure time is ~ 30,000 seconds.

appears to be powered by high-energy electrons accelerated in the high magnetic field of the spinning neutron star. As the electrons interact with the surrounding magnetic field, they produce the observed x-rays by synchrotron emission. No trace is seen of the external layers of the star presumably blown off in the explosion that created the pulsar. The two parts of the cover picture are time-phased x-ray photographs comprised of data folded with the 33-millisecond period of the Crab pulsar. Each photograph represents 10 percent of the period summed over many cycles. For the data centered on the main pulse (phase 0.0) the pulsar is very bright (overexposed) and dominates the picture, while for the data centered off the pulse (phase 0.6) only a very weak point component is found at the pulsar location (15). By using the data away from the main and secondary pulses, we can search for steady, blackbody emission from the surface of the rotating neutron star. Our preliminary analysis indicates that the weak point emission during the off phase corresponds to 1.4×10^{34} erg/sec in the energy range 0.1 to 4 keV, which corresponds to a temperature of about 2.5 million kelvins if we assume blackbody emission from a neutron star with radius 10 km. Even if the off-state emission is not from the neutron star surface, these data set an upper limit for blackbody emission and require rapid cooling for the Crab neutron star.

It is also quite interesting that no evidence has yet been found for pulsed or unpulsed emission from a neutron star at the center of any of the historic supernovas except for Crab and Vela. We are left wondering whether a supernova explosion must always result in the production of a pulsar or, if this is the case, whether more efficient cooling mechanisms than neutrino cooling may make the pulsar undetectable (16). Pion cooling has been suggested, for example, to explain the possibility that the neutron star may be so cold as to no longer be detectable at x-ray wavelengths.

Some of the historical supernova remnants, such as SN 1006, Tycho, and Cassiopeia A, have an appearance very different from the Crab, with the x-ray emission showing a well-developed shell structure. The HRI image of Cas A is shown in Fig. 4. We are able to recognize several regions of x-ray emission: (i) an emission shell lying outside the region of the optical filaments, due presumably to the interaction of the shock front with the interstellar material; (ii) a region associated with fast-moving knots from which material may be either evaporated due to heating by the shock or ablated due to the passage of these knots in the interstellar medium (the material of the knots was presumably ejected during the supernova explosion itself); (iii) a region of x-ray emission associated with stationary flocculi, possibly the remains of material ejected from a star prior to the explosion and heated to x-ray temperatures by the shock; and (iv) a region connected with radio emission, presumably from high-energy electrons spiraling in the local magnetic field and producing x-rays by nonthermal processes. From a preliminary analysis the total amount of mass involved in the supernova can be estimated as 10 to 30 solar masses (17).

Spectroscopic observations by the GSFC group with the solid-state spectrometer on the Einstein Observatory provide detailed information on the elemental composition and temperature of the material blown off in the explosion (18). A spectrum of a large region in the Cas A supernova shell is shown in Fig. 5. The data show substantial line emission consistent with transitions of helium-like ions of silicon, sulfur, and argon with abundances ranging from 1.7 to 6.9 times the solar abundances. The helium-like nature of the transitions limits the temperature in the emitting region to less than 10×10^6 K. The absence of excess iron group material suggests that at the time of the explosion the stellar core had not yet advanced to the stage of substantial burning of silicon group material to form iron group material. The data also support models in which the iron group material formed in the supernova explosion is located close to the center and is ejected less forcefully than the silicon group material. As a result, the more slowly expanding iron group material is located to the inside of the silicon-rich shell we observe and has not yet been reheated to x-ray–emitting temperatures by the inward-moving reverse shock (19).

Even from the preliminary account of the supernova observations made possible by the imaging techniques and high sensitivity of Einstein, it is apparent that spectrally resolved x-ray images of such explosive events may give us a unique tool for studying pre- as well as post-supernova conditions. Far from being a bizarre and rare event, supernovas are the factory where the heavy elements are manufactured and dispersed into the interstellar medium. The transfer of the kinetic energy of the explosions determines the temperature of the interstellar medium. It is clear that the x-ray contribution to our understanding of these events will noticeably advance our knowledge of stellar evolutionary processes.

X-ray Emission from Galaxies

Over the past 50 years, visible light and radio observations have identified characteristic properties that have been used to classify galaxies and to investigate their formation and development. For example, "normal" galaxies have

Fig. 5. Solid-state spectrometer observations of Cas A, showing the spectrum for the overall x-ray source in the energy band 0.9 to 4.5 keV. The lower dashed curve represents the continuum contribution from hydrogen, helium, carbon, nitrogen, oxygen, and neon components of the hot gas. The total emission (top curve) is comprised of this continuum plus line emission from magnesium, aluminum, silicon, sulfur, argon, calcium, and iron. The three largest peaks in the top curve are associated with silicon, sulfur, and argon. [Courtesy of S. S. Holt; this figure originally appeared in *Astrophysical Journal*]

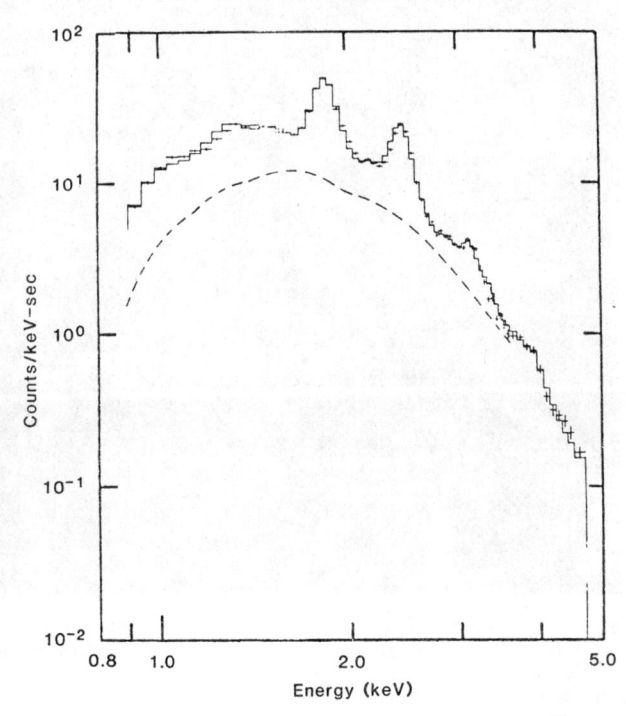

been described as elliptical, spiral, or irregular with several subclasses on the basis of their optical morphology. The presence of a very bright nucleus in a faint galaxy has been a primary signature for a Seyfert galaxy. Spectroscopic observations of broad hydrogen lines and broad forbidden lines (widths ≳ 1000 km/sec) have been used to characterize a subset of Seyfert galaxies as type 2, while Seyferts with hydrogen line widths of order 10,000 km/sec have been called type 1 (20). Similar spectra with very broad emission lines characterize the quasars (quasi-stellar objects, or QSO's), although surrounding faint nebulosities have, in general, not yet been detected for these objects, perhaps because they are considerably more distant than most Seyfert galaxies. Another class of galaxies recently identified and thought to be related to quasars are the BL Lacertae objects. BL Lacs are characterized by weak or absent emission lines, rapid optical variability, strong and variable optical polarization, and radio emission. Among the other categories of galaxies that have been established are those characterized by narrow emission lines, radio galaxies, and N-type galaxies.

This great variety of characteristics raises many questions about the formation and evolution of galaxies and about the relationships between different morphological types. When did galaxies first form? How do the gas content, stellar content, and morphological classification evolve as a function of time for different types of normal galaxies? Are type 1 Seyfert galaxies and normal galaxies remnants of quasars that were more luminous at an earlier time? In addition to these types of questions, a very basic unknown is the nature of the central energy source that powers the nuclei of normal and active galaxies, including quasars.

While still in a relatively early stage of development, our studies of the x-ray-emitting properties of galaxies can now be directed toward these questions. As described earlier, the x-ray emission from our own galaxy is comprised of several elements including mass-exchange binary systems, supernova remnants, and normal stars, as well as an overall diffuse component arising from hot interstellar gas. Our observations of nearby normal galaxies provide information on these same phenomena, but with the additional feature of having all of the sources for a given galaxy at essentially the same, known distance.

Thus the Einstein observations of 13 supernova remnants in the Large Magellanic Cloud (at a distance of 55,000 parsecs, or 180,000 light-years) have permitted our colleagues at Columbia University (21) to determine the x-ray luminosities and in several cases the remnant temperature and x-ray shell size. These observations have been combined with radio and optical data, and the Sedov solution for a symmetrical blast wave propagating into the interstellar medium has been used to solve for the age of the remnants, the initial energies in the supernova explosions, and the ambient densities of the interstellar medium. The Columbia group finds that the ages range from 1,000 to 20,000 years; the initial energies range from 0.8×10^{50} to 3.4×10^{50} ergs, similar to energies derived for supernovas in our galaxy; and the interstellar medium densities range from 5×10^{-2} to 10 cm^{-3}. They find that the densities are highly correlated with observed remnant diameters; this correlation is not consistent with a simple blast wave propagating into a homogeneous medium. A lower density interstellar medium studded with higher density clouds might fit the observed correlation between density and diameter in a reasonable manner.

Another example of a normal galaxy is provided by our twin, M31 or Andromeda. At a distance of ~ 700,000 parsecs (~ 2 million light-years), M31 occupies a region of about 2° on the sky, and several x-ray observations have to be combined in a mosaic to obtain an overall picture of the galaxy. At least 80 individual x-ray sources have been resolved in the combined HRI-IPC Einstein data (22), whereas previous observations were able to detect as a very faint source only the integrated emission from M31 as a whole. The individual sources observed by Einstein have luminosities of at least 9×10^{36} erg/sec and are probably mass-exchange binaries in most instances, although at least one supernova remnant has been identified. Individual normal stars would be too faint for us to detect, and the data have not been analyzed for extended interstellar emission. Many of the luminous x-ray sources are associated with the spiral arm structure of M31. Spiral arms are regions containing much gas and dust and are bright because many young, massive stars are being formed. Such conditions are conducive to the formation of massive binaries that eventually evolve into mass-exchange binary x-ray sources.

In the HRI image of the central region of M31 shown in Fig. 6, the outermost sources are probably associated with gas, dust, and spiral arm features. However, the 20 or so sources located within a central radius of 2 arc minutes may

Fig. 6. High-resolution imager exposure to the center of M31. North is to the top, east to the left. The many individual x-ray sources are apparent. On the basis of the observed x-ray luminosities, we believe that most of these sources are mass-exchange binary systems containing a collapsed star accreting gas from its companion.

well be a separate "inner bulge" population. These sources, as well as nine or ten probably associated with globular cluster systems in M31, may represent a different class of sources possibly connected with older, low-mass binary systems formed by a capture process. The average luminosity of the inner bulge sources in M31 is 4.5×10^{37} erg/sec, approximately twice the average luminosity of the spiral arm sources. More important, this inner bulge region, with a 400-parsec radius, contains about 1.5 percent of the mass of M31 and is responsible for ~ 33 percent of the 0.5- to 4.5-keV x-ray emission. It is also very much smaller than the inner bulge region of x-ray sources in our own galaxy. This concentration of x-ray-emitting systems in the center of M31 and the large x-ray output from a relatively small fraction of the galaxy's mass indicate that further x-ray studies of M31 and other nearby normal galaxies can teach us much more about the stellar evolutionary processes at work in different galaxies.

In addition to using observations of luminous binaries to study normal galaxies, we can apply observations of normal stars to study the stellar content of galaxies. For example, some models for elliptical galaxies invoke a population of M stars of low mass and low optical luminosity to account for a substantial fraction of the overall mass of the galaxy. For a galaxy containing 10^{11} to 10^{12} M stars we would expect an x-ray emission of 10^{38} to 10^{40} erg/sec from these stars alone. Einstein observations can detect such galaxies at least out to the Virgo cluster of galaxies (a distance of 20 million parsecs) and may provide the means of detecting the optically dark halos theorized to surround galaxies and comprise much of their mass.

The Einstein x-ray observations of the radio galaxy Centaurus A, located at a distance of 5 million parsecs, illustrate several different phenomena (23). A point x-ray source is associated with the nucleus of the galaxy. Extended x-ray emission has also been detected from a region 2 arc minutes in radius around the nucleus. This emission may be due to the superposition of presently unresolved luminous binaries or to a cloud of hot gas in the interior of the galaxy. Still farther away from the nucleus, x-ray emission is detected from the radio-emitting regions known as the inner lobes. Centaurus A shows two pairs of radio lobes aligned on opposite sides of the nucleus and thought to have been ejected in two separate explosive events. The data suggest that the x-ray emission from the inner lobes is produced by inverse Compton scattering of the radio-producing electrons off the photons responsible for the 3 K blackbody background. This then permits a first determination of the magnetic field for this region, which we calculate as 4×10^{-6} gauss.

Figure 7 is an isointensity contour plot of the HRI observations of Centaurus A, showing the pointlike source associated with the nucleus and an additional jetlike x-ray feature to the northeast of the nucleus. This x-ray jet is aligned with, but just within, a previously reported optical jet (24), and both are aligned in the direction of one of the inner radio lobes. Our calculations indicate that the x-ray emission in the jet can arise from thermal bremsstrahlung radiation, produced by the Coulomb acceleration of the hot electrons in the plasma. At the same time, the kinetic energy carried by the dense subrelativistic plasma comprising the jet is sufficient to power the inner radio lobe, which may resolve the question of the energy supply required by such long-lived radio-emitting regions.

In our description of x-ray emission from normal galaxies, we separate the galactic nucleus from the remainder of the galactic emission. It may well be that the emission from the nuclei of normal galaxies is closely related to the emission from the nuclei of active galaxies and quasars. Many astrophysicists believe that the nuclei of active galaxies are the sites of massive black holes and that the

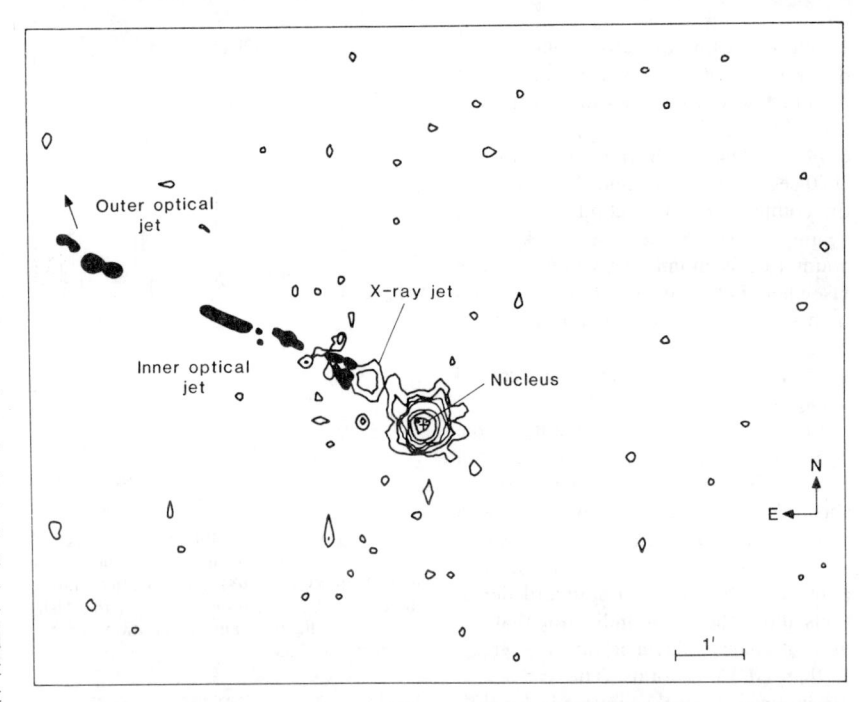

Fig. 7. Isointensity contour map of the HRI x-ray image obtained around the nucleus of the radio galaxy Centaurus A. The nucleus and x-ray jet toward the northeast are indicated, as is the optical jet. [This figure originally appeared in *Astrophysical Journal*]

emission from these objects is powered by the gravitational energy released by infalling matter (25). One way of providing the infalling material is through tidal disruption of stars that wander too close to the black hole. Normal galactic nuclei might contain "dormant" or "fossil" black holes, which are quiet because of the temporary or permanent absence of infalling material. X-ray luminosities for galactic nuclei range from 10^{35} erg/sec for our galactic nucleus to 10^{38} erg/sec for M31, 10^{42} to 10^{45} erg/sec for type 1 Seyfert galaxies, and 10^{43} to 10^{47} erg/sec for quasars. Narrow-emission-line galaxies, type 2 Seyfert galaxies, radio galaxies, N-type galaxies, and BL Lacs also fit within this range. Thus the x-ray luminosities may be useful in establishing specific relationships among these various classes of active galactic nuclei. In the context of the massive black hole model, the observed luminosity can be used to estimate a minimum black hole mass through the Eddington limit, which establishes a maximum rate for the infalling gas due to the pressure created by the outward-moving radiation. For a 10^{47} erg/sec source, a minimum black hole mass of 10^9 solar masses is required.

Additional evidence that the activity in galactic nuclei occurs in very small regions was obtained by monitoring source intensities as a function of time. It now appears that our x-ray observations are capable of probing the innermost (or smallest) regions of active galaxies and quasars. The data shown in Fig. 8 were obtained during a 2-day Einstein IPC observation of the Seyfert galaxy NGC 6814 (26). The data have been summed in 2000-second intervals and the count rate (in counts per 100 seconds) is plotted against time. Since the background count rate is monitored simultaneously elsewhere in the imaging detector (at less than 1 count per 100 seconds), we can be confident that the intensity variations seen in Fig. 8 are not due to background changes or instrumental effects. The error bars indicate the uncertainty in the count rate based on the actual number of counts. The mean for all of the data is shown by the dashed line. A chi-squared test indicates a negligible probability that the source intensity is constant. The data point at ~ 40 ksec is 5 standard deviations above the mean, indicating that significant changes occur in times no greater than 20,000 seconds. The average x-ray luminosity for this source is 4×10^{42} erg/sec and the observed variations indicate that a substantial fraction of this luminosity is produced in a region no more than 1/4 light-day in size. Similar variability has been observed in several type 1 Seyfert galaxies and quasars, although it has not been found in several others. Our findings of high luminosities and small x-ray–emitting regions, along with optical data showing high luminosities and rapid variability in some BL Lac objects and "violently variable" quasars, may be the most compelling evidence to date for the existence of massive black holes in active galactic nuclei.

X-ray Emission from Clusters of Galaxies

As is well known, stars are not uniformly or randomly dispersed in space, but are aggregated along with gas and dust to form galaxies. In a somewhat similar fashion, galaxies are not uniformly distributed, but tend to occur in groups or clusters with tens to thousands of members in regions of space typically a few million light-years across. In addition to their component galaxies, clusters also contain substantial amounts of hot gas, as was first conclusively demonstrated by the UHURU observations of extended x-ray emission from clusters (27). The Einstein observations of clusters have been directed toward improving our understanding of how the cluster environment affects the evolution of individual galaxies and how the clusters as a whole form and evolve.

Our studies of the development of individual galaxies in a cluster environment are perhaps best illustrated by the x-ray results obtained for a portion of the Virgo cluster of galaxies, shown in Fig. 9 (28). Here we have plotted the isointensity x-ray contours superimposed on a visible light photograph (from the Kitt Peak 4-m telescope). The two brightest galaxies, the ellipticals M86 and M84, are both relatively bright x-ray sources. X-ray emission is also observed from the tidally disrupted galaxy NGC 4438 to the northeast and from the narrow-emission-line galaxy NGC 4388 to the southwest. Since M84 is a radio galaxy, we initially hypothesized that the x-ray emission might be related to activity in the galaxy nucleus. Subsequent observations with the Einstein HRI, utilizing the arc-second resolution capability, demonstrated that the x-ray emission from M84 is diffuse and not primarily associated with the nucleus of the galaxy. Our present understanding of the x-ray emission from M84 is probably best discussed in the context of the M86 data.

The contours in Fig. 9 for M86 show that the x-ray emission is extended over several arc minutes with an apparent asymmetry to the northeast. A preliminary spectral analysis of this extended emission indicates that it is produced by hot gas with a temperature of $\sim 10 \times 10^6$ K. From the observed x-ray luminosity of 2×10^{41} erg/sec and the temperature and size, we can compute the central particle density as $\sim 4 \times 10^{-3}$ cm^{-3}. This leads to a mass estimate of several billion solar masses for the x-ray–emitting gas. This gas is thought to be generated by mass lost from stars and heated by supernova explosions within the galaxy itself. If the overall mass of M86 is 10^{12} solar masses, then this hot x-ray–emitting gas cannot be completely bound gravitationally to the galaxy. Recent calculations (29) suggest that the gas may be confined by the external pressure generated by the hotter, lower density gas that has been observed through its x-ray emission to fill the Virgo cluster. The cooling time for this hot, low-density intracluster gas is greater than the age of the universe, which rules it out as the source of the x-ray emission from M86.

The confinement of the gas associated with M86 by the hotter, lower density cluster gas must compete with ram pressure forces trying to strip the gas from the galaxy as it moves through the cluster. Since M86 is probably in an eccentric orbit, it has spent much of its time moving at relatively low velocity far from the center of the Virgo cluster. Thus a stripping-free time of up to 5 billion years has been available to generate the gas that we are now observing. As M86 moves at higher velocity through the cluster core, most of this gas associated with the galaxy will be decelerated and stripped away by the ram pressure of the intracluster gas. On the other hand, M84 is a well-bound member of the Virgo cluster core, spending most of its time in the denser central region of the cluster. In this case the gas produced in the galaxy is continually stripped from all but the innermost regions, which explains the much smaller size observed in x-rays for M84.

It is interesting to note that the intracluster gas exerts two opposing effects on the gas produced in these elliptical galaxies. On the one hand, the tenuous external medium provided by the cluster gas may be sufficient to confine the gas and prevent its being lost from the galaxy. On the other hand, when the galaxy moves through a sufficiently dense part of the intracluster gas at sufficiently high velocity, the gas will be stripped from the galaxy. It is likely that this gas stripped from the galaxies is subsequently heated to become part of the hot intracluster gas, which is further discussed below. By comparison, we expect that gas generated in isolated elliptical galaxies will be lost from these galaxies through a galactic wind in the absence of the confining pressure of the intracluster medium unless very massive dark halos are present to gravitationally bind the gas. As mentioned above, x-ray observations have the potential for detecting low-mass M stars, which may comprise such massive halos, as well as the hot gas that would be bound.

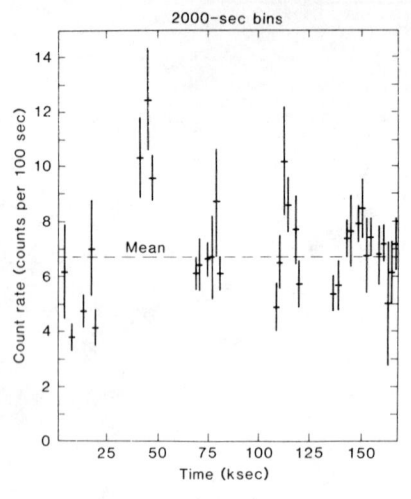

Fig. 8. X-ray intensity plotted in 2000-second intervals as a function of time for the type 1 Seyfert galaxy NGC 6814. The mean counting rate for the 2 days data is shown by the dashed line. Significant intensity variations are apparent in the data.

This capability for detecting hot gas makes x-ray observations a most sensitive probe of the intracluster medium and thereby of the gravitational potential of the cluster as a whole. Various scenarios have been developed to describe the formation and evolution of clusters of galaxies, and we can use our x-ray data to evaluate these different formulations. An example of a cluster believed to be in an early phase of its development is shown in Fig. 10 (30). This is the cluster Abell 1367 at a distance of about 120 million parsecs. The x-ray emission is several times 10^{43} erg/sec and the isointensity contours shown in Fig. 10 demonstrate the broad, highly clumped nature of the emission. The x-ray picture is superimposed on an optical photograph and some of the x-ray peaks are associated with bright galaxies in the cluster, much as was the case for the Virgo cluster galaxies. In both of these examples we believe that the x-ray emission is dominated by gas that originates in the individual galaxies and is still at least partially bound to the galaxies. The gravitational potential of these individual galaxies is apparently more important for containment than the gravitational potential of the entire cluster. The galaxies move about the cluster with relatively low velocities and the gas temperature is also relatively low—about 15×10^6 K for Abell 1367. In some instances Fig. 10 shows x-ray enhancements or clumps of hot gas where no bright optical galaxy is present. We are scheduling higher resolution x-ray observations for this cluster to pursue the possibility that the x-ray enhancements are indicators of optically dark, previously undetected mass concentrations.

As the cluster continues to evolve, gas generated in the galaxies will be stripped by processes such as we discussed above for M86. This may affect the distribution of morphological types of galaxies in the clusters, with fewer gas-rich spiral galaxies in more highly evolved clusters. With the passage of time the gas and galaxies will interact with each other and a smooth, centralized distribution will develop following the cluster gravitational potential. Figure 11 shows the x-ray isointensity contours superimposed on an optical photograph of the cluster Abell 85, which we believe to be in this more highly evolved state. The data show a smooth, centrally peaked x-ray distribution, indicating that the gas distribution is, in fact, dominated by the overall cluster gravitational potential. The x-ray temperature for this cluster is at least 100×10^6 K, and the relative velocities of the galaxies are also expected to be high. The optical data also show a very bright central dominant (cD) galaxy at the cluster center. The formation of such a massive, bright galaxy may be enhanced by the high density of galaxies and/or gas in the cluster core, although not all highly evolved clusters contain such a cD galaxy. Further cluster evolution may lead to a smooth but somewhat relaxed or less concentrated distribution of gas and galaxies, which will also be reflected by the x-ray morphology. Examples of such clusters may be the Coma cluster, Abell 2319, and Abell 2256.

These models suggest a trend toward increasing cluster x-ray luminosity and temperature as a function of time, since the amount of gas available is increasing and the strength of the central gravitational potential is also increasing with time. In addition, gas that has been pro-

Fig. 9. Isointensity IPC x-ray data superimposed on a Kitt Peak 4-m visible light photograph of a portion of the Virgo cluster of galaxies. Extended x-ray emission associated with galaxy M86 is apparent, as are sources associated with galaxies M84, NGC 4438, and NGC 4388. [This figure originally appeared in Astrophysical Journal]

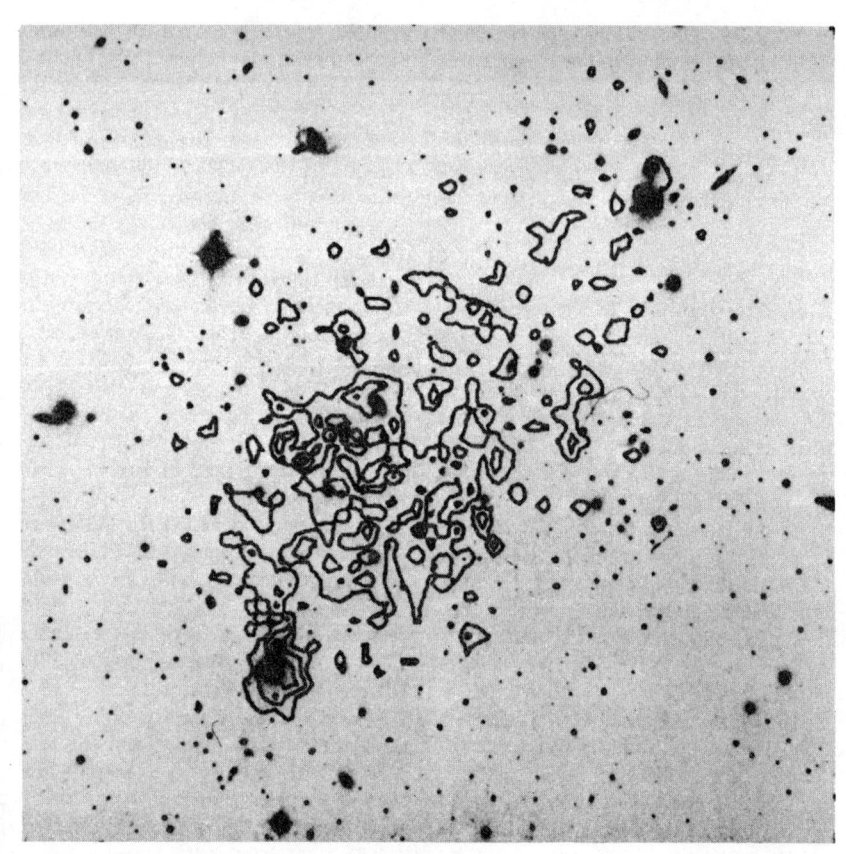

Fig. 10. Isointensity contour map of IPC x-ray data for cluster Abell 1367 superimposed on visible light photograph. Note the broad, highly clumped nature of the x-ray emission. [This figure originally appeared in Astrophysical Journal]

cessed through stars in galaxies and then released into the intracluster medium has been observed to be enriched in iron, with detectable iron emission lines in the x-ray spectra. This evolutionary picture can be tested in a statistical manner by observing the x-ray emission from a number of clusters at great distances (or large redshifts) and comparing it with data for relatively nearby clusters. On the average, clusters at large distances will appear earlier in their development than nearby clusters, since the radiation we observe now has taken millions or even billions of years to reach us. With Einstein we have initiated such an observing program. We have already detected some of the most distant known clusters out to a redshift (z) of 0.8, which corresponds to a distance of ~ 5000 million parsecs (*31*). Although we have not yet obtained sufficient data to conclusively test the picture, the data obtained so far are consistent with this evolutionary scenario. More sensitive observations with an x-ray observatory following Einstein will be required to accurately map the cluster gas distribution in very distant clusters and to determine the iron content and spatial distribution in these clusters. It is also possible that this next-generation x-ray telescope will be capable of determining the largest redshift at which clusters exist and thereby establishing the age of the universe at which the clusters first formed.

Since the mass of the x-ray–emitting gas in the clusters is usually at least as large as the mass corresponding to the visible galaxies in the clusters, we can see the importance of using x-ray observations to trace the gas and determine its mass distribution and total mass. One of the more puzzling questions in astronomy involves the so-called missing mass in clusters. Observations of galactic velocities indicate that the mass required to keep the clusters bound exceeds the mass corresponding to visible galaxies by as much as a factor of 10. One possible explanation was that the missing mass was in the form of gas. While the x-ray observations confirm the presence of considerable amounts of gas, there is still not enough in the core of the cluster to hold the cluster together. Substantially more gas may exist outside the cluster core although the contribution of such gas to the binding of the cluster is still the subject of some controversy. Another possibility is that the missing mass may exist in dark, massive halos of individual galaxies, a possibility that we discussed earlier. In each of these cases the x-ray data will be required to provide critical detailed information on the distribution of matter within the clusters.

Origin of the X-ray Background and Cosmological Research

The application of x-ray observations to the study of cosmological problems has been a tantalizing possibility since the very beginning of x-ray astronomy. The rocket flight that discovered extrasolar x-ray sources in 1962 (*32*) also discovered the existence of an apparently isotropic background of radiation in the energy range 2 to 8 keV. The results were immediately compared to the predictions of the hot-universe model of steady-state cosmology and were found to be incompatible. At energies below 1 keV, rocket experiments showed the presence of a structured galactic component of the background, while the results of the UHURU and other satellite observations made it clear that at higher energies the background was primarily extragalactic.

Once this was established, it was shown that a substantial fraction of the radiation must originate at redshifts greater than 1 (*33*). Even in the extreme assumption of a uniform distribution of sources in a Euclidean universe, this fraction is greater than 20 percent. Therefore, the study of the cosmic x-ray background may provide information about the early epochs of the universe, particularly in the range of redshifts intermediate between those accessible to optical observations ($z \leq 3$) and those from which the observed microwave background radiation appears to originate ($z \geq 1000$). It is in this intermediate range of redshifts that we may hope to study the transition from density enhancements in a largely homogeneous universe to the first stages of formation and evolution of stellar and/or galactic systems.

Until the advent of the Einstein Observatory we could not, however, use this powerful new tool of cosmological research because we did not understand the origin of the x-ray background radiation and we did not have the sensitivity to observe directly the x-ray emission of even the most powerful extragalactic objects at large redshifts. The great improvement in sensitivity as well as angular resolution brought about by the Einstein mission has given us the opportunity to attack the problem along two distinct, but complementary, lines. The

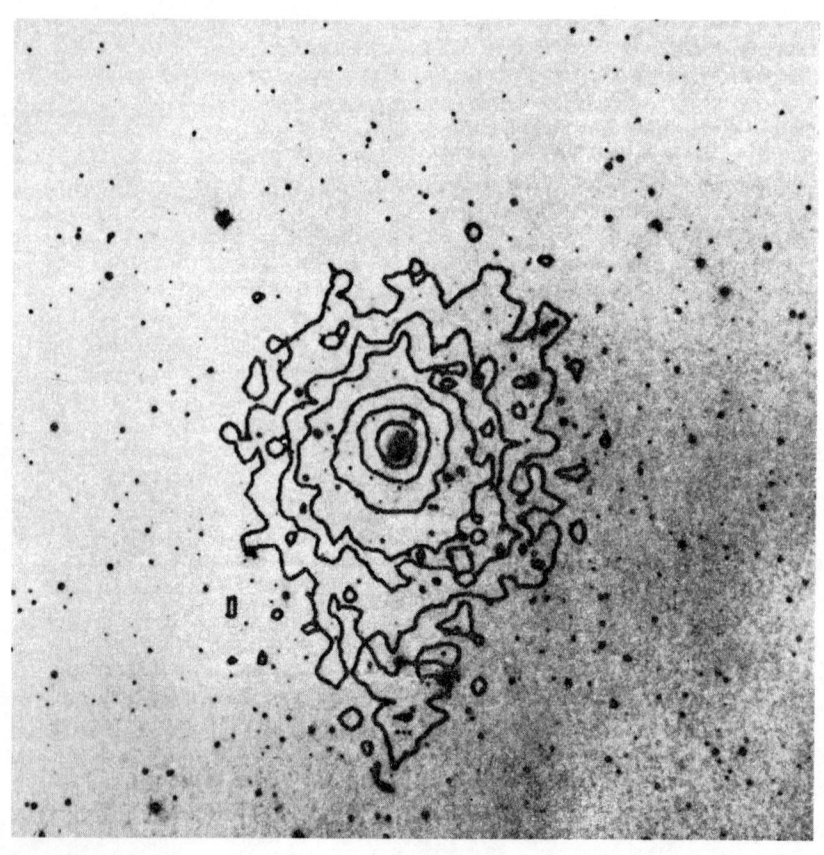

Fig. 11. Isointensity contour map of IPC x-ray data for cluster Abell 85 superimposed on visible light photograph. Note the bright, central galaxy and the highly peaked, smooth x-ray distribution. [This figure originally appeared in *Astrophysical Journal*]

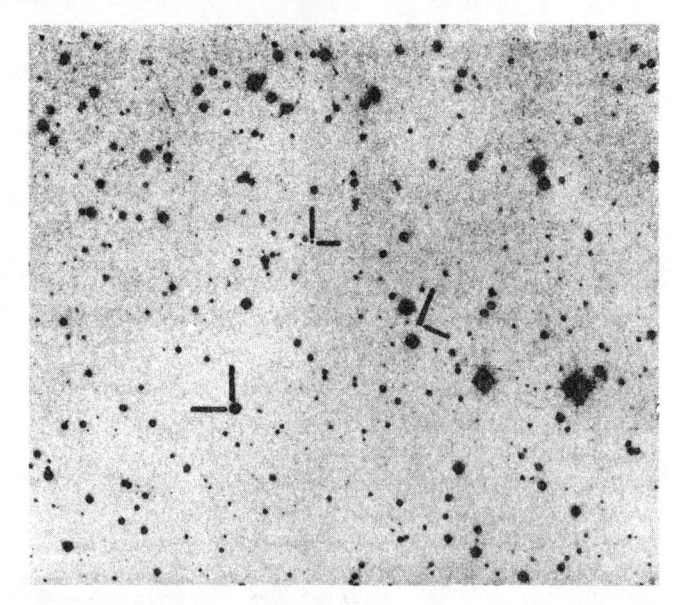

Fig. 12 (left). An HRI image obtained during Einstein deep x-ray survey of field in Eridanus. Three sources—two quasars and one star—are visible in the x-ray data.   Fig. 13 (right). Forty-eight-inch Schmidt plate showing the visible light photograph corresponding to the x-ray exposure of Fig. 12. The optical counterparts of the three x-ray sources are indicated.

first is to ascertain directly whether the x-ray background is due to a collection of unresolved individual sources or to truly diffuse processes. The second is to study the x-ray emission from individual classes of sources back to the largest redshifts at which they have been detected and to compare their summed contribution to the total background flux.

The first approach was needed because the data from earlier satellites did not provide sufficient constraints to decide the issue. The observed granularity of the background ($\sim$ 3 percent) and its spectrum could be reconciled with either of two hypotheses: that the background was due to the superposition of individual sources or that it was due to a single diffuse emission process. Although the totality of the background could not be explained in terms of any known class of extragalactic emitter without taking into consideration evolutionary effects, once such effects were taken into account, several classes of objects could yield a large fraction, if not all, of the background (34).

Among a number of plausible theories that were advanced, the view that the background was due to thermal bremsstrahlung emission from a hot intergalactic plasma appeared most attractive (35) and seemed to be supported by the results of HEAO-1 experiments (36). Boldt et al. (36) confirmed and refined previous work on the spectrum of the background and showed that to a high degree of accuracy it appeared to follow an exponential law, as would be predicted for thermal bremsstahlung emission from a gas with a temperature of 45 keV, or 500 $\times$ 10^6 K. For gas that was not excessively clumped, it could be shown that the mass associated with the gas could exceed the mass of every other component of the universe and, in fact, provide a non-negligible fraction of the mass required to slow down and eventually reverse the present expansion of the universe. In spite of the apparent attractiveness and simplicity of this explanation, several problems remained unresolved regarding the origin and heating mechanisms of the gas; in addition, such an explanation limited the contribution from discrete extragalactic sources, contrary to expectation (37).

The deep surveys carried out with the Einstein Observatory address this problem by extending the number-intensity relationship to intensities three orders of magnitude fainter than those detected by UHURU, or in other words by directly imaging the background (38). If the number of sources increased with decreasing intensity according to a 3/2 power law, as expected for a uniform distribution of sources in a Euclidean universe, we would observe several million sources in the sky or several tens of sources per square degree. At this source density, imaging is essential to avoid source confusion.

We select regions of the sky that have no known x-ray source or peculiar radio or optical features. This is to ensure that they represent as unbiased a sample of the sky as we can choose. We obtain radio and optical coverage of the regions at the very limit of ground-based telescopes and then cover the regions with a mosaic of x-ray exposures. Using automated detection algorithms to reveal the presence of sources, we cross-correlate the information from different detectors or exposures to ascertain the existence of these sources. We then attempt first-cut identifications to determine the galactic or extragalactic nature of the objects. An example is given in Figs. 12 and 13, where the x-ray image and optical photograph of a small region of the sky are shown. The three x-ray sources that can be seen are identified with visible light counterparts, marked by arrows on the 48-inch Schmidt plate. Two of the objects are newly discovered quasars and one a sunlike star, a few hundred light-years away. The quasars are at redshifts of 0.5 and 1.96, respectively.

This example is typical of the findings in the two deep-survey regions analyzed to date. Some of the objects (about one-third) turn out to be stars; the remainder are extragalactic, and of these several are identified with quasars. The survey of the number count of sources indicates that at least 30 percent of the extragalactic background is due to discrete sources, up to our current limit of sensitivity. It is not too daring to surmise that if one could image even fainter sources, perhaps all of the background could be resolved into individual discrete sources. The data are consistent with the view that the bulk of the sources may be quasars at redshifts greater than 1, in agreement with what we could expect from our knowledge of quasar evolutionary effects.

This view is very much strengthened by the complementary study of x-ray emission from quasars at large redshifts

(*39*). Only three quasars, among the nearest to us, had been detected before the launch of Einstein. The Einstein x-ray telescope has made it possible to extend the detection of quasar x-ray emission to the most distant known objects, up to redshifts of 3.5. Our studies show that the average ratio of x-ray to visible light luminosity of these objects, although varying by as much as 250 from object to object, remains fairly constant, independent of the visible light luminosity or the redshift of the object. Then the known increase in the number density of quasars in the past (or at large redshift) can account for all of the x-ray background. In fact, a simple extrapolation based on the observed population of quasars at faint optical magnitudes and on an assumed constant ratio of x-ray to visible luminosity can quickly exceed the observed background. This implies that at some point in the past the quasars have to become weaker in x-rays in proportion to visible light than they are at present, or that the quasar number density must eventually decrease, or both. This is the first important result on cosmological evolution to stem directly from the x-ray data.

We mentioned another example of cosmological research in progress—the search for evolutionary effects in clusters of galaxies. Complete studies of the x-ray properties of clusters at redshifts greater than 1 will probably require advanced x-ray facilities.

Finally, there appears to be within our grasp the prospect, not yet fully realized, of measuring some of the fundamental constants such as the rate of expansion and the age of the universe through a combination of microwave and x-ray measurements of distant clusters (*40*). The main difficulty, in this most important and far-reaching undertaking, is the measurement of the microwave portion of the effect.

These examples should demonstrate that x-ray astronomy has been established through the Einstein observations as a powerful tool for cosmological research. Future progress in these areas, as in many areas of astronomy, will require the combination of sensitivity radio, optical, and x-ray studies.

References and Notes

1. H. Wolter, *Ann. Phys.* **10**, 94 (1952).
2. R. Giacconi and B. Rossi, *J. Geophys. Res.* **64**, 773 (1960).
3. G. S. Vaiana, J. M. Davis, R. Giacconi, A. S. Krieger, J. K. Silk, A. F. Timothy, M. Zombeck, *Astrophys. J. Lett.* **185**, L47 (1973).
4. R. Giacconi et al., *ibid.* **230**, 540 (1979). (This paper furnishes a more detailed description of the Einstein instrumentation and references even more detailed descriptions of the individual instruments.)
5. S. E. Strom, K. M. Strom, G. L. Grasdalen, *Annu. Rev. Astron. Astrophys.* **13**, 187 (1975).
6. T. P. Snow and D. C. Morton, *Astrophys. J. Suppl.* **32**, 429 (1976).
7. H. Tananbaum, in *X- and Gamma-Ray Astronomy*, H. Bradt and R. Giacconi, Eds. (Reidel, Dordrecht, Netherlands, 1973), pp. 9–28.
8. R. Giacconi, in *Proceedings of the 16th Solvay Conference on Physics* (Editions de l'Université de Bruxelles, Brussels, 1974), pp. 27–72.
9. J. Grindlay et al., *Astrophys. J. Lett.* **205**, L127 (1976).
10. G. Garmire, "Low luminosity galactic x-ray sources," *Proceedings of the COSPAR Symposium, Innsbruck, Austria, 29 May to 10 June 1978.*
11. G. Vaiana, *Proceedings of the International Astronomical Union meeting, August 1979, Montreal* (in press).
12. F. R. Harnden et al., *Astrophys. J. Lett.* **234**, L51 (1979).
13. W. Ku and G. Chanan, *ibid.*, p. L59.
14. F. Seward, W. Forman, R. Giacconi, R. E. Griffiths, F. R. Harnden, Jr., C. Jones, J. P. Pye, *ibid.*, p. L55.
15. F. R. Harnden et al., *Bull. Am. Astron. Soc.* **11**, 789 (1980).
16. D. Helfand, G. Chanan, R. Novick, *Nature (London)* **283**, 337 (1980).
17. S. S. Murray, G. Fabbiano, A. C. Fabian, A. Epstein, R. Giacconi, *Astrophys. J. Lett.* **234**, L69 (1979).
18. R. H. Becker, S. S. Holt, B. W. Smith, N. E. White, E. A. Boldt, R. F. Mushotzky, P. J. Serlemitsos, *ibid.*, p. L73.
19. D. Arnett, private communication.
20. D. Weedman, *Annu. Rev. Astron. Astrophys.* **15**, 69 (1977).
21. K. S. Long and D. J. Helfand, *Astrophys. J. Lett.* **234**, L77 (1979).
22. L. Van Speybroeck, A. Epstein, W. Forman, R. Giacconi, C. Jones, W. Liller, L. Smarr, *ibid.*, p. L45.
23. E. J. Schreier, E. Feigelson, J. Delvaille, R. Giacconi, J. Grindlay, D. A. Schwartz, A. C. Fabian, *ibid.*, p. L39.
24. R. J. Dufour and S. van den Bergh, *ibid.* **226**, L73 (1978).
25. D. Lynden-Bell, *Nature (London)* **233**, 690 (1969).
26. M. Elvis, E. Feigelson, R. E. Griffiths, J. P. Henry, H. Tananbaum, in *Highlights of Astronomy* (Reidel, Dordrecht, Netherlands, in press).
27. H. Gursky, E. Kellogg, S. Murray, C. Leong, H. Tananbaum, R. Giacconi, *Astrophys. J. Lett.* **167**, L81 (1971).
28. W. Forman et al., *ibid.* **234**, L27 (1979).
29. A. C. Fabian, J. Schwarz, W. Forman, in preparation.
30. C. Jones, E. Mandel, J. Schwarz, W. Forman, S. S. Murray, F. R. Harnden, Jr., *Astrophys. J. Lett.* **234**, L21 (1979).
31. J. P. Henry, G. Branduardi, U. Briel, D. Fabricant, E. Feigelson, S. S. Murray, A. Soltan, H. Tananbaum, *ibid.*, p. L15.
32. R. Giacconi, H. Gursky, F. Paolini, B. Rossi, *Phys. Rev. Lett.* **9** (No. 11), 439 (1962).
33. D. Schwartz and H. Gursky, in *X-ray Astronomy*, R. Giacconi and H. Gursky, Eds. (Reidel, Dordrecht, Netherlands, 1974), pp. 359–388.
34. D. Schwartz, in *(COSPAR) X-Ray Astronomy*, W. Baity and L. E. Peterson, Eds. (Pergamon, New York, 1979), pp. 453–465.
35. G. Field, *Annu. Rev. Astron. Astrophys.* **10**, 227 (1972).
36. E. Boldt, F. Marshall, R. Mushotzky, S. Holt, R. Rothschild, P. Serlemitsos, in *COSPAR X-ray Astronomy*, W. A. Baity and L. E. Peterson, Eds., (Pergamon, New York, 1979), p. 443.
37. G. Field, in preparation.
38. R. Giacconi et al., *Astrophys. J. Lett.* **234**, L1 (1979).
39. H. Tananbaum et al., *ibid.*, p. L9.
40. J. Silk and S. D. M. White, *ibid.* **226**, L103 (1978).
41. We gratefully acknowledge the help of hundreds of engineers, scientists, students, program managers, computer programmers, technicians, and secretaries, without whose contributions the Einstein program would not have been possible. Although it is not possible to acknowledge individuals, we wish to take this opportunity to recognize the organizations to which they belong. The help and support of NASA Headquarters, the guidance by the management team led by F. Speer at Marshall Space Flight Center, the support by Kennedy Space Flight Center for launch, and the ongoing team efforts at Goddard Space Flight Center for mission operations have been essential. Also essential were the efforts of the industrial contractors: TRW for the spacecraft design and construction; AS & E for the experiment construction, integration, and testing; Perkin-Elmer for the fabrication of the x-ray mirror; Honeywell Electro-Optics Center for the construction and testing of the star trackers; Convair Division General Dynamics for the construction of the optical bench; LND for the preparation of proportional counters; Parker-Hannifin Corp., Lord Corp., Aeroflex Laboratories, Inc., R & K Precision Machine Co., and Astronautic Industries, Inc. for many of the mechanical assemblies; Spacetac, Inc. and Matrix Research and Development Corp. for low- and high-voltage power supplies; and, finally BASD (formerly Ball Brothers) for the construction of the cryostat. We wish to acknowledge the constant support given to this project by the management and staff of our own institutions: Harvard/Smithsonian Center for Astrophysics (R. Giacconi, Principal Investigator; Harvey Tananbaum, Scientific Program Manager; Leon Van Speybroeck, assumed scientific guidance for design and development of the grazing incidence telescope); Massachusetts Institute of Technology (George W. Clark, Principal Scientist); Goddard Space Flight Center (Stephen S. Holt, Principal Scientist); and Columbia University (Robert Novick, Principal Scientist). This research was sponsored under NASA contracts NAS8-30751, NAS8-30752, and NAS8-30753.

COVER

Time-phased x-ray images of the Crab Nebula and pulsar from the Einstein Observatory. The upper photograph, synchronized to the pulse peak, is dominated by the pulsar. The lower photograph, synchronized to the pulsar "off" state, is dominated by the extended nebular emission. The residual emission from the pulsar location is used to estimate a maximum temperature for the neutron star surface. See page 865. [F. R. Harnden, Jr., Harvard-Smithsonian Center for Astrophysics, Cambridge, Massachusetts]

THE ASTROPHYSICAL JOURNAL, 245:163-182, 1981 April 1
© 1981. The American Astronomical Society. All rights reserved. Printed in U.S.A.

Reprinted with permission from *The Astrophysical Journal*, 245 pp. 163-182, G.S. Vaiana, *et al.*, "Results from an Extensive *Einstein* Stellar Survey." © 1981 The American Astronomical Society.

RESULTS FROM AN EXTENSIVE *EINSTEIN* STELLAR SURVEY

G. S. VAIANA,[1] J. P. CASSINELLI,[2,4] G. FABBIANO, R. GIACCONI, L. GOLUB, P. GORENSTEIN, B. M. HAISCH,[3,4]
F. R. HARNDEN, JR., H. M. JOHNSON,[4,6] J. L. LINSKY,[3,4,5] C. W. MAXSON, R. MEWE,[4,7] R. ROSNER,
F. SEWARD, K. TOPKA, AND C. ZWAAN[4,7]

Harvard-Smithsonian Center for Astrophysics
Received 1980 April 2; accepted 1980 September 22

ABSTRACT

We report the preliminary results of the *Einstein* Observatory stellar X-ray survey. To date, 143 soft X-ray sources have been identified with stellar counterparts, leaving no doubt that stars in general constitute a pervasive class of low-luminosity galactic X-ray sources. We have detected stars along the entire main sequence, of all luminosity classes, pre–main sequence stars as well as very evolved stars. Early type OB stars have X-ray luminosities in the range $\sim 10^{31}$ to $\sim 10^{34}$ ergs s^{-1}; late type stars show a somewhat lower range of X-ray emission levels, from $\sim 10^{26}$ to $\sim 10^{31}$ ergs s^{-1}. Late type main-sequence stars show little dependence of X-ray emission levels upon stellar effective temperature; similarly, the observations suggest weak, if any, dependence of X-ray luminosity upon effective gravity. Instead, the data show a broad range of emission levels ($\sim$ three orders of magnitude) throughout the main sequence later than F0. Comparison of the data with published theories of acoustically heated coronae shows that these models are inadequate to explain our results. The data are consistent with magnetically dominated coronae, as in the solar case.

Subject headings: stars: coronae — X-rays: sources

I. INTRODUCTION

We report the preliminary results of an extensive, ongoing *Einstein* Observatory stellar survey. The observations discussed here involve the results of several distinct Harvard-Smithsonian Center for Astrophysics (CfA) observing programs (126 stars detected) as well as results provided by a number of collaborating Guest Observers (17 stars detected) and cover approximately the first 6 months of Observatory operation. These results are preliminary in nature; but because they provide a first overview of the levels of stellar X-ray emission throughout the H-R diagram, we felt that their intrinsic interest merited early dissemination.

The ongoing program has detected X-ray emission from stars and stellar systems throughout most of the H-R diagram (Fig. 1), and has established stars in general to be a class of low-luminosity ($< 10^{35}$ ergs s^{-1}) X-ray sources. The Observatory's high-efficiency imaging focal plane detectors and large collecting area optics combine to give a $\sim 10^3$ higher sensitivity for detection of point sources such as stars (Giacconi *et al.* 1979*b*) than previous surveys (cf. Vanderhill *et al.* 1975) and has therefore allowed detection of typical stellar X-ray emission levels. It can, therefore, now be asserted that X-ray

---

[1] Also Osservatorio Astronomico di Palermo, Italy.
[2] Washburn Observatory, University of Wisconsin, Madison.
[3] Joint Institute for Laboratory Astrophysics, University of Colorado and National Bureau of Standards.
[4] Guest Observer, *Einstein* Observatory.
[5] Staff Member, Quantum Physics Divison, National Bureau of Standards.
[6] Lockheed Palo Alto Research Laboratory.
[7] The Astronomical Institute, Utrecht, The Netherlands.

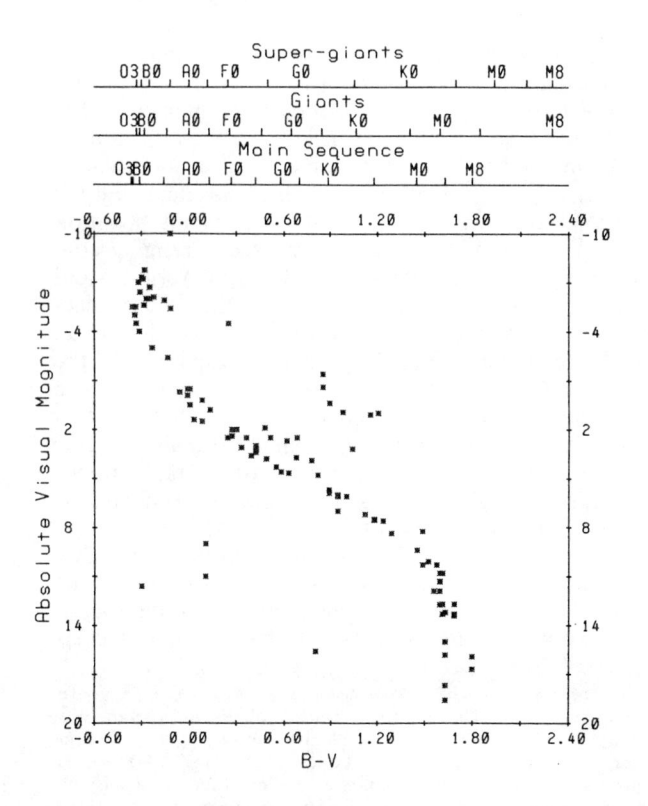

FIG. 1.—An H-R diagram for stars detected as soft X-ray sources by *Einstein*. The figure only includes that subset of Survey stars for which absolute visual magnitude could be determined either from the luminosity class or from parallax data; these stars are referred to as the "optically well-classified" sample.

emitting stars are the norm, not the exception; and it is now possible, for the first time, to test coronal theories throughout the H-R diagram. If, as we shall contend, stellar X-ray emission generally derives from hot stellar coronae, the *Einstein* Observatory results reported here provide the first *direct* observation of coronal activity throughout the H-R diagram; they thus extend previous, more limited X-ray observations (Catura, Acton, and Johnson 1975; Heise *et al.* 1975; Mewe *et al.* 1975; Vanderhill *et al.* 1975; Nugent and Garmire 1978; Cash *et al.* 1978; Topka *et al.* 1979; Walter *et al.* 1980*a*, *b*) and complement other observations in the radio, infrared, optical, and UV regions (Oster 1975; Zirin 1975; Wilson 1966, 1978; Stencel 1978; Dupree 1975; Evans, Jordan, and Wilson 1975; Vitz *et al.* 1976; Brown, Jordan, and Wilson 1979; Carpenter and Wing 1979; Haisch and Linsky 1976; Linsky and Haisch 1979; Hartmann *et al.* 1979; Cassinelli and Olson 1979).

Our paper is organized as follows: we briefly define the objectives and components of our stellar survey program and outline the data analysis (§ II); the detailed results are described in § III, and their implications for stellar coronal theory are discussed in § IV. The major results are summarized in § V.

II. THE *Einstein*/CfA STELLAR SURVEY

a) Objectives

The basic observational goal[8] of the CfA stellar survey is to establish the X-ray luminosity function and spectral characteristics for classes of stars throughout the H-R diagram. Corollary studies include the identification of relevant stellar characteristics (viz., age, multiplicity, rotation, effective temperature) which correlate with the presence of X-ray emission and its intensity; investigation of stellar variability on various time scales, ranging from flare transients to rotational and orbital periods; and determination of the stellar contribution to diffuse galactic soft X-ray emission from stellar clusters and galaxies.

These studies are readily within the capability of the *Einstein* Observatory, as $\sim 10^4$ stars are estimated to be accessible at flux levels comparable to that of stars already detected. Constraints upon the available observing time will limit observations to $\sim 10\%$ of that number by mission end, but this should suffice for our present survey purposes for most types of stars. However, at the present preliminary stage of the survey, in which roughly one-half of the observed stars are serendipitous discoveries with little corresponding optical data available, our aims are more modest. Our present analysis strategy is, first, to estimate *typical* emission levels via the Pointed Survey (which samples the nearest stars); second, to define the *upper range* of emission levels via the remaining serendipitous and magnitude-limited surveys (which are biased toward the most luminous sources). Definition of the *lower range* of emission levels is as yet relatively incomplete and must await completion of the ongoing Pointed Survey.

b) Organization

The data reported here were acquired as part of four independent observing programs which collectively comprise the *Einstein*/CfA stellar survey: (1) the *Pointed Survey*, consisting of pointed observations of a sample of nearby stars in each spectral type and luminosity class category; (2) the *8.5 Survey*, presently entailing a search of all imaging proportional counter (IPC) fields for sources associated with stars brighter than $V = 8.5$ (Topka 1980); (3) the *Deep Stellar Survey*, conducted as part of the *Einstein* Deep Surveys (Giacconi *et al.* 1979*b*), consisting of a search of Deep Survey fields for sources identified with stars; and (4) the *Serendipitous Survey*, consisting of a search of all CfA IPC and high-resolution imager (HRI) fields (other than Deep Survey fields) for serendipitous sources which can be identified with stars regardless of stellar optical magnitude. Each of these programs, biased by selection criteria used and presently in various stages of completion, contributes unique information regarding the X-ray luminosity function of the various types of stars.

c) Data Analysis

The details of the Observatory instrumentation and methods of data reduction have been described by Giacconi *et al.* (1979*a*). Many details of the general analysis techniques can also be found in Giacconi *et al.* (1979*b*), Harnden *et al.* (1979*a*), and Seward *et al.* (1979).

Rigorous spectral analysis is not yet possible for sources observed with the IPC, although it does provide pulse height spectra spanning the 0.2–4.0 keV passband. However, qualitative comparisons have been carried out for a subset of the stronger IPC sources which includes a substantial number of early-type stars and selected late-type stars. These comparisons of their pulse height spectra with pulse height spectra of the remaining IPC sources have yielded the qualitative result that all stellar sources show spectra consistent with temperatures of the order of or lower than $\sim 10^7$ K. In consequence, we have used conversion factors of 2×10^{-11} ergs per IPC count and 4×10^{-11} ergs per HRI count to derive the provisional 0.2–4.0 keV X-ray fluxes.[9]

---

[8] These objectives have evolved somewhat as the initial data have been interpreted. The original "pointed" observing program was designed to be conservative and exploratory in keeping with the firm knowledge available at that time; however, "surprises" which would open up new areas of investigation were allowed for by establishing several extensive search programs for serendipitous sources. Several of the original guest observer programs have also been particularly valuable in this latter regard by providing exploratory pointed observations which could not *a priori* have been justified within the time constraints on consortium programs. These observations have been distinguished from the CfA results in our tabular presentations.

[9] These conversions were derived by folding solar abundance thermal spectra through the IPC instrument response, with the conclusion that the IPC conversion factor varies by less than 20% over the source temperature range 0.25 to 1.0 keV if interstellar absorption is negligible (i.e., log $N_H \leq 19$). Comparison of HRI and IPC observations of the same stellar sources shows that the HRI conversion factor can vary substantially in this temperature range primarily for early type (OB) stars, for which this effect has been taken explicitly into account.

d) Tabulated Results

The number of *Einstein* X-ray sources identified with stars as part of the present program stands at 143. A list of these survey sources, together with their optical counterparts, is presented in Table 1. These data represent results of observations obtained from 1978 December to 1979 August,[10] excluding only fields with strong diffuse sources in the field of view or with very short exposure times. The X-ray source names indicate the X-ray source positions; positioning accuracy depends somewhat on source strength, typically ranging from $\sim 40''$ for IPC sources to $\sim 4''$ for HRI sources. Fluxes are determined after background subtraction.

Because the basic motivation of the Survey is to study the dependence of stellar X-ray emission upon stellar properties, Table 1 has been segregated by spectral type and luminosity class. In addition to the stellar effective temperature and surface gravity, other possibly relevant parameters include the stellar rotation rate, mean surface magnetic field strength, strength of various emission lines (viz., Ca II H and K), and multiplicity; where readily available, these data have been considered in the following discussion. We are presently conducting a long-range program of optical studies of all stars associated with *Einstein*/CfA stellar survey sources in collaboration with optical observers and more detailed analysis (including spectral fitting) of the X-ray data.

e) Luminosity and Flux Ratios

The most outstanding result of the survey is that all categories of stars—with the sole exception of very late type giants and supergiants—have been detected as soft X-ray emitters; this is illustrated by the H-R diagram shown in Figure 1, which shows all stars observed as X-ray sources, and whose spectral type and luminosity class are known (this sample of stars will henceforth be called the "optically well-classified" sample).

An important parameter for comparing X-ray emission levels from different types of stars is the ratio of X-ray to bolometric luminosity L_x/L_{bol}. In general this ratio can be computed only if both the spectral type and luminosity class of the star in question are available. Unfortunately, this is not the case for the majority of sources we have detected so far; many stars are relatively faint, and only color information is available. The available data will, however, allow us to compute—for all *Einstein* stellar sources—the X-ray flux to V-band flux ratio f_x/f_V. The ultimately desired ratio L_x/L_{bol} can then be derived if one knows (a) the extent of interstellar optical and X-ray extinction and (b) the proper bolometric correction.

Nevertheless, the f_x/f_V ratio does permit an immediate, qualitative insight into the variation of X-ray emission throughout the H-R diagram. Figures 2a–2c, which are plots of f_x/f_V versus spectral type show that X-ray emission is persistent along the entire main sequence, quite in contrast to expectations based upon traditional theories

of coronal heating (see, for example, Mewe 1979). Since, as we shall argue below, we are in most cases seeing *typical* X-ray emission levels (and not *just* the exceptionally luminous stars), our data call for a major reassessment of theories of stellar coronal formation.

III. DETAILED SURVEY RESULTS

A detailed analysis of the survey data leading to the construction of X-ray luminosity functions involves a comparison of data from each of the four distinct observing programs for each spectral type and luminosity class. The following discussion outlines the principal results of our preliminary analysis, separating the available sample of stars by spectral type and luminosity class; this separation is motivated by past theoretical expectations of levels of stellar X-ray emission and by organizational convenience, and should not be interpreted as a substantive categorization. The artificiality of this separation is particularly evident for very early type stars, where no evidence exists for luminosity class differences, and for very late-type dwarfs (dK and later), where levels of X-ray emission do not appear to be related to spectral type.

a) Early Type (O, B, and A) Main-Sequence Stars

The data for early main-sequence stars include a dozen stars of spectral types ranging from O3 to A3, with a significant gap in the early B range due to the fact that the Pointed Survey has not as yet (e.g., as of the cutoff date for this report) sampled any stars in this spectral range. From a theoretical point of view, these stars are unified by the fact that they are conventionally not expected to have a vigorous surface convection zone (and associated surface turbulence), and hence should not have the requisite conditions for acoustically heated coronae (de Loore 1970; Renzini et al. 1977); for this reason alone they are discussed together here. Considering the Pointed Survey stars only, the median X-ray luminosity varies dramatically from $\sim 10^{33}$ ergs s^{-1} to $\sim 10^{28}$ ergs s^{-1} (Figs. 3a and 4). If we consider the f_x/f_V ratio for these stars, shown in Figure 3b, we note that the emission levels of the sources found in the Serendipitous and 8.5 Surveys lie well within the upper range of the Pointed Survey sources (particularly in the O star case); because the former samples are strongly biased toward brighter sources, we can thus be confident that we have observed the upper range of the true luminosity function throughout this spectral range even in the nearby sample. We note here that the OB star results first reported by Harnden et al. (1979a) and Seward et al. (1979), which are included here, show X-ray luminosities somewhat larger than those reported recently by Long and White (1980).

Possible correlations of X-ray emission levels with multiplicity and stellar characteristics other than spectral type have been examined and are as yet unclear. The possible effects of multiplicity upon the existence and level of X-ray emission of OB stars have been discussed by Harnden et al. (1979), with the conclusion that no significant effect is discernible in our data; it is not as yet possible to arrive at a similar judgment for the late B and A stars.

---

[10] The cutoff date for inclusion is 1979 August for all but the Pointed Survey targets, for which some data up to 1979 December have been included.

TABLE 1

X-Ray Sources Detected in the *EINSTEIN* Stellar Survey[a]

X-ray Source Name	Star Name	Optical Position R.A.	Decl.	log F_x/F_v	log L_x (erg s^{-1})	Survey	log (ct s^{-1})
		Main Sequence - O Stars					
		h m s	° ' "				
1E 053249-0525.2	θ^1 Ori C	05 32 49.0	-05 25 16	-4.17	32.2	P	-0.94
1E 053255-0526.8	θ^2 Ori	05 32 55.4	-05 26 51	-4.94	31.3	P	-1.92
1E 104237-5928.4	HD 093205	10 42 37.4	-59 28 28	-3.81	33.0	P	-2.52
1E 104245-5931.0	CPD-59 2600	10 42 45.5	-59 31 08	-4.64	32.9	P	-3.0
1E 104248-5918.0	HD 093250	10 42 48.4	-59 18 07	-3.78	33.3	P	-2.4
1E 104309-5924.2	SAO 238431	10 43 09.2	-59 24 17	-4.73	...	P	-1.30
1E 2137.3+5714	HR 8281	21 37 24.3	+57 15 44	-4.2	31.9	8.5	-1.25
		Main Sequence - B Stars					
1E 0304.8+4045	β Per	03 04 54.3	+40 45 53	-4.04	30.7	P	+0.34
1E 0755.4-5251	χ Car	07 55 30.3	-52 50 30	-5.80	30.0 S	P	-1.96
		Main Sequence - A Stars					
1E 0634.8+1627	γ Gem	06 34 49.5	+16 26 36	-5.57	29.2	(E)	-1.12
1E 064255-1639.4	α C Ma A	06 42 55.3	-16 39 26	-7.60	26.9	P	-1.7
1E 110143+3830.6	51 U Ma	11 01 44.5	+38 30 40	-5.41	28.5	S	-2.89
1E 1155.4+3232	HR 4574	11 55 33.0	+32 33 09	-4.64	29.0 S	8.5	-2.00
1E 1329.9-4636	HD 117721	13 30 00.7	-46 36 16	-3.49	30.5 S	8.5	-1.48
1E 183515+3844.3	α Lyr	18 35 15.1	+38 44 18	-6.79 V	27.6 V	P	-1.89 V
		Main Sequence - F Stars					
		h m s	° ' "				
1E 0048.9-7125	HD 005028	00 49 01.4	-71 25 24	-3.79	29.5 S	8.5	-1.40
1E 0123.5+1854	ρ Psc	01 23 33.2	+18 54 46	-4.16	29.2	S	-1.09
1E 0124.1+3406	HD 008774	01 24 15.6	+34 07 06	-4.39	29.3 S	S	-1.68
1E 0150.2+2919	α Tri	01 50 13.5	+29 20 03	-4.31	29.5	P	-0.45
1E 0157.2-6148	α Hyi	01 57 12.8	-61 48 44	-5.91	27.9	P	-1.82
1E 0410.8+0734	46 Tau	04 10 51.3	+07 35 24	-4.22	29.1	P	-1.43
1E 0429.4+0518	HD 028736	04 29 25.1	+05 18 46	-4.39	28.9 S	8.5	-1.70
1E 0535.8-2842	ν^2 Col	05 35 47.3	-28 43 05	-4.57	28.7	8.5	-1.47
1E 1100.4+6155	HD 095638	11 00 17.1	+61 55 26	-4.00	29.1 S	8.5	-1.64
1E 1222.5+2548	HD 108102	12 22 31.8	+25 50 15	-3.23	29.9 S	8.5	-1.26
1E 1228.5+3141	HD 108944	12 28 32.3	+31 41 59	-3.92	29.0 S	S	-1.62
1E 1253.6+3835	α^1 C Vn α^2 C Vn	12 53 39.6 12 53 40.8	+38 35 00 +38 35 18	-4.65 T ...	29.0 T ...	P ...	-1.66 T ...
1E 1327.5-4621	SAO 224202	13 27 32.7	-46 20 33	-3.37	29.9 S	S	-1.82
1E 1335.8-2918	HR 5128	13 35 52.8	-29 18 24	-4.09	29.4 S	S	-1.20
1E 1755.8+1500	Z Her	17 55 51.2	+15 08 34	-3.27	30.3	P	-0.92

TABLE 1—Continued

X-ray Source Name	Star Name	Optical Position R.A.	Decl.	log F_x/F_v	log L_x (erg s^{-1})	Survey	log (ct s^{-1})
		Main Sequence - G Stars					
		h m s	° ' "				
1E 0305.5+4924	ι Per	03 05 30.5	+49 25 24	-5.35	27.6	P	-1.74
1E 0834.7+6512	π^1 U Ma	08 34 46.6	+65 11 47	-3.66	29.1	S	-0.68
1E 0930.1+7002	24 U Ma	09 30 05.5	+70 03 09	-3.5	30.0	(C)	-0.16
1E 1352.3+1838	η Boo	13 52 18.0	+18 38 41	-5.54	28.0	P	-1.39
1E 143556-6037.3	α Cen A	14 35 57.0	-60 37 28	-5.62	27.1	P	-0.83
1E 1744.4+2744	μ Her	17 44 29.2	+27 44 34	-5.48 T	27.6 T	P	-1.59
1E 2003.9-6620	δ Pav	20 03 56.4	-66 19 18	-5.55	27.1	(A)	-1.74
		Main Sequence - K Stars					
1E 0330.5-0937	ε Eri	03 30 32.4	-09 37 35	-3.86	28.3	(A)	-0.12
1E 041258-0745.9	40 Eri A	04 12 53.8	-07 45 25	-1.70	27.2	P	-1.77
1E 0431.0+0516	HD 028946	04 31 11.0	+05 16 57	-4.69	27.3	8.5	-2.64
1E 1332.1-0804	EQ Vir	13 32 06.0	-08 05 09	-2.07	29.4	P	-0.58
1E 1346.7+2713	DM+27 2296	13 46 46.4	+27 13 41	-4.68	27.0	S	...
				-4.50	27.0		-2.19
1E 143555-6037.6	α Cen B	14 35 55.5	-60 37 47	-4.73	27.5	P	-0.49
1E 1802.9+0229	70 Oph B	18 02 56.1	+02 30 02	-3.2 T	28.4 T	(A)	-0.35 T
1E 1810.5+6940	HD 167605	18 10 21.9	+69 00 00	-2.75	29.5 S	S	-1.16
1E 2009.6+3813	HD 192020	20 09 35.7	+38 14 58	-4.00	29.0 S	8.5	-2.25
1E 2104.8+3831	61 CYG	21 04 50.5	+38 31 35	-3.8 T	27.5 T	(A)	-0.94
		Main Sequence - M Stars					
		h m s	° ' "				
1E 0015.6+4344	+43 44 AB	00 15 38.9	+43 44 35	-3.3 T	27.1 T	(A)	-1.34 T
1E 0136.5-1813	UV Cet	01 36 31.4	-18 12 59	-0.75 T	27.5 T	P	-0.70 T
	L726-8			-0.95 T			
1E 041258-0745.9	40 Eri BC	04 12 53.8	-07 45 25	-5.00 T	27.8 T	P	-1.25 T
1E 0539.2+1228	ROSS 47	05 39 18.0	+12 28 32	-2.43	27.1	(A)	-1.84
1E 0731.4+3158	YY Gem	07 31 25.8	+31 58 47	-1.74	29.6	S *	-0.14
1E 1053.9+0718	CN Leo	10 53 58.5	+07 17 53	-0.88	27.1	P	-1.07
1E 1103.0+4346	DM+44 2051	11 02 47.5	+43 47 30	-3.34	27.2	(A)	-1.32 T
	WX U Ma	11 02 50	+43 47 10	-1.02	27.2		
1E 1255.3+3529	DM+36 2322	12 55 17.0	+35 29 32	...	...	S	...
	G 164-31			-0.92 T	28.9 T	...	-0.96
1E 1425.9-6228	Proxima Cen	14 26 03.0	-62 27 42	-1.24	27.4	(D)	-0.28
1E 1652.7-0815	WOLF 630	16 52 46.8	-08 15 09	-1.30 T	29.3 T	(A)	+0.35
1E 1744.4+2744	μ Her	17 44 29.2	+27 44 34	-5.48 T	27.6 T	P	-1.59
1E 1755.3+0438	Barnard's	17 55 21.5	+04 38 25	-3.18	26.1	P	-1.77
1E 191429+0505.5	GL 752 A	19 14 28.0	+05 05 08	-3.4	27.1	P	-1.96
1E 2226.1+5726	Kruger 60	22 26 10.0	+57 26 38	-2.44 T	27.4 T	(A)	-1.17 T
	DO Cep						
1E 2329.3+1939	EQ Peg A	23 29 21.0	+19 39 41	-1.28 T	28.8 T	P	-0.20 T
	EQ Peg B			-0.47 T			

* YY Gem: Serendipitous detection; target was α Gem, which was not seen. YY Gem, separation 72", was also a Columbia Astrophysical Laboratory target.

TABLE 1—Continued

X-ray Source Name	Star Name	Optical Position R.A.	Decl.	log F_x/F_v	log L_x (erg s^{-1})	Survey	log (ct s^{-1})
\multicolumn{8}{c}{Giants & Supergiants - O Stars}
1E 052926+0020.0	δ Ori A	05 29 27.0	-00 20 04	-4.57	32.5	P	-0.63
1E 1041.9-5917	HD 093129A	10 42 00.9	-59 17 05	-3.84	33.6	P	-2.28
1E 1042.6-5905	HD 093249	10 42 46.8	-59 05 37	-3.82	33.0	P	-1.96
1E 1043.6-5909	HD 093403	10 43 46.7	-59 08 39	-4.18	33.0	P	-1.53
1E 203035+4108.1	Cyg OB2-5	20 30 35.2	+41 08 03	-3.86	33.8	P	-2.02
1E 203122+4104.8	Cyg OB2-9	20 31 22.7	+41 04 51	-3.98	33.5	P	-2.28
1E 203127+4108.5	Cyg OB2-8A	20 31 27.1	+41 08 33	-3.2	34.3	P	-1.85
\multicolumn{8}{c}{Giants & Supergiants - B Stars}
1E 0515.1-0654	τ Ori	05 15 10.5	-06 53 49	-5.41	30.2 S	P	-1.62
1E 0533.6-0113	ε Ori	05 33 40.5	-01 13 56	-4.97	32.3	(B)	-0.42
1E 0545.4-0940	κ Ori	05 45 23.0	-09 41 09	-5.40	31.8	(B)	-0.99
1E 1758.0+0257	67 Oph	17 58 08.4	+02 55 56	-5.47	31.3	(B)	-1.82
1E 1943.4+4501	δ Cyg	19 43 24.8	+45 00 30	-5.74	29.1	P	-1.68
1E 203053+4104.2	Cyg OB2-12	20 30 53.2	+41 04 12	-4.65	34.0	P	-2.25
\multicolumn{8}{c}{Giants & Supergiants - A Stars}
1E 0850.9+1401	HD 075976	08 51 00.2	+14 01 15	-3.89	30.6 S	8.5	-1.51
1E 1732.6+1200	α Oph	17 32 36.8	+12 35 35	-5.69	28.6	P	-1.35
\multicolumn{8}{c}{Giants & Supergiants - F Stars}
1E 0622.8-5239	α Car	06 22 50.5	-52 40 03	-6.34	30.0	P	-0.82
\multicolumn{8}{c}{Giants & Supergiants - G Stars}
1E 021432+5717.2	χ Per	02 14 32.2	+57 17 10	-4.58	30.3	P	-2.05
1E 051259+4556.7	α Aur	05 12 59.7	+45 56 46	-4.53	30.3	P	+0.37
1E 0526.0-2048	β Lep	05 26 06.1	-20 47 56	-5.9	29.3	P	-1.82
1E 1623.4+6137	η Dra	16 23 18.4	+61 37 38	-6.3 T	28.0 T	P	-2.15 T
\multicolumn{8}{c}{Giants & Supergiants - K Stars}
1E 1256.2+3833	DM+39 2586	12 56 14.6	+38 32 59	-2.69	31.5 S	S	-0.96
1E 1541.8+0633	α Ser	15 41 48.4	+06 34 55	-6.45	27.8	P	-2.28
1E 1646.9-3412	ε Sco	16 46 53.7	-34 12 23	-6.44	27.8	P	-2.13
\multicolumn{8}{c}{No Luminosity Class - O Stars}
1E 203120+4103.0	Cyg OB2-22	20 31 20.8	+41 03 01	-5.88	31.3	P	-2.70
1E 203135+4058.9	Cyg OB2-E	20 31 35.4	+40 58 55	...	...	P	-2.77
\multicolumn{8}{c}{No Luminosity Class - B Stars}
1E 1640.4+6225	HD 151067	16 40 32.5	+62 24 08	-4.06	...	8.5	-1.72
\multicolumn{8}{c}{No Luminosity Class - A Stars}
1E 1208.2+4009	HD 105824	12 08 16.0	+40 10 10	-4.67	...	8.5	-2.36
1E 1650.5-3020	HD 152287	16 50 31.6	-30 19 34	-3.41	...	8.5	-1.42

168

TABLE 1—Continued

X-ray Source Name	Star Name	Optical Position R.A.	Decl.	log F_x/F_v	log L_x (erg s^{-1})	Survey	log (ct s^{-1})
		No Luminosity Class - F Stars					
		h m s	° ′ ″				
1E 0039.2+4024	HD 003914	00 39 17.5	+40 24 53	-4.82	...	8.5	-2.42
1E 0411.8+1035	HD 026781	04 11 49.2	+10 34 36	-3.93	...	S	-1.55
1E 0425.8+6456	HD 028122	04 25 49.7	+64 56 33	-4.16	...	8.5	-2.16
1E 043947-1620.5	...	04 39 47.0	-16 20 30	-2.77	...	D	-3.04
1E 0535.7-2838	HD 037484	05 35 42.0	-28 39 16	-4.25	...	8.5	-1.94
1E 0536.6-2850	HD 037627	05 36 35.9	-28 51 50	-4.16	...	8.5	-2.16
1E 110157+3830.9	HD 095976	11 01 57.2	+38 31 00	-4.01	...	S	-2.05
1E 1208.6+3924	HD 105881	12 08 38.2	+39 24 30	-4.03	...	8.5	-2.00
1E 1224.9+1001	BD+10 2425	12 25 00.7	+10 02 01	-2.97	...	S	-1.96
1E 1309.7+3221	HD 114723	13 09 41.1	+32 21 01	-4.13	...	8.5	-1.80
1E 1424.3+1638	HD 126695	14 24 21.9	+16 38 17	-3.22	...	8.5	-1.28
1E 171205+7111.8	...	17 12 05.0	+71 11 54	-3.72	...	D	-3.00
1E 171312+7111.1	SAO 008737	17 13 12.6	+71 11 07	-4.09	...	D	-2.53
1E 1742.4-2823	HD 161247	17 42 27.4	-28 23 16	-4.10	...	8.5	-2.13
1E 1743.3-2853	SAO 185730	17 43 28.3	-28 52 48	-3.10	...	S	-1.57
1E 1743.9-2809	BN Sgr	17 43 55.8	-28 07 56	-3.92	...	S	-2.38
1E 2138.7+5721	HD 206482	21 38 47.9	+57 21 15	-3.87	...	8.5	-1.54
1E 2154.9+0354	HD 208632	21 54 59.4	+03 55 09	-4.87	...	8.5	-2.44
1E 2333.9+2023	HD 221972	23 33 56.8	+20 23 20	-3.97	...	8.5	-1.85
		No Luminosity Class - G Stars					
		h m s	° ′ ″				
1E 0134.3+2027	SAO 074827	01 34 24.8	+20 26 45	-2.99	...	S	-1.33
1E 0410.0+1029	SAO 093816	04 10 02.6	+10 28 50	-3.49	...	S	-1.82
1E 0412.3+0717	SAO 111689	04 12 21.5	+07 17 24	-3.65	...	S	-1.72
1E 0429.1+6432	HD 028495	04 29 10.7	+64 31 41	-3.06	...	S	-1.13
1E 043754-1633.2	...	04 37 54.0	-16 33 12	-2.96	...	D	-3.03
1E 043838-1641.0	...	04 38 38.0	-16 41 00	-2.92	...	D	-3.00
1E 073019+6547.0	SAO 014241	07 30 19.4	+65 47 00	-3.34	...	S	-1.92
1E 0851.7+1426	HD 076081	08 51 41.2	+14 26 07	-3.57	...	8.5	-1.55
1E 1225.3+0910	SAO 119414	12 25 21.2	+09 10 30	-4.25	...	S	-2.53
1E 1226.6+3139	HD 108693	12 26 38.7	+31 40 01	-4.35	...	8.5	-2.30
1E 1330.6-0811	HD 117860	13 30 34.0	-08 11 14	-3.66	...	S	-1.33
1E 1532.9+0917	SAO 121078	15 32 55.9	+09 18 01	-3.41	...	S	-1.60
1E 1549.8+2023	SAO 084044	15 49 50.9	+20 23 50	-2.81	...	S	-1.19
1E 170606+7107.1	...	17 06 06.0	+71 07 12	-1.77	...	D	-2.95
1E 170957+7100.1	...	17 09 57.0	+71 00 06	-3.50	...	D	-3.34
1E 1854.7+0116	BD+1 3828	18 54 47.1	+01 16 31	-3.06	...	S	-1.36
1E 2203.4-0536	HD 209779	22 03 28.2	-05 36 06	-3.83	...	8.5	-1.62

169

TABLE 1—*Continued*

X-ray Source Name	Star Name	Optical Position R.A.	Decl.	log F_X/F_V	log L_X (erg s^{-1})	Survey	log (ct s^{-1})
No Luminosity Class - K Stars							
		h m s	° ′ ″				
1E 043916-1622.5	...	04 39 16.0	-16 22 30	-2.26	...	D	-3.17
1E 1053.8+0738	HD 094765	10 53 54.9	+07 39 21	-4.05	...	8.5	-1.77
1E 1238.9+1921	HD 110350	12 38 54.7	+19 20 57	-3.45	...	8.5	-1.31
1E 1527.4+1600	HD 138157	15 27 26.1	+16 21 46	-3.64	...	8.5	-1.35
1E 1615.4+3500	SAO 065201	16 15 28.6	+35 00 59	-3.51	...	S	-1.72
1E 1900.0+7037	HD 177620	18 59 58.7	+70 37 33	-3.70	...	8.5	-1.80
Pre-Main Sequence Stars							
1E 053228-0525.1	KM Ori	05 32 28.3	-05 25 07	-2.30	30.9	S	-2.17
1E 053250-0524.6	MT Ori	05 32 50.3	-05 24 38	-2.45	31.1	S	-2.00
1E 053256-0532.6	V358 Ori	05 32 56.2	-05 32 38	-2.38	30.8	S	-2.25
1E 053314-0530.0	AN Ori	05 33 14.2	-05 30 04	-2.58	31.0	S	-2.05
Dwarf Novae							
1E 0752.1+2208	U Gem	07 52 07.8	+22 08 09	-0.54	30.1	P	
1E 1815.0+4948	AM Her	18 15 00	+49 48 00	+1.1	32.6	P	
1E 2140.7+4321	SS Cyg	21 40 44.4	+43 21 22	+0.23	32.2	P	

U Gem: 0.5 - 4.5 keV (IPC), low state
AM Her: 0.1 - 3.5 keV (OGS+MPC), high state (normal)
SS Cyg: 0.4 - 6 keV (OGS+MPC;IPC), low state

White Dwarf Detections							
		h m s	° ′ ″				
1E 064255-1639.3	α C Ma B	06 42 55.9	-16 39 19	-1.21	28.8	P	+0.24
1E 131400+2921.7	HZ 43	13 14 00.3	+29 21 47	+0.60 T	31.6 T	P	+0.46
Other Stars							
		h m s	° ′ ″				
1E 104306-5925.2	η Car	10 43 06.8	-59 25 15	-3.47	33.5	P	-1.30
1E 1458.2-4122	SAO 225377	14 58 15.1	-41 22 15	-3.59	...	S	-1.96
1E 203043+4103.9	Cyg OB2-C2	20 30 43.1	+41 03 55	...	32.5	P	-2.38

[a] The table includes all detections and corresponding optical counterparts of the survey programs described in § II, as well as of collaborating guest investigators, sorted by stellar type and Right Ascension. The particular program or observer is identified under the "Survey" heading (P: pointed survey, 8.5: V=8.5 magnitude-limited survey, S: serendipitous survey, D: deep stellar survey). The focal plane instrument used for primary detection is specified by the precision of the Einstein source name identification (HRI sources have five significant digits in the Declination identifier; see § II in text).

Guest observers (in "Survey" column):

    (A) H. Johnson
    (B) J. Cassinelli
    (C) C. Zwaan & R. Mewe
    (D) B. Haisch & J. Linsky
    (E) A. Fabian, Inst. of Astron., Cambridge, U. K.

L_X: Letter "S" after luminosity indicates that the distance was determined by spectroscopic parallax

L_X or F_X/F_V: Letter "T" after value indicates that the total X-ray flux is used (assigned to indicated component for F_X/F_V)

Letter "V" after an entry indicates X-ray variability.

RESULTS FROM EINSTEIN SURVEY

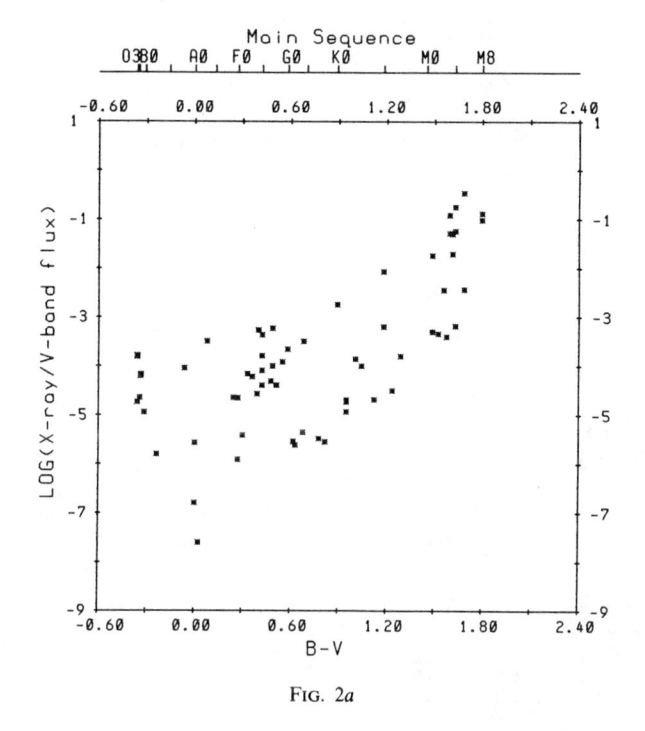

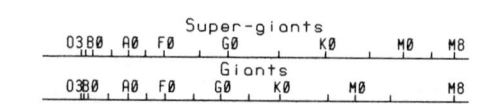

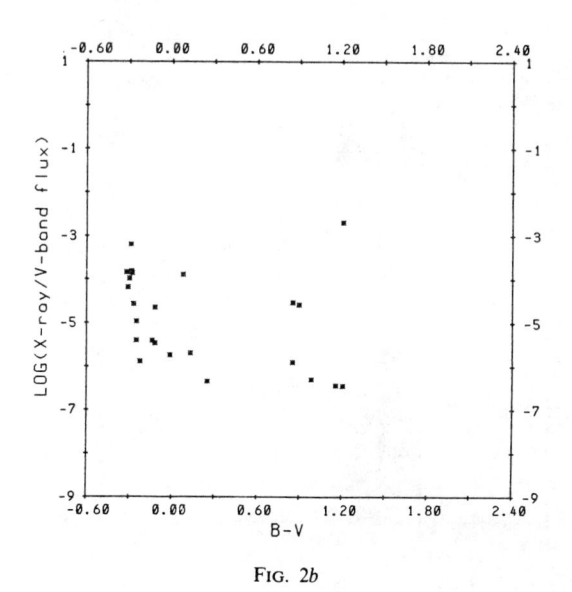

FIG. 2a

FIG. 2b

FIG. 2c

FIG. 2.—The ratio of soft X-ray to V-band fluxes for stars detected in the Survey; stars are divided into: (a) main sequence (luminosity classes IV, V, and VI); (b) giants and supergiants (luminosity classes I, II, and III), and (c) stars whose luminosity class has not as yet been determined. In order to compare the latter group with the others, we have used V-band fluxes, which are readily available, rather than bolometric fluxes which cannot be calculated for group (c).

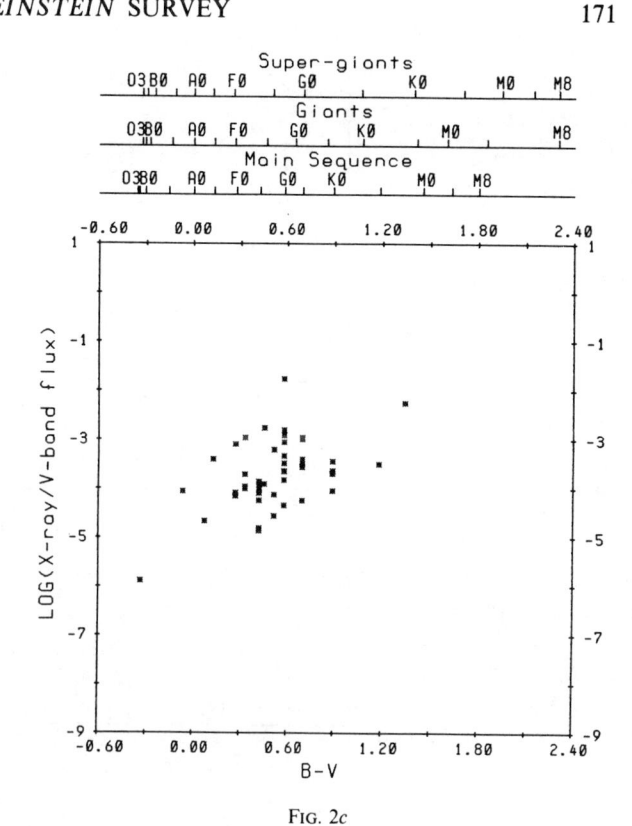

b) Late Type (F, G, K, M) Main-Sequence Stars

i) F Stars

The nearby sample of optically well-characterized dF stars shows them to be clustered about an X-ray luminosity of $\sim 10^{29}$ ergs s^{-1} (Fig. 3a), corresponding to an f_x/f_V ratio of $\sim 10^{-4.5}$ (Fig. 3b). F stars from the Serendipitous and 8.5 Surveys, in which F stars are the dominant stellar type detected, have an f_x/f_V ratio distribution which overlaps that of the nearby (Pointed Survey) sample, but with a bias toward somewhat larger levels of X-ray emission. The substantial overlap in the two distributions suggests that the stars of these two surveys are dominantly the same as those of the Pointed Survey, i.e., main-sequence stars. This possibility is consistent with the frequency distribution of the observed sources as a function of V-band magnitude; that is, the optically fainter stars are systematically intrinsically brighter in X-rays, as would be expected if we are seeing the upper range of the true dF X-ray luminosity distribution as we look at optically fainter main-sequence stars. Furthermore, although the number of dF stars per $(1°)^2$ in a random field is still increasing at $V = 8.5$ (Allen 1973), the actual number of detected X-ray sources drops dramatically above this limiting magnitude, consistent with "running out" of sources for exposures with an average limiting sensitivity of $\sim 10^{-13}$ ergs cm^{-2} s^{-1} at Earth. The brightest sources in the 8.5 Survey therefore represent the brightest sources of the dF luminosity function.

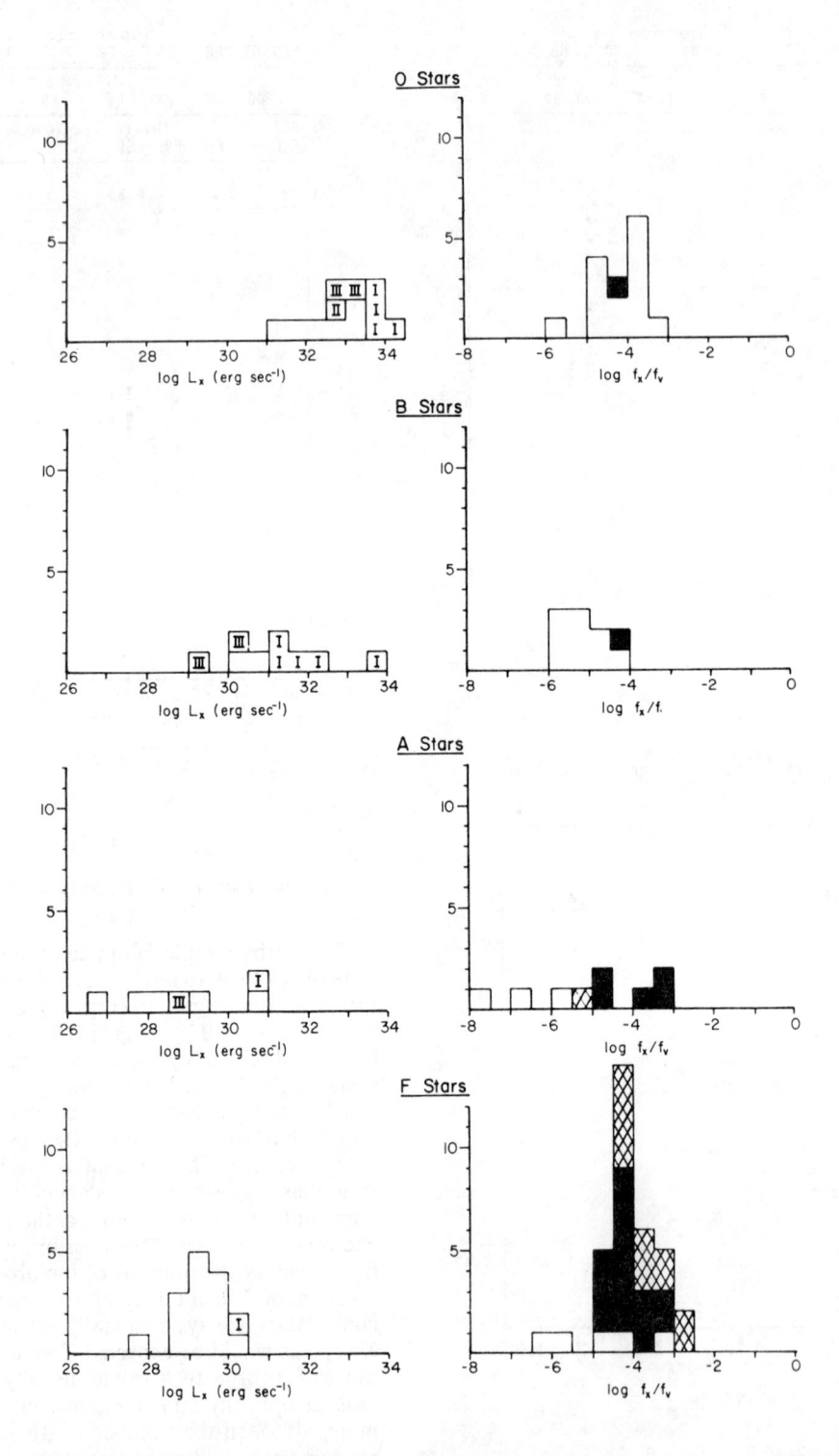

FIG. 3.—(a) *Left-hand columns:* histograms showing the distributions of X-ray luminosities L_x for stars of spectral type O through M; labels I, II, III indicate those luminosity classes and blanks indicate main-sequence stars (IV, V, or VI). (b) *Right-hand column:* histograms showing the distribution of the ratio X-ray flux/V-band flux f_x/f_V for the stars of Fig. 3a, but including also those stars whose luminosity classes are not known. The histograms are subdivided according to the surveys which yielded the data; blank indicates Pointed Survey, black indicates 8.5 Survey, and crosshatch indicates Serendipitous Survey (see § II for descriptions of the various surveys). Also indicated for G stars are the f_x/f_V ratios which would be seen if the Sun's corona consisted entirely of coronal holes (CH), quiet corona (QS), active regions (AR), or X-ray flares (F), respectively.

RESULTS FROM *EINSTEIN* SURVEY

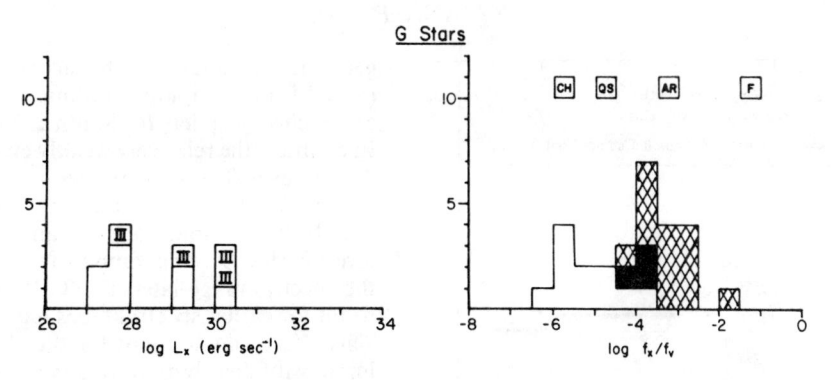

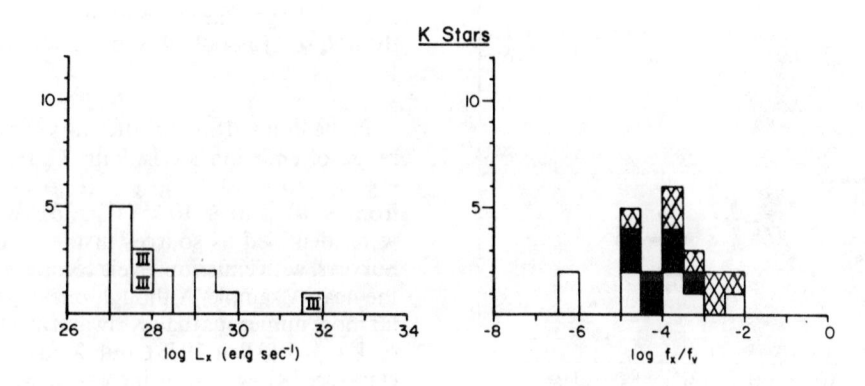

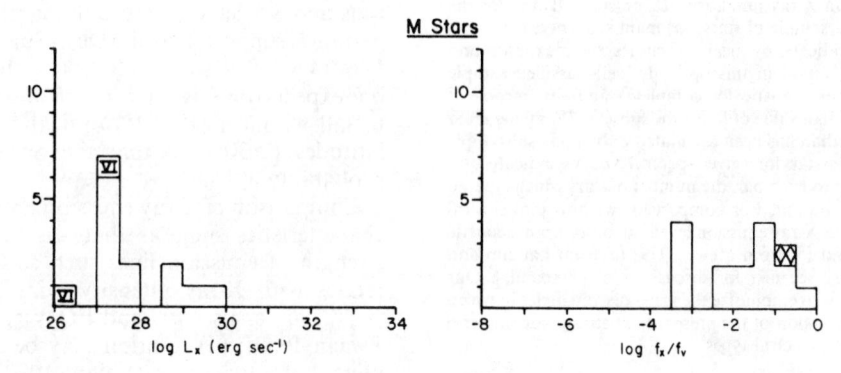

FIG. 3—Continued

Here we note that the RS CVn stars in our sample, which we have chosen not to segregate, show emission levels comparable to the brightest of the non-RS CVn stars (either F, G, or K).

ii) *G Stars*

The nearby (Pointed Survey) G stars show a fairly wide range of emission levels which, if combined with the known level of solar soft X-ray emission levels (see Vaiana and Rosner 1978), spans over four orders of magnitude; the X-ray luminosity ranges from $\sim 10^{26}$ ergs s^{-1} to $\sim 10^{30}$ ergs s^{-1}, with corresponding f_x/f_V values ranging from $10^{-6.5}$ to $-10^{-3.6}$. The optically unclassified G stars of the 8.5 and Serendipitous surveys have characteristics similar to those of the F stars of the same surveys: they are uniformly fainter optically and tend to be intrinsically brighter in X-rays than G stars of the nearby sample (Fig. 3). Study of the proper motions of 8.5 survey stars shows that, under the assumption that they are giants, a means space velocity in excess of 100 km s^{-1} is obtained; under the assumption that they are dwarfs, a mean space velocity of ~ 10 km s^{-1} is obtained (Topka 1980). As the latter result is far more plausible (cf. Allen 1973), the presumption is therefore very strong that—as in the case of F stars—the optically unclassified sample consists dominantly of dwarfs.

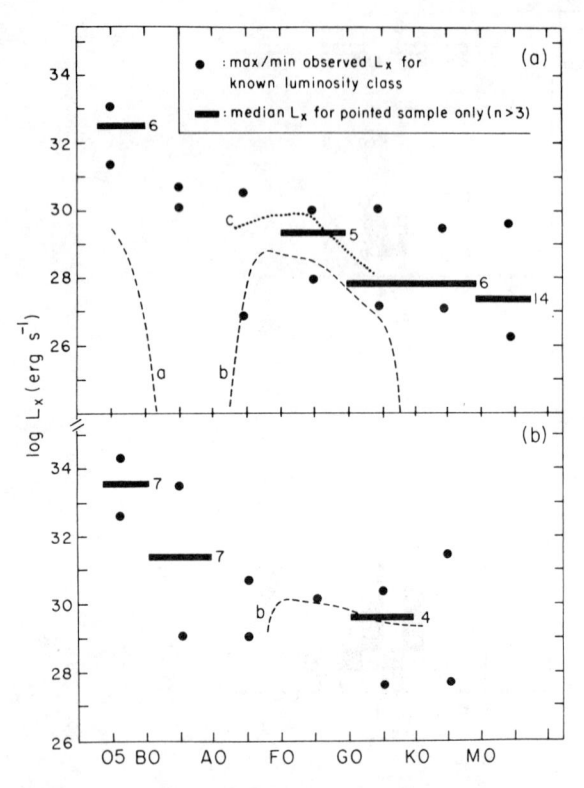

FIG. 4.—Variation in X-ray luminosity L_x vs. spectral type for the optically well-classified sample of stars: (a) main sequence, (b) giants and supergiants. We indicate, by means of circles, the maximum and minimum value of L_x found in this optically well-classified sample (which is by no means statistically complete) and, by means of horizontal bars, the median value of L_x for the subset of Pointed-survey stars (see § II). The median has been calculated only if this subsample contains more than three stars for a given spectral type; we indicate, by a small numeral adjacent to each bar, the number of stars which entered into the median computation. For comparison we also plot several theoretical predictions of X-ray emission levels, all based upon acoustic coronal heating [(a) and (b) from Mewe 1979; (c) from Landini and Monsignori-Fossi 1973; see text for discussion of passbands]. Our primary intention here is to emphasize the gross discrepancies between such theories and observation of the present (statistically incomplete) sample at early and late spectral types.

In spite of the wide range of X-ray emission levels of dG stars, there appears to be evidence that the true luminosity function does change as one goes from F to G stars. This tentative conclusion rests upon the observation that the X-ray luminosity functions of nearby dF and dG stars peak at distinctly different luminosity values (even though they were observed with comparable sensitivities; cf. Fig. 3), and cannot be due to a magnitude-dependent selection effect since the optical magnitude distribution of the two samples is similar; this conclusion is supported by a statistical analysis of the 8.5 Survey results (Topka 1980; Topka *et al.* 1979). It appears that the true X-ray luminosity function of dG stars extends to significantly lower luminosities than that of the dF stars.

Finally, we note that a possible hint to the underlying cause responsible for the broad spread in dG X-ray emission levels (as well as in that of late type dwarfs in general) is contained in the fact that the brightest dwarf (π^1 UMa) is a rapidly rotating, single star with a very active chromosphere (cf. Smith 1978; Linsky *et al.* 1979). In contrast, the relatively weakly emitting Sun and α Cen A are fairly slow rotators, with comparatively inactive chromospheres (Ayres and Linsky 1980a). This association between rapid rotation and high X-ray emission levels for the *Einstein* sample of stars persists throughout the spectral range later than G (Vaiana 1980); a similar result for earlier spectral types has not been established. This effect may account for the skew of the dG X-ray luminosity distribution, relative to that of the dF stars, toward lower emission levels (because F dwarfs have, as a class, larger mean rotation rates than G and later dwarfs; see Tassoul 1978 for review and references).

iii) K Stars

In the Pointed Survey, dK stars show a typically broad range of emission levels, with L_x ranging from $\sim 10^{27}$ ergs s^{-1} to $\sim 10^{29}$ ergs s^{-1} (Fig. 3a) and f_x/f_V ranging from $\sim 10^{-5}$ to $\sim 10^{-2}$ (Fig. 3b). A number of K stars were identified as sources in the Serendipitous and 8.5 Surveys, with emission levels comparable to that found in the nearby sample. Although one expects K giants to be far more numerous than K dwarfs at a limiting magnitude of $V = 8.5$ (Allen 1973), our X-ray data on nearby gK stars (see below)—showing a somewhat lower f_x/f_V ratio for giants—suggest that the stars seen in the Serendipitous and 8.5 Surveys are dominantly dwarfs. This suggestion is supported by the Deep Survey observations of a $V = 14.6$ K8–9 star, with $f_x/f_V \sim 10^{-2.5}$; at $V \sim 14.6$, one expects roughly three dwarfs and essentially no giants to fall within a ($1° \times 1°$) optical field at high galactic latitudes ($\gtrsim 30°$), so that the observed star is in all probability a dwarf.

Comparison of X-ray emission levels with other stellar characteristics suggests that—as for dG stars—(i) the strength of emission lines such as Ca II H and K correlates with X-ray intensity (viz., α Cen B and DM +272296 versus ϵ Eri and EQ Vir; also see Mewe and Zwaan 1980); (ii) rotation may be an important determinant of activity level (Vaiana 1980; cf. EQ Vir, which is known to be a rapid rotator [Anderson, Schiffer, and Bopp 1977] and is the X-ray brightest dK star). We note that the high fraction of multiple star systems in the dK sample makes it difficult to uniquely assign emission levels to the various components; for the nearest systems (viz., α Cen A, B), the HRI can resolve the individual components (Golub *et al.* 1979) and has been and will continue to be used to resolve this problem whenever possible.

iv) M Stars

The most striking result of observations of dM stars is that *all* show emission levels comparable to that of earlier spectral type dwarfs, with L_x ranging from $\sim 10^{26}$ ergs s^{-1} to $\sim 10^{29}$ ergs s^{-1} (Fig. 3a), corresponding to f_x/f_V ranging from $\sim 10^{-3}$ to $\sim 10^{-0.7}$ (Fig. 3b). Because of their intrinsic faintness, dM stars with $V < 8.5$ are extremely rare; it is hence not surprising that none were found in the 8.5 survey. However, extrapolating these observations to exposures with highest sensitivity, one would expect to detect a significant number of dM stars as X-ray sources in deep survey fields, with optical counterparts whose V magnitude exceeds ~ 20 (Rosner et al. 1979a); the observations are as yet inconclusive on this point.

There is some evidence in the data that the level of X-ray emission for dM stars, as for the other late-type dwarfs, correlates with: (i) multiplicity (and hence, for close binaries, with rotation) and (ii) emission-line behavior (Rosner et al. 1979a). Thus the least luminous sources (e.g., CN Leo and Barnard's star) are the only unambiguously single stars in our sample. Similarly, the X-ray brightest stars tend to be classified as active stars in general (e.g., as dMe, spot, and/or flare stars; viz., YY Gem, EQ Peg), whereas the weakest X-ray emitting stars tend toward less dramatic activity (viz., Barnard's star).

v) Summary for Late Type Dwarfs

We conclude that, if one examines the data for dwarf stars later than F in toto, a remarkable similarity in X-ray properties emerges: the median X-ray emission levels of the Pointed sample, as well as the range of emission levels of all optically well classified stars, change relatively little as the spectral type varies from G to late M. Since the stellar surface area experiences a drastic reduction in this spectral range (an almost 100-fold change), these stars seem to obey a kind of total X-ray flux conservation law, such that the average X-ray surface brightness increases as the stellar radius decreases. The major change in X-ray luminosity for these types of stars appears to occur at the dF/dG spectral type boundary; this change (a skewing of the X-ray luminosity function to lower emission levels) coincides roughly with the sharp decrease in mean rotation rates of field stars observed in this spectral type range.

c) Evolved Stars

i) Early Type Giants and Supergiants

The relatively large number of evolved early-type stars found to be soft X-ray sources by the *Einstein* stellar survey (Fig. 1, Table 1), as is the case for their main-sequence counterparts, is primarily a reflection of their strong spatial correlation: the discovery that stars in OB associations are indeed X-ray sources Harnden et al. 1979b; Seward et al. 1979; Long and White 1980) ensured that many such stars would be observed.

Examination of the observed X-ray luminosity distribution of these stars (Fig. 3) shows that: (a) O giants and supergiants have high emission levels, somewhat in excess of those of main-sequence stars of the same spectral type and consistent with the proportionality between X-ray and visual luminosity reported by Harnden et al. (1979a, b; see also Rosner et al. 1979b; Cassinelli et al. 1979; Long and White 1980); (b) B and A spectral type evolved stars have distinctly lower emission levels than O giants and supergiants, thus showing the same behavior with spectral type as their dwarf counterparts.

ii) Late Type Giants and Supergiants

The Pointed Survey has detected one F supergiant, four G giants, and two K giants (Table 1). We have also looked at K and M giants and M supergiants, but have thus far obtained only upper limits; there is some support, therefore, for the suggestion that there is decreasing prevalance of hot coronae toward evolved stars of later spectral type (Linsky and Haisch 1979).

The detections show these late-type evolved stars to be very modest soft X-ray sources if their X-ray emission is compared to their optical output (or if their mean surface X-ray flux is computed), with $f_x/f_V \lesssim 10^{-5}$; for example, the corresponding surface flux for Canopus (F0 I) is comparable to solar coronal hole values. The two exceptions (Capella and Algol) are complex multiple-component systems for which unique assignment of emission levels is not possible. We note, however, that the total X-ray luminosity of the detected sources can be substantial; the F supergiant Canopus, for example, is as bright in X-rays as the brightest of the single late-type main-sequence stars.

Perhaps most striking of our observations of these stars are the strong upper bounds upon the mean surface flux we can place upon some very late giants and supergiants. We have upper bounds for ϵ Sco (K2 III-IV; $f_x/f_V \lesssim 10^{-6.6}$), β Peg (M2 II; $f_x/f_V \lesssim 10^{-5.9}$), α Ori (M2 Iab; $f_x/f_V \lesssim 10^{-6.8}$) and α Sco (M1 Ib, $f_x/f_V \lesssim 10^{-7.1}$). For α Sco the corresponding upper bound upon the mean X-ray surface flux is $\sim 10^{0.9}$ ergs cm^{-2} s^{-1}, characteristic of solar coronal holes (Maxson and Vaiana 1977), the least active portion of the solar corona. Any corona on α Sco must be rather feeble; but because inhomogeneity cannot be excluded, it is still possible for α Sco to have significant *localized* coronal emission, that is, the upper bound upon its total X-ray luminosity is $10^{29.3}$ ergs s^{-1}, quite comparable to that of the brighter late-type dwarfs. In summary, our data indicate that the X-ray surface flux (but not the luminosity) varies with effective gravity, with late supergiants consistently showing the lowest surface fluxes (or upper bounds) at any given spectral type.

d) Pre–Main-Sequence Stars

Early *Einstein* IPC observations of the Orion complex strongly suggested that pre–main-sequence stars are also soft X-ray sources (Ku and Chanan 1979). These observations have been confirmed by extensive HRI observations reported by Chanan et al. (1979) and by early HRI observations carried out as part of the present survey and reported here; the much higher spatial resolution of the HRI leaves no doubt as to the optical identification (see the four sources listed in Table 1). The sources observed in the CfA pointing, corresponding in the optical to Orion nebular variables (Kukarkin 1968), have relatively large values of f_x/f_V, lying at the upper range of f_x/f_V ratios observed for main-sequence stars of similar spectral type. For an assumed distance of 400 pc, these sources have an X-ray luminosity of $\sim 10^{30.6}$–$10^{30.9}$ ergs s^{-1}. Our survey of main-sequence stars reveals only one single late-type star (24 UMa) with comparable luminosity; 24 UMa is a giant-subgiant chosen because of its high Ca II emission (see Table 1).

We caution that the above narrow range of inferred luminosities for these young stars is in all probability a selection effect. We have obtained a number of upper limits at the emission level of the weakest detected source. However, Chanan et al. (1979) have reported a far larger number of detections, based on much more extensive observations than reported here. The implication is that the limiting sensitivity of our shorter HRI observation allowed only a small number (e.g., the brightest) of the Orion nebular variables in the field to be seen.

e) White Dwarfs

Because of the intrinsic optical faintness of white dwarfs, the 8.5 Surveys could not reasonably be expected to supplement our Pointed observations of the nearest such stars. Instead, a computer-based catalog of white dwarfs was used to search all IPC Survey fields for X-ray emission from white dwarfs falling serendipitously into these fields (Topka et al. 1979). The catalogs upon which this search was based were copies of those due to Luyten (1969) and Gliese (1969), which are not in any sense magnitude-limited;[11] at present we do not regard this search for serendipitous white dwarf X-ray sources as statistically complete. Our search provided upper bounds for 34 white dwarfs: none was detected as an X-ray source in the CfA survey (see Table 2).

The white dwarf portion of the Pointed Survey has been similarly unfruitful, pointing at 11 white dwarfs, and detecting only the two previously known sources, HZ 43 (Hearn et al. 1976; Lampton et al. 1976; Heise and Huizenga 1980) and Sirius (Mewe et al. 1975). In the second case, however, our HRI observation definitively resolves the question of which component of the Sirius system is the emitter: *both* Sirius A and B are X-ray sources, the white dwarf being by far the *dominant* contributor to the total system X-ray flux. We note that because the Pointed Survey concentrated upon nearby white dwarfs, the stronger (relative to the Serendipitous results) upper bounds obtained are to be expected (Table 2).

Considering our white dwarf data (detections and upper bounds) *in toto*, several results emerge. First, a substantial number of our upper bounds fall below the flux values derived for the detected sources. Comparing Tables 1 and 2, we see that a number of nearby white dwarfs have upper bounds on L_x at or below the weakest detected white dwarf (Sirius B, $L_x \sim 10^{28.8}$ ergs s^{-1}). We are therefore forced to conclude that *if* white dwarfs constitute a class of X-ray sources, their X-ray luminosity distribution must be broad and their median luminosity far below our detection levels.

---

[11] As pointed out by Gliese (1969), the catalog also contains high-velocity dM and other stars; it therefore provides an additional means of searching for serendipitous X-ray sources associated with nondegenerate stars fainter than $V = 8.5$, but not in any statistically complete sense.

A second striking characteristic is that the detected white dwarfs have large X-ray to optical emission ratios with values of f_x/f_V in excess of the values attained by any of the nondegenerate stars. Under the assumption that the X-ray emitting volume overlies the white dwarf surface (as in a coronal model), we have computed the mean surface X-ray flux $\mathscr{F}_x$ for these stars, and find that it is many orders of magnitude in excess of that observed for any nondegenerate star. For example, we compute an $\mathscr{F}_x \sim 10^{10}$ ergs cm^{-2} s^{-1} for Sirius B, assuming its radius to be $\sim 10^{-2} R_\odot$; to place this figure in perspective, the mean solar surface X-ray flux under the assumption that the entire solar surface is covered by small (C) flares would be $\sim 3 \times 10^8$ ergs cm^{-2} s^{-1} (Vaiana and Rosner 1978). If coronal, the process leading to X-ray emission from these white dwarfs would then be indeed spectacular; the extremely large value of $\mathscr{F}_x$ in fact casts some doubt upon a simple coronal interpretation, and the relatively low estimated effective temperature of Sirius B seems to exclude the possibility of direct (thermal) surface emission. This suggests that in this case the X-ray source surface area is substantially larger than the white dwarf surface area (as would be the case if the emission derived from an accretion disk).

f) Cataclysmic Variables

Cataclysmic variables have long been considered to be a separate class of objects (at least as far as X-ray emission is concerned) from the "normal" stars; we have included them here for the sake of completeness from the point of view of low-luminosity galactic X-ray sources. If, as is presently believed, the main source of energy in these systems is the release of the gravitational energy of the matter onto the white dwarf and an accretion disk, they are quite different from the other stellar systems discussed in the present survey paper.

As part of the CfA program, we have looked at three such objects: U Gem, SS Cyg, and AM Her. The results of a detailed study of *Einstein*, *IUE*, and optical simultaneous observations and their implications for the model of these systems are reported by Fabbiano et al. (1980); here we note that their X-ray luminosity is of the order of 10^{30}–10^{32} ergs s^{-1}, with $f_x/f_V \sim 10^{-0.5}$–$10^{1.1}$. This places them at the upper end of the late type star X-ray luminosity distribution and at the extreme upper end of the f_x/f_V distribution for all stellar systems. We note that the values of L_x and f_x/f_V determined for U Gem, SS Cyg, and AM Her are similar to those of HZ 43 and Sirius, both binary systems containing a white dwarf (see Margon et al. 1976 for discussions of binary nature of HZ 43); it should be noted, however, that the spectral characteristics of the latter two sources differ substantially from those of the cataclysmic variables.

IV. DISCUSSION

Our fundamental conclusion, only partially illustrated by Figure 4, is that *stars in general—virtually irrespective of spectral type, luminosity class, age, or any other distingu-*

TABLE 2A

SELECTED UPPER LIMITS[a]

Star Name	Spectral Type	Visual Magnitude	Upper Limit log F_x/F_v	Upper Limit log L_x (erg s^{-1})	Survey
66 Oph	B2 Ve	4.84	−5.2	30.6 S	P
HR 2142	B2 IV-Vne	5.23	−5.2	30.7 S	P
139 Tau	B1 Ib	4.78	−5.3	31.7 S	(B)
o² C Ma	B3 Ia	3.04	−6.1	31.2 S	(B)
56 Ari	Ap	5.60	−5.07	29.3	P
o Vir	A2p	4.93	−5.15	29.1	P
γ Tra	A0 Vp	2.89	−6.42	28.1	P
HD 014590	Am	5.46	−5.11	28.7 S	8.5
HD 255690	A1m	6.86	−4.26	30.0 S	8.5
ζ U Ma	A2 V + Am	3.96 V	−5.56	28.3	P
78 Vir	Ap	4.94	−5.3	28.5 S	P
α Lyr [*]	A0 V	0.04	−7.4 [*]	27.0 [*]	P
τ Cet	G8 VI	3.50	−5.8	26.8	(A)
82 Eri	G5 V	4.26	−5.6	27.0	(A)
α Ari	K2 III	2.00	−6.7	27.8	P
α U Ma	K0 II-III	1.80	−6.6	28.3	P
α Ori	M2 Iab	0.80	−6.8	30.2	P
α Sco	M1 Ib	1.08	−7.1	29.4	P
β Peg	M2 II	2.56	−5.9	29.3	P

[a] Ordered by spectral type; limits selected to include only those which add significant information to Table 1.

[*] α Lyr, though not seen with the IPC, was detected with the HRI (see Table 1).

TABLE 2B

WHITE DWARF UPPER LIMITS[b]

Star Name	Spectral Type	Visual Magnitude	Upper Limit log F_x/F_v	Upper Limit log L_x (erg s^{-1})	Survey
GR 267	DA	15.0	−0.96	28.9	P
EG 005	DG	12.37	−2.38	26.5	P
EG 011	DA	12.84	−2.17	27.7	P
EG 248	DG	14.60	−1.42	27.4	P
EG 046	DA?	12.9	−2.08	28.3	P
EG 054	DF + M	13.00	−2.35	26.8	P
EG 099	DA	12.31	−2.23	27.9	P
EG 100	DK	14.71	−1.33	27.1	P
GR 372	DKp?	14.15	−1.58	26.9	P
EG 129	DA	13.18	−2.11	27.6	P
EG 131	DA +dM5	12.36 12.75	−2.36 −2.21	27.3	P
EG 144	DA	12.88	−2.17	27.9	P
ROSS 627	DF	14.24	−2.25	27.5	S
BPM 94172	DA	13.01	−2.33	27.3	S
LP 658-2	DK	14.52	−1.52	26.8	(A)

[b] Only the most significant white dwarf upper limits are given. Less significant values also exist for a number of other white dwarfs.

Guest observers (in "Survey" column):

(A) H. M. Johnson
(B) J. Cassinelli

ishing feature—constitute a class of X-ray sources; only a few types of stars have not as yet been detected as X-ray sources (very late giants and late-type supergiants), and for these the upper bounds on luminosity are relatively high. Much of the evidence for this conclusion is schematically summarized by the curves in Figure 4, which show the variation of the median observed soft X-ray luminosity L_x with spectral type for main sequence, giant, and supergiant stars, thus also displaying the variation of L_x with effective gravity for fixed spectral type. In the present section, we address the questions of the source of this X-ray emission and its pervasiveness.

a) "Coronal" versus Alternative Interpretations

A number of mechanisms have been proposed to account for possible X-ray emission associated with *particular* types of stars; these processes can involve local plasma heating in the vicinity of the stellar surface (the "coronal model" [de Loore 1970; Renzini *et al.* 1977; Landini and Monsignori-Fossi 1973; Hearn 1975; Mullan 1976; Cassinelli and Olsen 1978; Thomas 1975; Bisnovatyi-Kogan and Lamzin 1979; Rosner and Vaiana 1979; Lucy and White 1980]), plasma heating via mass accretion (Gorenstein and Tucker 1976 and references therein), or shocking of strong stellar winds (Weaver *et al.* 1977; Cooke, Fabian, and Pringle 1978); the latter two processes have received particular attention in the context of X-ray emission from early-type stars.

Consider first the early-type stars. The initial *Einstein* data on X-ray emission from early-type (OB) stars in Cygnus OB2 and η Car, which first established such stars as a distinct class of low-luminosity X-ray sources (Harnden *et al.* 1979*a*; Seward *et al.* 1979; see also Cassinelli *et al.* 1979; Harnden *et al.* 1979*b*; Stewart *et al.* 1979; Long and White 1980), provide strong evidence for a "coronal" interpretation alone. A significant aspect of the argument presented by Harnden *et al.* was the demonstration that X-ray emission properties were not strongly dependent upon the characteristics of the O star involved, other than through its optical luminosity[12] (also Rosner *et al.* 1979*b* and Long and White 1980); in particular the presence of binary components does not appear to be an essential element in determining the existence of X-ray emitting plasma. A comparable argument can be advanced against models which involve wind collision with the interstellar medium (which are most strongly excluded by source temperatures and the absence of spatial source extension in HRI images; see Harnden *et al.* 1979*a, b*), contrary to the conclusions of Ku and Chanan (1979).

These arguments remain in force over the entire H-R diagram: that is, examination of Table 1 shows no correlation between *existence* of stellar X-ray emission and existence of binary components,[13] and HRI images of the nearest stars give no evidence for source extension. This, of course, does not mean that X-ray producing processes unique to binary or multiple star systems are not operative in some cases and that for particular systems (viz., compact close binaries, dwarf novae) such processes cannot dominate "coronal" processes, and also lead to the existence of hot plasma. Nevertheless, we conclude that the pervasive nature of X-ray emission is in all probability due to the presence of *coronae as a general stellar phenomenon*, with other processes playing a dominant role only in rare instances. The *Einstein* Stellar Survey has therefore settled in the negative—on the basis of direct observations—the long-debated question of whether stellar coronae were characteristic *only* of a restricted class of late-type stars.

b) Comparison of Data with Standard Coronal Models

If we accept the above argument on the coronal origin of most of the observed stellar X-ray emission, we may ask whether the data presented here—which for the first time display X-ray emission levels throughout the H-R diagram—are in accord with published models of stellar coronal emission. We focus upon three representative such theories: those of Landini and Monsignori-Fossi (1973), Gorenstein and Tucker 1976), and Mewe (1979); the latter is an application and refinement of Hearn's (1975) minimum flux coronal theory (see also Mullan 1976). These three models assume that coronal heating is due to the shock dissipation of a uniform flux of acoustic modes,[14] that the hot coronal plasma responsible for the observed emission is constrained from escape only by stellar gravity, and that the mean coronal radiative loss for a given star can meaningfully be represented by calculations based upon homogeneous model atmospheres.

These assumptions can be tested for the solar corona where, unlike stellar coronae, the X-ray emitting plasma can be spatially resolved. However, none of these three assumptions can be easily reconciled even with solar coronal observations (see reviews by Vaiana and Rosner 1978; and Wentzel 1978). In order to test whether the above models, together with the underlying assumptions, gain any credence in the more general stellar context, we have plotted as a function of spectral type for all three models the predicted X-ray luminosity L_x in Figure 4.[15]

---

[12] Note that the particular dependence upon optical luminosity found by Harnden *et al.* obtains *only* for early-type stars.

[13] On the other hand, the *level* of X-ray emission may well depend strongly upon the presence or absence of binary components, viz., RS CVn stars (Walter *et al.* 1980*a*) and dM stars (§ III above), presumably because of tidal coupling and enhanced stellar rotation rates in close binaries (cf. Tassoul 1978).

[14] Note, however, that in Hearn's (1975) formulation, no specific heating process is assumed.

[15] The instrumental passbands used in these predictions differ somewhat from that of *Einstein*; a detailed analysis of thermal spectra shows, however, that these predicted luminosities require relatively little correction (generally less than 0.5 in the log) in order to allow meaningful comparison with the observed values of L_x (shown in Fig. 4) over the temperature range $10^6 < T < 10^7$ K.

The comparison is dominated by two striking contradictions. First, stars of late B and early A spectral type as well as a subset of late-type stars (late F to early G) have levels of X-ray emission substantially, or even many orders of magnitude, in excess of predictions. Second, dwarf stars later than ~ G5 remain at a luminosity level of log L_x ~ 28 ± 1.5 with increasing $B - V$, completely in conflict with the theoretical expectation that L_x should steeply decrease with increasing $B - V$. The only (qualitative) correspondence between theory and data occurs in a narrow spectral range from late A to early G for main-sequence stars, but even here the broad range of emission levels encountered remains unexplained.

Some of these results confirm early studies which noted the need for revising standard coronal theories.[16] The key difficulties can be brought into sharper focus by comparing the theoretically predicted acoustic wave flux f_a (ergs cm^{-2} s^{-1}) available to heat the corona above the stellar surface with the X-ray surface flux f_x (ergs cm^{-2} s^{-1}) derived from our data; such a comparison is presented in Figure 5 and reveals two distinct facets of contradiction.

i) There is a *quantitative* discrepancy between the calculated mechanical wave flux available for coronal heating and the observed X-ray flux. Although uncertainties in the theory (due to difficulties in theories for convection, sound generation in a turbulently convective fluid, and the thermalization of acoustic flux) may account for some of the discrepancies (see Jordan 1973 and Toomre *et al.* 1976), cases such as early A dwarfs and late-type dwarfs require corrections of many orders of magnitude; it seems unlikely that the *quantitative* differences can be resolved by considering more sophisticated acoustic heating theories in the context of revised convection theories and acoustic conversion and propagation models.

ii) There is a *qualitative* discrepancy between the observed functional dependence of X-ray flux on spectral type and the theoretical predictions. This discrepancy manifests itself in two ways. There is a large spread in the

---

[16] For example, Walter *et al.* (1980a) review observations of RS CVn stars (principally from *HEAO 1*), which show anomalously high coronal activity levels, given both the relatively high ($\gtrsim 10^7$ K) observed coronal temperatures and relatively low surface gravity of the stars involved. Walter *et al.* point out that the large observed coronal emission measure can only be understood if the X-ray emitting plasma is confined to the stellar surface by forces other than gravity, such as Lorentz forces exerted by surface magnetic fields (see Noci 1973; Pneuman 1973; Rosner, Tucker, and Vaiana 1978). This particular point has been further amplified by spectroscopic studies of Capella using the Solid State Spectrometer of *Einstein* (Holt *et al.* 1979; see also Swank *et al.* 1979 for similar data on the RS CVn star UX Ari) which show that magnetic confinement of coronal plasma seems unavoidable if the high-temperature component of Capella's corona is to be understood. We also note that earlier studies of stellar *chromospheric* activity similarly concluded that standard models invoking acoustic heating were inadequate to account for the observations of late type dwarf Ca II emission (viz., Blanco *et al.* 1974) or the apparent absence of a dependence of Mg II surface flux on stellar gravity (Linsky and Ayres 1978; Basri and Linsky 1979). Furthermore, a direct search by *OSO 8* investigators for the acoustic flux required to heat the upper chromosphere and corona appears to exclude its existence (cf. Athay and White 1979 and references therein).

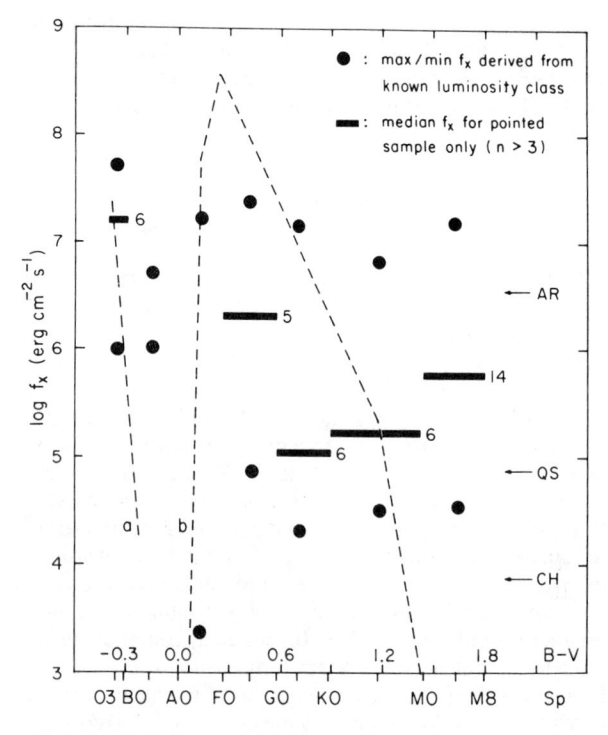

FIG. 5.—Variation of derived X-ray surface flux f_x vs. spectral type for main-sequence stars only. Circles indicate the maximum and minimum observed f_x for main-sequence stars in the optically well-classified sample; bars indicate the median f_x for the Pointed sample of main-sequence stars only (calculated only if this sample contained more than three stars; this number is indicated next to each bar).

For comparison we plot the variation in total available acoustic surface flux f_a as a function of spectral type predicted by: (*a*) Hearn (1972, 1973) and (*b*) Renzini *et al.* (1977); more recent calculations by Ulmschneider and collaborators suggest that acoustic flux levels at very late spectral types may have been severely underestimated by Renzini *et al.* (Ulmschneider, private communication), but such revisions do not appear to suffice to eliminate the discrepancies between theory and observations. We note that f_a places a very strict upper bound on f_x if acoustic heating dominates (as such heating must also account for chromospheric, etc., losses, which are not shown here); as discussed in the text, the vastly different qualitative variation of f_a and f_x appears to exclude acoustic heating as a viable universal coronal heating mechanism.

As a guidepost we have also indicated in the right-hand margin the typical values of soft X-ray surface flux for various solar features (AR = active region, QS = Quiet Sun, CH = coronal hole) taken from Vaiana and Rosner (1978); we note that the range of observed stellar surface fluxes corresponds fairly well to that of the inhomogeneous solar corona.

X-ray luminosity function for any fixed spectral type along the main sequence later than spectral type A, suggesting that coronal activity levels are determined by additional parameters (e.g., stellar rotation [Ayres and Linsky 1980b; Rosner and Vaiana 1979] and confining magnetic fields [Rosner, Tucker, and Vaiana 1978; Walter *et al.* 1980a, b]). There is also an observed increase in f_x with $B - V$ for late-type dwarfs, contrary to the expected sharp drop seen in f_a. This decrease of f_a with decreasing stellar mass along the main sequence appears to be an essential (and unavoidable) element of all acoustic heating models; it is a consequence of the strong dependence of the acoustic flux upon the magnitude of the turbulent velocity at the stellar surface, and the decrease of this flow speed with decreasing stellar mass (and bolometric luminosity; see reviews by Jordan 1973 and Kippenhahn 1973).

c) A Possible Alternative: Magnetic Field Dominated Coronae

One possible alternative to canonical acoustic coronal heating models which we have been pursuing is that stellar magnetic fields play the dominant role in determining the level of quiescent coronal emission by (1) channeling free energy to the corona, (2) allowing nongravitational confinement of hot plasma, and (3) actively participating in the coronal heating process (Tucker 1973; Rosner et al. 1978; Rosner and Vaiana 1979; Golub et al. 1980, 1981; Linsky 1980; Vaiana 1980; Walter et al. 1980a). In this scenario, the modulation of the surface magnetic flux level (for example, by variations in interior stellar convection and stellar rotation in late-type stars) and the level of stressing of surface magnetic fields (by surface turbulence) together determine the variation of the X-ray luminosity function in the H-R diagram (e.g., variation in both its mean and its spread). Thus coronal activity may be fixed by (i) convection and rotation-driven magnetic dynamo activity and surface convective turbulence together producing magnetically confined coronae in late-type stars; (ii) surface turbulence (driven by rotational, radiative, or pulsational stellar surface instabilities) acting directly on a magnetically confined atmosphere, or on the remnant surface flux of primordial stellar magnetic fields which in turn heats the outer confined atmospheres of early type stars;[17] and (iii) turbulent convection acting on dynamo-generated or primordial magnetic fields in pre–main sequence stars (viz., T Tauri and nebular variables). This conceptual framework is primarily designed to focus and motivate further work, and of course is as yet speculative; it should be seen as a useful departure point for forthcoming observational and theoretical studies of stellar coronae.

It may still be argued that several distinct processes give rise to stellar coronae, and that the standard acoustic heating theories survive only for stars in the spectral range $\sim$ A5–$\sim$ G5, with other (perhaps as yet unknown) processes responsible for coronae on other types of stars. This line of reasoning, which attempts to minimize the necessary adjustments, must, however, contend with the results of recent solar coronal research, which have shown that solar coronal X-ray emission is largely modulated by solar surface magnetic fields (see review by Vaiana and Rosner 1978); that is to say, straightforward acoustic heating theory is inadequate to explain observations of the corona of even the G2 V star for which it was originally developed. These difficulties cannot be simply alleviated by, for example, resorting to coupling pure acoustic modes to magnetic modes unless one can demonstrate that the strong dependence of acoustic mode generation upon the magnitude of the turbulent flow velocity can be eliminated; yet the qualitative agreement between acoustically based coronal theories and the data for dwarfs in the spectral range $\sim$ A5–$\sim$ G5 suggests that stellar surface turbulence must be a necessary ingredient in any sensible theory.

V. CONCLUSIONS

The *Einstein* stellar survey reported here has established that stars *in general* constitute a new class of low-luminosity galactic X-ray sources. X-ray emission appears to be characteristic of stars from youth (e.g., pre–main sequence) to old age (late type giants and supergiants), crossing virtually all spectral type and luminosity class boundaries, and is in all probability associated with the presence of stellar coronae. The preliminary *Einstein* results thus settle the long controversy surrounding the pervasiveness of the coronal phenomenon in the affirmative, raise a multitude of new questions regarding the formation process responsible for stellar coronae, and clearly contradict standard theories of coronal X-ray emission.

Previous observational studies had shown certain types of stars to be persistent low-luminosity X-ray sources (principally the RS CVn stars, Walter et al. 1980a), and had suggested that other classes of stars may also be X-ray sources (for example, A stars [Topka et al. 1979; Cash, Snow, and Charles 1979; den Boggende et al. 1978]); but these data were not sufficient to establish whether the observed X-ray sources are exceptional stars or represented the norm. Optical and, more recently, UV observations of stellar emission characteristic of chromospheres and solar-like transition regions had led to the inference that certain types of stars (principally the "active" late type dwarfs and giants) might well have hot coronae (see Wilson 1966; Blanco et al. 1974; Linsky 1977; Hartmann et al. 1979; Linsky and Haisch 1979). These relatively indirect results have been both confirmed and vastly extended by the ongoing work described here to stars of all activity levels.

The *Einstein* Stellar Survey will continue to explore the coronal phenomenon, focusing in particular upon (a) further investigation of the X-ray luminosity functions for the various stellar categories, shown here in preliminary form, (b) establishing stellar X-ray spectral characteristics in the H-R diagram; (c) correlating stellar X-ray emission with other stellar characteristics, including age, rotation, surface magnetic fields (viz., Ap stars), and possibly related emission phenomena in the radio, optical, and UV; (d) placing limits upon the stellar contribution to the diffuse soft X-ray background and to the integrated emission from stellar clusters and galaxies.

---

[17] Recently Lucy and White (1980) have suggested an alternative possibility, which involves wind instabilities and consequent shock-heating in the expanding atmosphere of early type stars, and does not invoke magnetic fields. Following their scenario, one would require a transition between early and late type stellar coronae which reflects the relative dominance of plasma heating by local (wind) instabilities and photospherically generated, propagating modes.

This work has been supported at the CfA in part by NASA under grants NAS8-30751, NAG-8302, and NSG 7176, by the Langley Abbot Program of the Smithsonian and by the CNR of Italy. H.M.J.'s part was done under contract NAS8-33332 between NASA and Lockheed; J.P.C.'s, through NASA grant NAS8-33340 with the University of Wisconsin. The work of J.L.L. and B.M.H. was supported by NASA grants NAS5-23274 and NAS8-33333 to the University of Colorado.

REFERENCES

Allen, C. W. 1973, *Astrophysical Quantities* (London: Athlone Press).
Anderson, C. M., Schiffer, F. H., and Bopp, B. W. 1977, *Ap. J.*, **216**, 42.
Athay, R. G., and White, O. R. 1979, *Ap. J.*, **229**, 1147.
Ayres, T. R., and Linsky, J. L. 1980a, *Ap. J.*, **235**, 76.
———. 1980b, *Ap. J.*, **241**, 279.
Basri, G. S., and Linsky, J. L. 1979, *Ap. J.*, **234**, 1023.
Bisnovatyi-Kogan, G. S., and Lamzin, S. A. 1979, Moscow Space Institute preprint 531.
Blanco, C., Catalano, S., Marilli, E., and Rodono', M. 1974, *Astr. Ap.*, **33**, 257.
Brown, A., Jordan, C., and Wilson, R. 1979, *Proc. Symposium on the First Year of IUE*, in press.
Carpenter, K. G., and Wing, R. F. 1979, *Bull. AAS*, **11**, 419.
Cash, W., Bowyer, S., Charles, P. A., Lampton, M., Garmire, G., and Riegler, G. 1978, *Ap. J. (Letters)*, **223**, L21.
Cash, W., Snow, T. P., Jr., and Charles, P. 1979, preprint.
Cassinelli, J. P., and Olson, G. L. 1979, *Ap. J.*, **229**, 304.
Cassinelli, J. P., Waldron, W. L., Vaiana, G. S., and Rosner, R. 1979, *Bull. AAS*, **11**, 775.
Catura, R. C., Acton, L. W., and Johnson, H. M. 1975, *Ap. J. (Letters)*, **196**, L47.
Chanan, G. A., Ku, W., Simon, M., and Charles, P. A. 1979, *Bull. AAS*, **11**, 623.
Cooke, B. A., Fabian, A. C., and Pringle, J. E. 1978, *Nature*, **273**, 645.
Cordova, F. A., and Mason, K. 1980, preprint.
Cordova, F. A., Nugent, J. J., Klein, S. R., and Garmire, G. P. 1980, *M.N.R.A.S.*, **190**, 87.
de Loore, C. 1970, *Ap. Space Sci.*, **6**, 60.
den Boggende, A. J. F., Mewe, R., Heise, J., Brinkman, A. C., Gronenschild, E. H. B. M., and Schrijver, J. 1978, *Astr. Ap.*, **67**, L29.
Dupree, A. K., 1975, *Ap. J. (Letters)*, **200**, L27.
Evans, R. G., Jordan, C., and Wilson, R. 1975, *M.N.R.A.S.*, **172**, 585.
Fabbiano, G., Gursky, H., Schwartz, D. A., Schwarz, J., Bradt, H. V., and Doxsey, R. E. 1978, *Nature*, **275**, 721.
Giacconi, R., et al. 1979a, *Ap. J.*, **230**, 540.
Giacconi, R., et al. 1979b, *Ap. J. (Letters)*, **234**, L1.
Gliese, W. 1969, *Catalogue of Nearby Stars* (Karlsruhe: Braun).
Golub, L., Harnden, F. R., Jr., Rosner, R., Topka, K., and Vaiana, G. S. 1979, *Bull. AAS*, **11**, 775.
Golub, L., Maxson, C. W., Rosner, R., Serio, S., and Vaiana, G. S. 1980, *Ap. J.*, **238**, 343.
Golub, L., Rosner, R., Vaiana, G. S., and Weiss, N. O. 1981, *Ap. J.*, in press.
Gorenstein, P., and Tucker, W. H. 1976, *Ann. Rev. Astr. Ap.*, **14**, 373.
Gottlieb, D. M. 1978, *Ap. J. Suppl.*, **38**, 287.
Haisch, B. M., and Linsky, J. L. 1976, *Ap. J. (Letters)*, **205**, L39.
Haisch, B. M., Linsky, J. L., Harnden, F. R., Jr., Rosner, R., Seward, F. D., and Vaiana, G. S. 1980, *Ap. J. (Letters)*, **242**, L99.
Harnden, F. R., Jr., et al. 1979a, *Ap. J. (Letters)*, **234**, L51.
Harnden, F. R., Jr., Golub, L., Rosner, R., Seward, F., Topka, K., and Vaiana, G. S. 1979b, *Bull. AAS*, **11**, 775.
Hartmann, L., Davis, R., Dupree, A. K., Raymond, J. P. C., Schmidtke, P. C., and Wing, R. F. 1979, *Ap. J. (Letters)*, **233**, L69.
Hearn, A. G. 1972, *Astr. Ap.*, **19**, 417.
———. 1973, *Astr. Ap.*, **23**, 97.
———. 1975, *Astr. Ap.*, **40**, 355.
Hearn, D. R., et al. 1976, *Ap. J. (Letters)*, **203**, L21.
Heise, J., and Huizenga, H. 1980, preprint.
Heise, J., et al. 1975, *Ap. J. (Letters)*, **202**, L73.
Holt, S. S., White, N. E., Becker, R. H., Boldt, E. A., Mushotzky, R. F., Serlemitsos, P. J., and Smith, B. W. 1979, *Ap. J. (Letters)*, **234**, L65.
Jordan, S. D. 1973, in *Stellar Chromospheres*, ed. S. D. Jordan and E. H. Avrett (NASA SP-317), p. 181.
Kippenhahn, R. 1973, in *Stellar Chromospheres*, ed. S. D. Jordan and E. H. Avrett (NASA SP-317), p. 265.
Ku, W., and Chanan, G. A. 1979, *Ap. J. (Letters)*, **234**, L59.
Kukarkin, B. V., et al. 1968, *General Catalogue of Variable Stars* (Moscow: Publishing House of the Academy of Sciences, USSR).
Lampton, M., Margon, B., Paresce, F., Stern, R., and Bowyer, S. 1976, *Ap. J. (Letters)*, **203**, L71.
Landini, M., and Monsignori-Fossi, B. C. 1973, *Astr. Ap.*, **25**, 9.
Linsky, J. L. 1977, in *The Solar Output and Its Variation*, ed. O. R. White (Boulder: Colorado Associated Universities Press), p. 477.
———. 1980, invited presentation, HEAD/AAS Meeting on X-ray Astronomy, Cambridge, MA, 1980 Jan.
Linsky, J. L., and Ayres, T. R. 1978, *Ap. J.*, **220**, 619.
Linsky, J. L., and Haisch, B. M. 1979, *Ap. J. (Letters)*, **229**, L27.
Linsky, J. L., Worden, S. P., McClintock, W., and Robertson, R. M. 1979, *Ap. J. Suppl.*, **41**, 47.
Long, K. S., and White, R. L. 1980, *Ap. J. (Letters)*, **239**, L65.
Lucy, L. B., and White, R. L. 1980, *Ap. J.*, **241**, 300.
Luyten, W. 1969, *Proper Motion Survey with the 48-Inch Schmidt* (Minneapolis: University of Minnesota).
Margon, B., Liebert, J., Gatewood, G., Lampton, M., Spinrad, H., and Bowyer, S. 1976, *Ap. J.*, **209**, 525.
Maxson, C. W., and Vaiana, G. S. 1977, *Ap. J.*, **215**, 919.
Mewe, R. 1979, *Space Sci. Rev.*, **24**, 101.
Mewe, R., Heise, J., Gronenschild, E. H. B. M., Brinkman, A. C., Schrijver, J., and den Boggende, A. J. F. 1975, *Ap. J. (Letters)*, **202**, L67.
Mewe, R., and Zwaan, K. 1980, Proc. Cambridge Cool Star Symposium (1980 Jan.).
Mullan, D. J. 1976, *Ap. J.*, **209**, 171.
Noci, G. 1973, *Solar Phys.*, **28**, 403.
Nugent, J., and Garmire, G. 1978, *Ap. J. (Letters)*, **226**, L83.
Oster, L. 1975, in *Problems in Stellar Atmospheres and Envelopes*, ed. B. Baschek, W. H. Kegel, and G. Traving (New York: Springer), p. 901.
Osterbrock, D. E. 1961, *Ap. J.*, **134**, 347.
Pneuman, G. W. 1973, *Solar Phys.*, **28**, 247.
Renzini, A., Cacciari, C., Ulmschneider, P., and Schmitz, F. 1977, *Astr. Ap.*, **61**, 39.
Ricketts, M. J., King, A. R., and Raine, D. J. 1979, *M.N.R.A.S.*, **186**, 233.
Robinson, E. L. 1976, *Ann. Rev. Astr. Ap.*, **14**, 119.
Rosner, R., Giacconi, R., Golub, L., Harnden, F. R., Jr., Topka, K., and Vaiana, G. S. 1979a, *Bull. AAS*, **11**, 776.
Rosner, R., Grindlay, G., Harnden, R. F., Jr., Seward, F., and Vaiana, G. S. 1979b, *Bull. AAS*, **11**, 446.
Rosner, R., Tucker, W. H., and Vaiana, G. S. 1978, *Ap. J.* **220**, 643.
Rosner, R., and Vaiana, G. S. 1979, in *Proc. Intl. School of Astrophysics*, Erice, ed. G. Setti, and R. Giacconi (in press).
Seward, F. D., et al. 1979, *Ap. J. (Letters)*, **234**, L55.
Smith, M. A. 1978, *Ap. J.*, **224**, 584.
Stencel, R. 1978, *Ap. J. (Letters)*, **223**, L37.
Stewart, G. C., Fabian, A. C., Cook, M., and Pringle, J. E. 1979, *Bull. AAS*, **11**, 775.
Swank, J. H. 1979, in *Proc. IAU Colloquium No. 53*, ed. H. M. Van Horn and V. Weicemann (Rochester: University of Rochester).
Swank, J. H., Becker, R. H., Boldt, E. A., Holt, S. S., Mushotzky, R. F., Serlemitsos, P. J., and White, N. E. 1979, *Bull. AAS*, **11**, 782.
Tassoul, J.-L. 1978, *Theory of Rotating Stars* (Princeton: Princeton University Press).
Thomas, R. N. 1975, *J. Chem. Phys.*, **32**, 259.
Toomre, J., Zahn, J.-P., Latour, J., and Spiegel, E. A. 1976, *Ap. J.*, **207**, 545.
Topka, K. 1980, thesis, Harvard University.

Topka, K., Fabricant, D., Harnden, F. R., Jr., Gorenstein, P., and Rosner, R. 1979, *Ap. J.*, **229**, 661.
Topka, K., Golub, L., Harnden, F. R., Jr., Gorenstein, P., Rosner, R., and Vaiana, G. S. 1979, *Bull. AAS*, **11**, 781.
Tucker, W. H. 1973, *Ap. J.*, **186**, 285.
Tuohy, I. R., Lamb, F. K., Garmire, G. P., and Mason, K. O. 1979, *Ap. J. (Letters)*, **226**, L17.
Vaiana, G. S. 1980, invited presentation, HEAD/AAS Meeting on X-ray Astronomy, Cambridge, MA, 1980 Jan.
Vaiana, G. S., and Rosner, R. 1978, *Ann. Rev. Astr. Ap.*, **16**, 393.
Vaiana, G. S., et al. 1979, *Bull. AAS*, **11**, 446.
Vanderhill, M. L., Borken, R. J., Bunner, A. N., Burstein, P. H., and Kraushaar, W. L. 1975, *Ap. J. (Letters)*, **197**, L19.
Vitz, R. C., Weiser, H., Moos, H. W., Weinstein, A., and Warden, E. S. 1976, *Ap. J. (Letters)*, **205**, L35.
Walter, F. M., Cash, W., Charles, P. A., and Bowyer, C. S. 1980a, *Ap. J.*, **236**, 212.
Walter, F. M., Linsky, J. L., Bowyer, S., and Garmire, G. 1980b, *Ap. J. (Letters)*, **236**, L137.
Warner, B. 1976, in *IAU Symposium 73, The Structure and Evolution of Close Binary Systems*, ed. P. Eggleton, S. Mitton and J. Whelan (Dordrecht: Reidel).
Weaver, R., McCray, R., Castor, J., Shapiro, P., and Moore, R. 1977, *Ap. J.*, **218**, 377.
Wentzel, D. G. 1978, *Rev. Geophys. Space Phys.*, **16**, 757.
Wilson, O. C. 1966, *Science*, **151**, 1487.
———. 1978, *Ap. J.*, **226**, 379.
Zirin, H. 1975, *Ap. J. (Letters)*, **199**, L63.

J. CASSINELLI: Washburn Observatory, University of Wisconsin, 475 N. Charter St., Madison, WI 53706

G. FABBIANO, R. GIACCONI, L. GOLUB, P. GORENSTEIN, F. R. HARNDEN, JR., C. MAXSON, R. ROSNER, F. SEWARD, K. TOPKA and G. S. VAIANA: Center for Astrophysics, 60 Garden St., Cambridge, MA 02138

B. HAISCH and H. M. JOHNSON: Lockheed Missiles & Space Co., Inc., Hanover St., Pálo Alto, CA 94304

J. LINSKY: JILA, University of Colorado, Boulder, CO 80302

R. MEWE and C. ZWAAN: The Astronomical Institute, Utrecht, The Netherlands

Reprinted with permission from *Pulsars*, pp. 353-356, edited by W. Sieber and R. Wielebinski. R.L. Kelley and S. Rappaport, "Masses of Neutron Stars in X-Ray Binary Systems." Copyright © 1981 by the IAU.

MASSES OF NEUTRON STARS IN X-RAY BINARY SYSTEMS

R.L. Kelley and S. Rappaport
Department of Physics and Center for Space Research
Massachusetts Institute of Technology

The masses of 6 neutron stars have now been established through studies of binary X-ray and radio pulsars. All of the masses are found to be consistent with, but not necessarily constrained to, the range 1.2-1.6 $M_\odot$. In this talk we discuss the methods and assumptions used in determining the masses of neutron stars in binary X-ray pulsar systems. For other recent reviews of this subject, the reader is referred to Bahcall (1978), Rappaport and Joss (1981), and references therein. Neutron-star parameters may also be obtained from studies of X-ray bursts that result from thermonuclear flashes near the surface of an accreting neutron star (see Joss 1980 and references therein), which we will not discuss here.

There are currently about 18 known binary X-ray pulsars, with pulse periods ranging from 0.7 s to 835 s. Compelling evidence that these objects are indeed magnetic neutron stars comes from studies of their pulse period histories. The long-term secular decrease in pulse period, first observed in Her X-1 and Cen X-3 (Giacconi 1974; Gursky and Schreier 1975; Schreier and Fabbiano 1976), is found in at least 8 other binary X-ray pulsars (see, e.g., Rappaport and Joss 1981). Both the magnitude and sign of this "spin-up" behavior are in excellent agreement with theoretical predictions of the torques exerted on a highly magnetized, accreting neutron star (see, e.g., Pringle and Rees 1972; Lamb, Pethick and Pines 1973; Rappaport and Joss 1977; Mason 1977; Lamb 1981). Further direct evidence for the presence of magnetic fields of the order of several times 10^{12} G is provided by the identification of possible cyclotron line features in the spectra of two of the X-ray pulsars (Trümper et al. 1978; Wheaton et al. 1979).

In general, the determination of the masses of neutron stars in binary systems requires a knowledge of the orbital elements of the system. The neutron-star orbit can be measured by tracking the pulse arrival times over a suitably long time interval (see, e.g., Schreier et al. 1972). The X-ray mass function is then determined from the orbital parameters:

$$f(M) = \frac{4\pi^2}{G} \frac{(a_x \sin i)^3}{P_{orb}^2} = \frac{M_c \sin^3 i}{(1+q)^2} \quad ,$$

where $a_x \sin i$ is the semi-major axis of the neutron-star orbit, P_{orb} is the orbital period, M_c is the mass of the companion, i is the orbital inclination angle, and q is the ratio of the neutron-star mass to the companion mass. The orbits, and hence the mass functions, of 7 X-ray pulsars have now been measured in this manner (see Kelley, Rappaport, and Petre 1980 for references). For systems in which the optical companion can be identified, it may be possible to obtain the Doppler velocity of the companion, K_c, from spectral studies of strong photospheric absorption lines. This information can be combined with the parameters obtained from the X-ray observations to deduce the mass ratio q. Provided that the orbital inclination can also be determined, the neutron-star mass is obtained from the mass function.

If the source exhibits X-ray eclipses, it is possible to estimate the orbital inclination angle as follows. The photospheric radius of the companion star in units of the orbital separation (R_c/a) can be simply related, by geometry, to the inclination angle and the eclipse angle, θ_e. On the other hand, the size of the critical potential lobe in units of the orbital separation (R_L/a), which sets an upper limit to the size of the companion, depends only on q and the rotation rate of the companion star (see, e.g., Avni 1976; Rappaport 1979). Studies of the ellipsoidal light variations displayed by several X-ray binaries (see, e.g., Bahcall 1978) indicate that the companion stars nearly fill their critical potential lobes (i.e., $\beta \equiv R_c/R_L \gtrsim 0.9$; Avni and Bahcall 1975a,b). The rotation rates of the companion stars are known only approximately, but the ones that exhibit ellipsoidal light variations appear to have rotation rates that lie in the range $0 \lesssim \Omega \lesssim 1.5$, where $\Omega = P_{orb}/P_{rot}$ (see, e.g., Conti 1978). Furthermore, the effects of tidal dissipation in the companion star are expected to force these systems into approximately synchronous rotation (i.e., $\Omega \sim 1$; see, e.g., Lecar, Wheeler and McKee 1976).

With the above set of relations and assumptions, the orbital inclination can be estimated and the most probable neutron-star mass obtained. In practice, however, it is difficult to propagate the experimental uncertainties in $a_x \sin i$, K_c and θ_e, and the experimental and theoretical uncertainties in β and Ω. The probability distributions have therefore been calculated by a Monte Carlo technique (Rappaport, Joss and Stothers 1980) in which $a_x \sin i$, K_c, θ_e are chosen for each source from the appropriate experimental probability distributions, and β and Ω are chosen from uniform distributions in the ranges $0.9 \leq \beta \leq 1.0$ and $0 \leq \Omega \leq 1.5$. The resulting probability distributions for 4 neutron-star masses are shown in Figure 1a.

These distributions can be integrated to find the mass limits for a desired confidence level. In Figure 1b we show the 95% confidence limits for the 4 neutron-star masses determined in this way. The inner limits on the X-ray pulsar masses in Figure 1b were computed for the less conservative case where $\beta = \Omega = 1$, which corresponds to Roche geometry with the companion star filling its Roche lobe. Also shown are the mass limits for Her X-1 and the binary radio pulsar PSR1913+16, added for completeness.

MASSES OF NEUTRON STARS IN X-RAY BINARY SYSTEMS 355

For the Her X-1 system the high degree of X-ray heating of the companion star, HZ Her, has prevented a reliable determination of the companion's orbital velocity. However, studies of the optical pulsations, which are apparently due to the reprocessing of the X-ray pulsations that impinge on the photosphere of the companion, have resulted in a determination of the mass of Her X-1 (Middleditch and Nelson 1976; Bahcall and Chester 1977). The mass of PSR1913+16 has been obtained through the measurement of general relativistic effects in that binary system (Taylor, Fowler and McCulloch 1979; Taylor 1981).

The measured neutron-star masses in Figure 1b are consistent with the range of masses (shaded region in Fig. 1b) that might be expected if the neutron stars were formed during the collapse of the degenerate cores of highly evolved stars (Arnett and Schramm 1973; Iben 1974), or from the collapse of accreting degenerate dwarfs in close binary stellar systems (see, e.g., Canal and Schatzman 1976). The allowed range of neutron-star masses is consistent with, but does not yet significantly constrain, neutron-star models based on conventional many-body nuclear and high-energy physics (see, e.g., Arnett and Bowers 1977; Fechner and Joss 1978; and references therein).

In the future we anticipate that the discovery of more binary X-ray pulsar systems will significantly add to the statistical sample of neutron-star masses. Improved X-ray and optical measurements should further reduce the uncertainties in neutron-star masses by a factor of at least several over present results.

The authors are grateful to P.C. Joss for helpful discussions. This work was supported in part by NASA contract NAS5-24441.

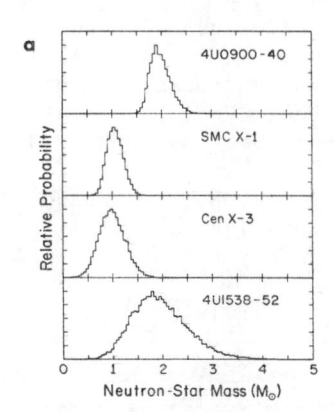

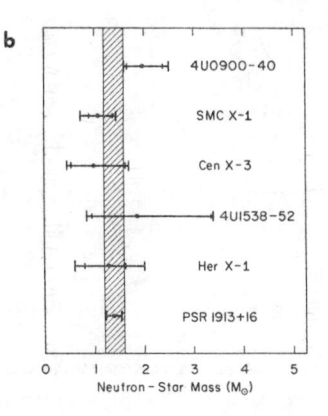

Fig. 1a) Probability distributions of the neutron-star masses for four binary X-ray pulsars. (b) Integrated 95% confidence limits on the masses of neutron stars (from Rappaport and Joss 1981). The mass of Her X-1 is from Bahcall and Chester (1977). The inner mass limits are for the special case where the companion fills its Roche lobe. The mass of the radio pulsar PSR1913+16 is from Taylor et al. (1979).

REFERENCES

Arnett, W.D. and Schramm, D.N.: 1973, Astrophys. J. Letters 184, p. L47.
Arnett, W.D. and Bowers, R.L.: 1977, Astrophys. J. Suppl. 33, p. 415.
Avni, Y.: 1976, Astrophys. J. 209, p. 574.
Avni, Y. and Bahcall, J.N.: 1975a, Astrophys. J. 197, p. 675.
Avni, Y. and Bahcall, J.N.: 1975b, Astrophys. J. Letters 202, p. L131.
Bahcall, J.: 1978, Ann. Rev. Astron. Astrophys. 16, p. 241.
Bahcall, J.N. and Chester, T.J.: 1977, Astrophys. J. Letters 215, p. L21.
Canal, R. and Schatzman, E.: 1976, Astron. Astrophys. 46, p. 229.
Conti, P.S.: 1978, Astron. Astrophys. 63, p. 225.
Fechner, W.B. and Joss, P.C.: 1978, Nature 274, p. 347.
Giacconi, R.: 1974, in "Astrophysics and Gravitation", Proc. of the 16th International Solvay Congress, Université de Bruxelles, p. 27.
Gursky, H. and Schreier, E.: 1975, in "Neutron Stars, Black Holes and Binary X-Ray Sources", eds.: H. Gursky and R. Ruffini, D. Reidel Publ. Co., Dordrecht, p. 175.
Iben, I.: 1974, Ann. Rev. Astron. Astrophys. 12, p. 215.
Joss, P.C.: 1981, in "Fundamental Problems in the Theory of Stellar Evolution", Proc. IAU Symp. No. 93, D. Reidel Publ. Co., Dordrecht, p. 207.
Kelley, R., Rappaport, S., and Petre, R.: 1980, Astrophys. J. 238, p. 699.
Lamb, F.K., Pethick, C.J., and Pines, D.: 1973, Astrophys. J. 184, p. 271.
Lamb, F.K.: 1981, this volume.
Lecar, M., Wheeler, J.C., and McKee, C.F.: 1976, Astrophys. J. 205, p. 556.
Mason, K.O.: 1977, Mon. Not. R. Astron. Soc. 178, p. 81P.
Middleditch, J. and Nelson, J.: 1976, Astrophys. J. 208, p. 567.
Pringle, J.E. and Rees, M.J.: 1972, Astron. Astrophys. 21, p. 1.
Rappaport, S. and Joss, P.C.: 1977, Nature 266, p. 683.
Rappaport, S.: 1979, in Proceedings of the NATO Advanced Study Institute on Galactic X-Ray Sources, Cape Sounion, Greece, June 1, 1979.
Rappaport, S. and Joss, P.C.: 1981, in "X-Ray Astronomy", Proc. HEAD-AAS Meeting, ed.: R. Giacconi, D. Reidel Publ. Co., Dordrecht, p. 123.
Rappaport, S., Joss, P.C., and Stothers, R.: 1980, Astrophys. J. 235, p. 570.
Schreier, E., Levinson, R., Gursky, H., Kellogg, E., Tananbaum, H., and Giacconi, R.: 1972, Astrophys. J. Letters 172, p. L79.
Schreier, E.J. and Fabbiano, G.: 1976, in "X-Ray Binaries", NASA SP-389, p. 197.
Taylor, J.H., Fowler, L.A., and McCulloch, P.M.: 1979, Nature 277, p. 437.
Taylor, J.H.: 1981, this volume.
Trümper, J., Pietsch, W., Reppin, C., Voges, W., Staubert, R., and Kendziorra, E.: 1978, Astrophys. J. Letters 219, p. L105.
Wheaton, W.A. et al.: 1979, Nature 282, p. 240.

Reprinted with permission from *Comments on Astrophysics*, 7 (5) pp. 151-160, D.M. Eardley, A.P. Lightman, N.I. Shakura, S.L. Shapiro, and R.A. Sunyaev, "A Status Report on Cygnus X-1." © 1978 Gordon and Breach Science Publishers Ltd.

A Status Report on Cygnus X-1

I. Cygnus X-1 : A Black Hole?

We would be happy if Cygnus X-1 were a black hole. However, in honesty, we are quite uncertain. Despite vigorous searches for black holes in nature, Cygnus X-1 is the only good candidate. It is just this uniqueness, together with the vast range of peculiarities in astrophysics, which requires caution in judging the nature of Cygnus X-1. Yet it is important to consolidate our understanding of Cygnus X-1, not only for its particular candidacy for a black hole, but also because of its similarities (e.g. chaotic and variable temporal behavior) to other discrete sources, notably active galactic nuclei. In this brief review, we shall be more interested in indicating focal areas of future research, than in giving an exhaustive summary of the existing literature. Unfortunately, there have been few major advances in theory or observations since 1973.

What evidence do we have that Cygnus X-1 is a black hole? Despite the many detailed models of gas flow and X-ray production discussed below, by far the strongest argument in favor of a black hole is that Cygnus X-1 is a compact object and has a mass larger than that allowed for a neutron star or white dwarf. Evidence for the former feature is associated with the observation of very rapid time variability in X-rays, discussed in Section IV. Evidence for the latter follows from a union of various types of observations of the Cyg X-1 star system. The visible star in the Cyg X-1 system is the single-line spectroscopic binary HDE 226868, a normal supergiant of spectral type O9.7 I ab.[1] The identification of Cyg X-1 with HDE 226868 was made by the simultaneous observations of the sudden appearance of a very accurately located radio source and a sudden spectral change in the X-ray source[2] in 1971. From the observed orbital parameters of the system,[3] it can be deduced that $a_1 \sin i = 7.9 R_\odot$ and $M^3 \sin^3 i/(M_1 + M_2)^2 = 0.21 M_\odot$. From the ellipsoidal light variations of HDE 226868 first discovered by Lyutiy and later interpreted and studied by Lyutiy, Sunyaev, Cherepaschchuk,[4] and by Avni and Bahcall,[5] it can be deduced that the mass of the unseen secondary satisfies $8 \lesssim M_2/M_\odot \lesssim 11$. Even without the latter observations, purely geometrical considerations[6] lead to a lower limit $M_2 \gtrsim 5 M_\odot$. In addition to the above arguments, it has been noted that Cygnus X-1 differs noticeably in its spectrum and chaotic time variability from other X-ray sources, believed to be accreting neutron stars.[2]

Devil's advocates point out that it is an unproved assumption that the compact X-ray source in the Cygnus X-1 system is identical with the high mass binary companion of the supergiant HDE 226868. Examples are given[7,8] of triple star systems in which an X-ray emitting neutron star orbits one or both members of a massive binary system of normal stars. In the former case the neutron star would perhaps have to be unmagnetized because of the absence of regular X-ray pulsations. In the latter case of the extended orbit, the gas supply to the neutron star would be extremely meager, and the energy source may well be rotation rather than gravitational accretion: the neutron star could be a 'young pulsar'. It is important to calculate the stability of such systems over a timescale $\gtrsim 10^6$ years; tidal dissipation may be highly important in the case of a close orbit. It is also important to search for spectroscopic and photometric evidence for a secondary, normal early type star. In favorable cases, composite line profiles with 1% photometric spectroscopy could reveal the presence of such a secondary.[5] Observational upper limits for the magnitude of modulation of the 5.6 day periodicity of the Cyg X-1–HDE 226868 system[9] have begun to place constraints upon triple system models, but such models remain possible. All in all, the triple system models are beset with difficulties, but as long as one is dealing with a unique source, they cannot be ruled out on grounds of implausibility alone. Kemp has recently reported that a 39 day periodicity exists in the optical polarization from the system.[10] This conceivably could be the other period of a triple system, but it is also reminiscent of and conceivably related to the ill-understood 35 day periodicity in the Her X-1 + HZ Her binary system.

But let us turn from these exotic possibilities to the more likely hypothesis that the system contains a black hole.

II. The Geometry of the Accretion Flow

The mode of mass transfer in the Cyg X-1 system is probably intermediate between the two extreme cases of Roche (or tidal) lobe-overflow, and a strong stellar wind. Following the discussion of van den Heuvel[11] and Lamers et al.,[12] we note that typical mass loss rates for 15–30 $M_\odot$ supergiants overflowing their Roche lobes are $\dot{M} \sim 10^{-3} M_\odot$ yr^{-1}. A significant fraction of this material will be captured by the compact secondary, will form an accretion disk, and will be converted into radiation at $\sim 10\%$ efficiency. Because the resulting accretion rate far exceeds the Eddington value, $\dot{M}_E = 10^{-7} M_\odot$ yr$^{-1}(M_x/10 M_\odot)$, most of the matter may

152

not be consumed by the black hole but will be blown out of the disk plane by radiative pressure. For such 'supercritical accretion' around a black hole, the emitted X-rays may be absorbed by an opaque ambient cloud of undigested gas and re-emitted as optical photons.[13] Even in the absence of such an opaque cloud the resulting luminosity would approach the critical Eddington value ($L_E = 1.3\ 10^{39}(M_x/10M_\odot)$ erg s^{-1}). If approximately half of that flux is emitted above 2 keV in the disk, this gives a luminosity an order of magnitude larger than is observed.

On the other hand there is no spectroscopic evidence for a spherical stellar wind strong enough to power the observed X-ray source, $\dot{M}_w \sim 10^{-5}$-$10^{-6} M$ yr^{-1} (cf. Davidsen and Ostriker[14]). The theoretical fits to the optical light curves[5,15] show that the companion nearly fills its Roche lobe (filling factor $\gtrsim 95\%$), however, so that the wind may strongly be enhanced by the low surface gravity of the primary below the Lagrange point L_1 and may possess significant angular momentum as it flows through a 'broadened nozzle' toward the compact star.[16]

It is not clear whether or not the accreted gas from the wind flows spherically onto the X-ray source or forms an accretion disk. Spherically symmetric laminar flow will not reproduce the observed X-ray spectrum (see Ref. 17 for a review of spherical accretion); turbulent accretion,[18,19] with magnetic reconnection, may generate the required spectrum[20] via Comptonization at $T_e \sim 10^9$ K, but detailed calculations are lacking for the efficiency (which must be high, $\sim 10\%$) and spectrum.

Calculations of the angular momentum transported in the accreted gas from a spherical wind[19,21] indicate that a disk in Cyg X-1 may not be very large. One can express the outer radius of the disk r_D in the form

$$r_D \simeq 700 r_I (1 + v_x^2/v_w^2)^{-4} (L/10^{38}\ \text{erg s}^{-1})^2 (M_w/10^{-6} M\ \text{yr}^{-1})^{-2}$$

where $r_I = 6GM/c^2$ is the innermost stable circular orbit around a Schwarzschild black hole. If one then requires[12] $\dot{M}_w \lesssim 10^{-5} M$ yr^{-1}, and the source is in the 'high state', $L \sim 10^{38}$ erg s^{-1}, one obtains $r_D/r_I \gtrsim 5$, indicating the existence of a disk. No firm conclusions can be drawn when the source is in the 'low state'.

If a large disk is present, nearly filling the Roche lobe of the compact star, it should be partially eclipsed by the companion near phase 0.0, and there is likely to be an observable signature in the light curve. No such effect has been reported. Taking all into account, it seems likely that there exists a small disk in this system.

Independent of the precise geometry of the flow, it seems almost certain that the observed hard X-rays originate from a hot ($T_e \gtrsim 5 \times 10^8$ K) optically thin plasma near the compact source (see Section III). There are many ways in which the gas near the hole can achieve high temperatures. If the flow is disk-like the geometry may consist of:

1) A cool thin outer region surrounding a hot bloated inner region[22,23] (e.g. $r \lesssim 10$-$20\ GM_x/c^2$).

2) A cool disk surrounded by a hot corona[24, 25] from which there may emanate a wind. (But convection in a cool disk inner region may not transport more than 25% of the total flux in the vertical direction.[26] This gives an upper limit to the ratio of hard X-rays to soft X-rays in standard disks.)

3) A cool thin disk subjected to instability-driven random outbursts of gas[27] which extend far above the disk midplane and travel inwards toward the black hole like 'sky-scrapers on wheels'.

4) A hot disk by itself. If the flow is spherical, highly turbulent flow characterized by reconnection, an equipartition magnetic field may convert kinetic energy into thermal energy efficiently and produce the required hot gas. It may turn out in all cases that the ion temperature exceeds the electron temperature in the plasma. There is no direct observational evidence for the existence of a cool disk, with the exception of the intense, soft X-ray flux observed during the high state ($L \sim 10^{38}$ erg s^{-1}, $E \lesssim 10$ keV).

Observations to date cannot distinguish between models, though polarization measurement may provide clues in the near future. The observations of Weisskopf et al.[28] of a 3% linear polarization at 2.6 keV at the 2σ level is not inconsistent with any of the disk models described above. If confirmed, it would rule out a perfectly spherical inner region. Knowledge of the polarization as a function of energy would provide more specific information regarding the disk geometry[29-32] but this experiment appears to be years away. Such a positive polarization measurement could confirm the presence of a well-behaved disk, but a negative result could not rule out disk-like inflow, since instabilities, turbulence, flares, or corrugations could so distort the outer scattering layers that no net polarization remains. Faraday rotation in the surface layers of the disk could also depolarize the X-rays, if the magnetic field is sufficiently strong.[33] The absence of strong linear and/or circular polarization ($\gtrsim 20\%$) from the compact source would indicate that if the source is really a neutron star, it does not possess a strong magnetic field.[29, 34]

III. The Emission Mechanism

Cygnus X-1 always shows a hard component (out to $h\nu \gtrsim 100$ keV) in its spectrum. It was shown that such a feature can arise as a hard tail in the standard disk model[13, 35] through 'Comptonization', or inverse Compton scattering[36, 37] by hot ($T_e \gtrsim 5 \times 10^8$) electrons. The source spends about 90% of the time in its 'low state', during which it shows nothing but this hard component, with a single power-law spectrum over the range $h\nu = 1\text{-}100$ keV,

$$I_\nu \propto \nu^{-\alpha}.$$

Here $\alpha = 0.5\text{-}1$ for averaging times greater than a few seconds.

It used to be common belief that such a power-law spectrum must arise from a power-law distribution of relativistic electrons, by either synchrotron radiation or inverse Compton scattering. However, recent calculations of 'Comptonization' show that a simple power-law spectrum can also arise by the mechanism of repeated inverse Compton scattering of soft photons in a finite, optically thin volume of plasma containing a thermal distribution of electrons at either a non-relativistic, semi-relativistic or relativistic electron temperature. This mechanism is really very similar to the Fermi mechanism for the acceleration of cosmic rays. The semi-relativistic and relativistic case was studied by Pozdnyakov, Sobol', and Sunyaev[38] using Monte Carlo techniques, and the non-relativistic case by Shapiro, Lightman and Eardley[22, 39] and by Katz,[40] using the Kompanéets equation. In all cases there arises an upper cutoff at $h\nu = (2-3)kT_e$. (Contrary to previous statements there does not arise an extra knee at $h\nu \sim 1/3 m_e c^2 \sim$ 150 keV at the onset of the Klein–Nishina scattering regime.) Figure 1 (from Ref. 38) shows a Monte Carlo calculation of the spectrum in a semi-relativistic case $kT_e = 0.5 m_e c^2 = 260$ keV for several values of Thomson optical depth τ. The value $\tau = 1$ yields $\alpha \sim 0.5$. Most model-builders feel that Cyg X-1 cools by this mechanism.

The value of T_e in Cyg X-1 is a very important parameter which must be determined by observation, through the identification of the upper cutoff. There are some observational suggestions of a knee in the spectrum[41] at $h\nu \simeq 125\text{-}150$ keV, but observations are difficult near and above this energy, and the existence and strength of this feature are in doubt. If this knee is real, then the average value of T_e in the hot plasma must be near 5×10^8 K. On the other hand, if the power-law spectrum extends to $E \gtrsim 500$ keV, as is suggested by other observations,[41] then T_e must be correspondingly greater. According to the disk models we discussed in Section II, we should have $T_e \simeq 5 \times 10^8 \div 2 \times 10^9$ K, but other models could give much higher values: rapid thermalization of gravitational energy near the black hole could give $T_e \sim 10^{11}$ K; particle acceleration could give even higher values. The positive observation of the cutoff for this source would be an important result.

The number of emission regions, independent in the sense that one region does not share photons with any other during the process of repeated inverse Compton scattering, is another important parameter of the source. If only one independent emission region exists, then the spectrum must be a single power-law (with cutoff, of course) at any one instant of time. However, the spectrum can vary on the thermal timescale, which can be as short as 1 ms. Canizares and Oda[42] recently reported that the spectral shape over $h\nu \simeq 2\text{-}20$ keV does in fact vary on timescales of ~ 1 s. This can mean that there exist a number of separate emission regions which come and go, as in the unstable disk model; or it can mean that there is a single emission region with rapidly variable parameters, as in the hot disk model.

155

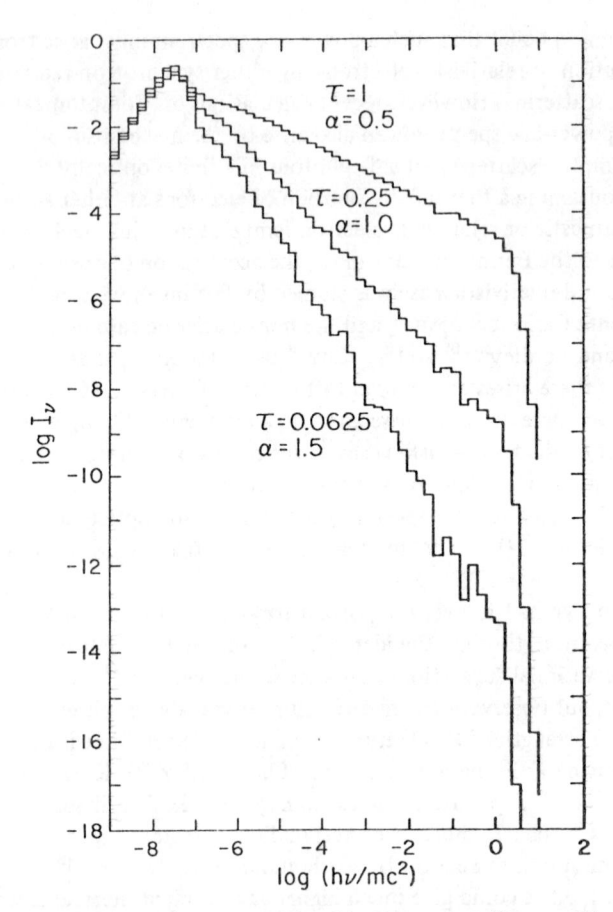

FIGURE 1 Inverse Compton spectra computed by Monte Carlo[38] for weakly relativistic electron temperature $T_e = 0.5 m_e c^2$, and for clouds of three different optical depths τ. Spectra are power-law with index $\alpha \simeq 0.5$-1.5, and with cutoff at $\sim 3kT_e$. Soft photons are injected in an adopted black-body spectrum with $kT_r = 10^{-8} m_e c^2$. The photons harden by repeated inverse Compton scattering from the hot electrons, and simultaneously leak out of the cloud.

According to the idea that the emission mechanism is inverse Compton scattering there should exist a separate, very soft spectral component somewhere below $h\nu = 1$ keV, which has not been observed. This 'soft photon source' could lie as low as $h\nu \sim 0.01$ keV where it would be completely unobservable. In the unstable disk model, the soft photons are produced in the cool region between the hot 'skyscrapers'.

Cygnus X-1 undergoes 'flares' or 'transitions to a high state',[2, 43-46] during which a new, intense, very soft spectral component appears below 10 keV. The spectral intensity above 10 keV seems to decrease,[44] so that soft and hard X-ray intensities are anti-correlated. The high state sets in on a timescale of days, and persists for on the order of a month; the source has spent ~ 10% of its time in this state over the last 5 years. It is very important for model-building to determine the true total luminosity in high state. Under the assumption that all of the observed low-energy cutoff at E = 2–3 keV is due to a constant absorption at a column density $N_H \simeq 7 \times 10^{21}$ H atom cm^{-2}, the total luminosity must increase by a factor of ~ 5, to ~ 10^{38} erg s^{-1}, in high state.[45] However, if the cutoff is intrinsic, there may be little or no increase in total luminosity. We do not know the underlying cause of these remarkable transitions, but the timescales are so long that changes in accretion rate from the companion star must surely be involved.

The soft component of this spectrum is direct and impressive evidence for the existence of an extended, rather cool, and very luminous body near the black hole in 'high state', which is naturally identified as the accretion disk.[47] Under the assumption that this body cannot emit more efficiently than a black body, we obtain $T \lesssim 6 \times 10^6$ K, $R \gtrsim 50$ km $\simeq 3GM/c^2$ for it. Therefore, the optically thick standard disk model, or something very much like it, must be partially present during 'high state'. In this connection we note that the observed timescales for transition to and decay of 'high state' are consistent with theoretical calculations for time-evolution of a standard accretion disk.[48] The persistence of the hard spectral component shows that some hot plasma is still present even in 'high state', perhaps in the form of unstable 'hot spots', of a corona, or of an optically thin region of the disk. The long-term transitions from high and low states on timescales of months and years might be associated with fluctuations in an accreted stellar wind[21,49] or changes in the optical depth of gas at the outer edge of an accretion disk,[50] which in turn, drive transitions in the mass flow near the hole.

IV. Time Variability

Cygnus X-1 is a highly variable X-ray source, with intensity fluctuations observed on timescales extending all the way from milliseconds to months and years (for a review, see Ref. 51). We can distinguish at least four theoretical ideas about the chaotic, short-term variability. The orbital period[52] of a 'hot spot' around the black hole may show up at timescales of 1–10 ms. More generally, there may be many hot spots, many locally unstable regions, all contributing at once at many periods, to yield chaotic variations.[52,53] In the unstable disk model,[27] organized temporal behavior in timescales up to ~ 1–10 s can occur due to the growth and inward propagation of unstable 'batches' of matter at the outer boundary of the radiation-pressure-dominated region: faster time-variations down to the orbital period will be superposed, because all spatial wavelengths down to the disk thickness are unstable. Finally, Compton reverberation[54] may play an important role on timescales of 1–20 ms. Theoretical understanding of all these processes is qualitative at best. The shortest intensity fluctuations, on timescales of $\lesssim 1$ ms, were reported by Rothschild et al.,[55] but their statistical significance has been questioned.[56] Recent observations of Canizares et al.[57] confirm the existence of a characteristic timescale on the order of 20 ms. The duration of these timescales is comparable with orbital periods at 10–50 Schwarzschild radii $\sim(2 \times 10^7$–$10^8)$ cm from a black hole of mass $10 M_\odot$. If nearly periodic, such fluctuations could indicate a disk around a black hole.[52] Chaotic fluctuations on these timescales could result from accretion-flow instabilities e.g. the optically thin thermal instability of Pringle et al.,[58] the optically thick thermal instability of Shakura and Sunyaev,[27] or the secular instability of Lightman and Eardley,[53] also at 10–50 Schwarzschild radii. In optically thin or radiation-pressure dominated disks with high viscosity, the thermal, radial drift, and dynamical timescales near the black hole are all comparable.

Several different investigators, e.g. Terrell,[59] have demonstrated that the time variability on timescales extending from 10^{-2}–1 s may be well modeled by random shot noise with individual shots of duration 0.2–0.5 s, and repetition rates of 10–100 $s^{-1} f^2$, where f is the fraction of the signal due to shots. Characteristic timescales of 0.1–1 s would suggest an emitting region up to ten times larger than that discussed above and would thereby diminish the strength of the argument for a highly compact central mass. Theoretical calculations of the origin of the observed shot noise are desperately needed. A promising suggestion[27] is that these fluctuations may arise from instability-driven batches created at the gas pressure-radiation pressure boundary of an accretion disk.

Recently, Canizares and Oda[42] have observed flares from Cygnus X-1 of duration 1–10 s, exhibiting both spectral and intensity fluctuations. These flares can only be explained by 'correlated shots' and suggest processes with ultimate causes at large distances from the black hole. It is important for understanding of the emission spectrum (see Section III) to determine the spectral shape on the shortest possible timescales.

V. Outlook

Cygnus X-1 seems to be the nearest actively accreting black hole. We are still far from understanding all details of its behavior, but it provides an excellent opportunity to study hydrodynamic and radiative processes near a black hole. Such processes may well be dominant in active galactic nuclei and quasars, should these even more intriguing objects turn out to involve supermassive black holes. There is a great need for the discovery of further objects such as Cyg X-1 in our galaxy. Circinus X-1 appears to be a good candidate because it shows strong, rapid time-variations[60, 61] on timescales down to $\lesssim$ 10 ms. The advent of new X-ray satellites, notably the HEAOs, promises further progress in this field.

D.M. EARDLEY, A.P. LIGHTMAN, N.I. SHAKURA,
S.L. SHAPIRO and R.A. SUNYAEV
Yale University, Harvard-Smithsonian Center for Astrophysics,
Sternberg Astronomical Institute, Cornell University, and
Space Research Institute of the U.S.S.R. Academy of Sciences

Acknowledgements

This report was written at the Joint U.S.A.-U.S.S.R. Workshop on Cosmic X-Ray Sources held in Protvino, U.S.S.R., Summer 1977, and sponsored by the U.S.S.R. Academy of Sciences and by the National Academy of Sciences of the U.S.A.

References

1. N.R. Walborn, Astrophys. J. Lett. **179**, L123 (1973).
2. H. Tananbaum, H. Gursky, E. Kellogg, R. Giacconi and C. Jones, Astrophys. J. Lett. **177**, L5 (1972).
3. C.T. Bolton, Nature Phys. Sci. **240**, 124 (1972).
4. V.M. Lyutiy, R.A. Sunyaev and A.M. Cherepashchuk, Astron. Zh. **50**, 3 (1973) [Sov. Astron. AJ **17**, 1 (1973)].
5. Y. Avni and J. N. Bahcall, Astrophys. J. **197**, 675 (1975).
6. B. Paczynski, Astron. Astrophys. **34**, 161 (1974).
7. J.N. Bahcall, F.J. Dyson, J.I. Katz and B. Paczynski, Astrophys. J. Lett. **189**, L17 (1974).
8. A.C. Fabian, J.E. Pringle and J.A. Whelan, Nature **247**, 351 (1974).
9. H.A. Abt, P. Hintzen and S.G. Levy, Astrophys. J. **213**, 815 (1977).
10. J.C. Kemp, submitted to Astrophys. J. (1977).
11. E.P.J. van den Heuvel, Astrophys. J. Lett. **198**, L109 (1975).
12. H.J.G.L.M. Lamers, E.P.J. van den Heuvel and J.A. Petterson, Astron. Astrophys. **49**, 327 (1976).
13. N.I. Shakura and R.A. Sunyaev, Astron. Astrophys. **24**, 337 (1976).
14. K. Davidsen and J.P. Ostriker, Astrophys. J. **179**, 585 (1973).
15. N.P. Bocharev, E.A. Karitskaja and N.I. Shakura, Pis'ma Astron. Zh. **1** (1975) [Sov. Astron. Lett. **1** (1975)].
16. M. Basko and R.A. Sunyaev, Astrophys. Space Sci. **23**, 71 (1973) [Ibid, **23**, 117 (1973)].
17. A.P. Lightman, M.J. Rees and S.L. Shapiro, Proc. Varenna Summer School on Physics of Neutron Stars and Black Holes (1975).
18. V.F. Shvartzman, Astron. Zh. **48**, 479 (1971) [Sov. Astron. AJ **15**, 377 (1971)].
19. A.F. Illarionov and R.A. Sunyaev, Astron. Astrophys. **39**, 185 (1975).
20. P. Mészáros, Nature **258**, 583 (1975).
21. S.L. Shapiro and A.P. Lightman, Astrophys. J. **204**, 555 (1976).
22. S.L. Shapiro, A.P. Lightman and D.M. Eardley, Astrophys. J. **204**, 187 (1976).
23. K.S. Thorne and R.H. Price, Astrophys. J. Lett. **195**, L101 (1975).
24. E.P. Liang and R.H. Price, University of Utah preprint, (1977); J.P. Ostriker, private communication (1975).

25. G.S. Bisnovatyi-Kogan and S.I. Blinnikov, Astron. Astrophys. **59**, 111 (1977).
26. N.I. Shakura, R.A. Sunyaev and S.S. Zilitinkevich, Space Research Institute preprint (1977).
27. N.I. Shakura and R.A. Sunyaev, M.N.R.A.S. **175**, 613 (1976).
28. M.C. Weisskopf, E.H. Silver, H.L. Kestenbaum, K.S. Long, R. Novick and R.S. Wolf, Astrophys. J. Lett. **215**, L65 (1977).
29. M.J. Rees, M.N.R.A.S. **171**, 457 (1975).
30. A.P. Lightman and S.L. Shapiro, Astrophys. J.Lett. **198**, L73 (1975).
31. A.P. Lightman and S.L. Shapiro, Astrophys. J. **203**, 701 (1976).
32. P.A. Connors and R.F. Stark, Nature **269**, 128 (1977).
33. Yu. N. Gnedin and N.A. Silant'ev, Astron. Zh. **53**, 338 (1976). [Sov. Astron. AJ **20**, 192 (1976)].
34. Yu. N. Gendin and R.A. Sunyaev, Astron. Astrophys. **36**, 379 (1974).
35. J.E. Pringle and M.J. Rees, Astr. and Ap. **21**, 1 (1972).
36. A.F. Illarionov and R.A. Sunyaev, Astron. Zh. **49**, 58 (1972) [Sov. Astron. AJ **16**, 45 (1972)].
37. J.E. Felten and M.J. Rees, Astron. Astrophys. **17**, 226 (1972).
38. L.A. Pozdnyakov, I.M. Sobol', and R.A. Sunyaev, Pis'ma Astr. Zh. (1976) [Sov. Astron. Lett. (1976)].
39. D.M. Eardley, Symposium on X-Ray Binaries, ed. Y. Kondo and E. Boldt (NASA SP-389, Greenbelt, MD, 1976).
40. J.I. Katz, Astrophys. J. **206**, 910 (1976).
41. P.C. Agrawal, G.S. Gokhale, V.S. Iyengar, P.K. Kunte, R.K. Manchanda and B.V. Sreekantan, Astrophys. Space Sci. **18**, 408 (1972).
42. C.R. Canizares and M. Oda, Astrophys. J. Lett. **214**, L119 (1977).
43. P.W. Sanford, J.C. Ives, S.J. Bell-Burnell, K.O. Mason, and P. Murdin, Nature **256**, 109 (1975).
44. M.J. Coe, A.R. Engel, J.J. Quenby, Nature **259**, 544 (1976).
45. J. Heise *et al.* Nature **256**, 107 (1975).
46. S.S. Holt, E.A. Boldt, L.J. Kaluzienski and P.J. Serlemitsos, Nature **256**, 108 (1975).
47. D.M. Eardley and A.P. Lightman, Nature **262**, 196 (1976).
48. A.P. Lightman, Astrophys. J. **194**, 419 (1974).
49. A.F. Illarionov and R.A. Sunyaev, Pis'ma Astron. Zh. (1976) [Sov. Astron. Lett. (1976)].
50. S. Ichimaru, Astrophys. J. **214**, 840 (1977).
51. M. Oda, K. Doi, Y. Ogawara, K. Takagishi and M. Wada, Astrophys. Space Sci **42**, 223 (1976).
52. R.A. Sunyaev, Astron. Zh. **49**, 1153 (1972) [Sov. Astron. AJ **16**, 941 (1973)].
53. A.P. Lightman and D.M. Eardley, Astrophys. J. Lett. **187**, L1 (1974).
54. C.R. Canizares, Astrophys. J. Lett **207**, L101 (1976).
55. R.E. Rothschild, E.A. Boldt, S.S. Holt and P.J. Serlemitsos, Astrophys. J. Lett. **189**, L13 (1974).
56. M.C. Weisskopf and P.G. Sutherland, Columbia University Astrophysics Laboratory preprint (1977).
57. C.R. Canizares, B. Laufer and F. Primini, Bull. Amer. Astron. Soc. **8**, 439 (1976).
58. J.E. Pringle, M.J. Rees, and A.G. Pacholczyk, Astron. Astrophys. **29**, 179 (1973).
59. N.J. Terrell, Astrophys. J. Lett. **174**, L35 (1972).
60. W. Forman, C. Jones and H. Tananbaum, Astrophys. J. **208**, 849 (1976).
61. A. Toor, Astrophys. J. Lett. **215**. L57 (1977).

THE ASTROPHYSICAL JOURNAL, 224:46–52, 1978 August 15
© 1978. The American Astronomical Society. All rights reserved. Printed in U.S.A.

Reprinted with permission from *The Astrophysical Journal*, **224** pp. 46-52, L. Cominsky, C. Jones, W. Forman, and H. Tananbaum, "Transient X-Ray Sources in the Galactic Plane." © 1978 The American Astronomical Society.

TRANSIENT X-RAY SOURCES IN THE GALACTIC PLANE

L. COMINSKY,[*] C. JONES,[†] W. FORMAN, AND H. TANANBAUM
Harvard-Smithsonian Center for Astrophysics
Received 1977 September 19; accepted 1978 February 24

ABSTRACT

Uhuru observations of the galactic plane indicate the presence of four X-ray sources not previously characterized as transient: MX 0836−42 (Markert *et al.*), A1918+14 (Seward *et al.*), and 4U 1730−22 and 4U 1807−10 (Forman *et al.*). X-ray light curves as well as positional and spectral information are presented for these sources and for 4U 1908+00, a recurrent transient source. The frequency, duration, and intensity of galactic plane transients during the *Uhuru* lifetime are discussed. Transient X-ray sources appear to be divided into two classes based primarily on an observed bimodal spectral temperature distribution.

Subject headings: interstellar: molecules — molecular processes — X-rays: general

I. INTRODUCTION

Centaurus X-2 was the first X-ray source to be described as "transient" (Chodil *et al.* 1968 and references therein). Since those early observations, many sources have been detected with similar X-ray light curves (see Table 1 for references). These transient X-ray sources have been observed with many types of detectors over different energy ranges and for various time intervals. However, for all transient sources the X-ray emission was detected at an intensity well above the survey limit for an interval of length greater than ~1 day which was short compared with the total amount of time the region was observed. This is the simplest observational definition of transient and may include sources of a type previously described as "highly variable." Transient outbursts have occurred only once for most sources; but as observations continue to be made, more "recurrent transients," i.e., sources with multiple outbursts, are reported (Jones *et al.* 1976; Kaluzienski *et al.* 1977a; Clark and Li 1977; Ricker and Primini 1977).

In this paper, we present *Uhuru* observations of four sources which satisfy the above definition of transient and of one source which is a recurrent transient, as well as a summary of all existing observations. The frequency of occurrence, intensity distribution, and spectral characteristics of transient sources will be discussed and compared with those of similar but less sensitive surveys (Kaluzienski *et al.* 1977b). Transient sources are observed to be divided into two classes based primarily on spectral temperatures (Maraschi *et al.* 1976; Kaluzienski *et al.* 1977b), and the relationship between these classes and existing models is discussed.

II. OBSERVATIONS

A summary of the observations for the 24 low-galactic-latitude ($|b^{II}| < 7°$) transient and recurrent

[*] Currently at Department of Physics and Center for Space Research, MIT.
[†] Junior Fellow, Harvard Society of Fellows.

transient X-ray sources is given in Table 1. Several transient events at higher latitudes have also been reported (Barnden and Francey 1969; Evans, Belian, and Conner 1970; Harries *et al.* 1971; Shukla and Wilson 1971; Conner, Evans, and Belian 1969; Rappaport *et al.* 1976; Cooke 1976; Ricketts, Cooke, and Pounds 1976; Forman *et al.* 1978) but will not be discussed in this paper, as they may represent distinctly different phenomena. All the galactic plane transients were detected at intensities well above the survey limit for a time interval $\tau_{\min}$ in days (Table 1, col. [4]) which is short (less than 50%) compared with the total amount of time the region was observed. For a source that was continuously monitored and for which exponential decay from the peak was observed, τ_e, the e-folding time constant in days, is given in column (5). Column (6) lists the maximum equivalent *Uhuru* intensity (2–6 keV) reported for each source. Column (7) gives the best 3 σ upper limit to persistent emission from the transient source from either the experiment which detected the outburst or the *Uhuru* lifetime based on the 4U catalog (Forman *et al.* 1978). $S_{HI/UL}$ (col. [8]) is the lower limit on the maximum range in intensity for each source, calculated from columns (6) and (7). Column (9) lists the published exponential temperature spectral parameter (Jones 1977) or the equivalent for non-*Uhuru* transients. The five sources at the bottom of Table 1 were later detected at intensities comparable to or greater than the original primary maximum and thus may be classified as recurrent transients.

The *Uhuru* X-ray observatory scanned the galactic plane frequently during 2.5 years following its launch on 1970 December 12. During this time, nine transient and three recurrent transient X-ray sources were observed within 7° of the galactic equator ($|b^{II}| \leq 7°$). The *Uhuru* observations of seven of these sources have been previously reported (Kellogg *et al.* 1971; Matilsky *et al.* 1972; Forman, Jones, and Tananbaum 1976a, b; Jones *et al.* 1976; Tananbaum *et al.* 1976). In this paper we present the data for the remaining

TRANSIENT X-RAY SOURCES

TABLE 1
Observations of Transient and Recurrent Transient X-Ray Sources

Source (1)	l^{II} (2)	b^{II} (3)	τ_{min} (4)	τ_e (5)	Observed High (6)	Best Upper Limits (7)	$S_{HI/UL}$ (8)	Exp. Temp. (keV) (9)	References (10)
4U 0115+63	125.92	+1.03	>50	~14	70	<5	>14	>15	7
A0620−00	209.96	−6.54	>100	~25	~50000	<5	>10000	<3	13–15
MX 0836−42	261.93	−0.97	>50	...	55	<2	>27	6 ± 2	19
A1118−61	292.6	−1.1	44	~7	~75	<5	>15	>15	20–21
Cen X-2	310.2	0	>44	~20	~6500	<5	>1300	<3.5	22, 37
Centaurus transient	313	0	>30	...	~10	<5	>2	>15	23
A1524−62	320.2	−4.49	>100	~60	~900	<2	>450	5 ± 1.5	24
4U 1543−47	330.93	+5.36	>110	~10	2000	<10	>200	<3	29–30
4U 1730−22	4.47	+5.89	>60	~30	125	<4	>31	4 ± 1	1
4U 1735−28	359.57	+1.56	>2	...	565	<40	>14	4.8 ± 9	34
A1742−28	359.94	−0.04	>18	...	~2300	<40	>57	3 ± 1	36
A1743−29	359.6	−0.42	...	...	~150	<40	>3	...	35
A1745−36	354.1	−4.2	...	...	~250	<10	>25	...	32
MX 1746−20	7.7	+3.8	>45	...	150	<7.5	>20	4 ± 0.5	3, 4
MX 1803−24	6.1	−1.9	...	...	~1000	<2	>500	...	2
4U 1807−10	18.6	+3.93	...	...	10	<2	>5	...	1
4U 1901+03	37.21	−1.39	>80	...	87	<2	>43	>15	7
A1918+14	49.7	+0.45	>13	...	45	<5	>9	6 ± 2	8
Cep X-4	98.96	+3.4	>15	...	~75	3	>25	>15	9
A0535+26	181.5	−2.65	54	~19	~2000	2.4	>833	>15	10–12, 25
MX 0656−07	220.17	−1.66	>15	...	~80	<3	>27	5 ± 2	16–18
4U 1608−52	330.91	−0.84	...	...	~1100	<3	>367	Variable	26–28
4U 1630−47	336.9	+0.28	>50	...	220	<10	>22	2 ± 0.5	31
4U 1908+00	35.67	−4.00	>12	...	400	<10	>40	3.5 ± 0.5	1, 5, 6, 33

REFERENCES.—(1) Forman *et al.* 1978. (2) Jernigan 1976. (3) Forman, Jones, and Tananbaum 1976b. (4) Markert *et al.* 1975. (5) Kaluzienski *et al.* 1977a. (6) Watson 1976. (7) Forman, Jones, and Tananbaum 1976a. (8) Seward *et al.* 1976. (9) Ulmer *et al.* 1973. (10) Rosenberg *et al.* 1975. (11) Coe *et al.* 1975. (12) Kaluzienski *et al.* 1975a. (13) Elvis *et al.* 1975. (14) Kaluzienski *et al.* 1977b. (15) Matilsky *et al.* 1976. (16) Clark 1975. (17) Carpenter *et al.* 1975. (18) Kaluzienski *et al.* 1976. (19) Markert *et al.* 1977. (20) Eyles *et al.* 1975a. (21) Ives, Sanford, and Bell-Burnell 1975. (22) Chodil *et al.* 1968. (23) Wheaton, Baity, and Peterson 1975. (24) Kaluzienski *et al.* 1975b. (25) Ricker and Primini 1977. (26) Tananbaum *et al.* 1976. (27) Li 1976. (28) Clark and Li 1977. (29) Matilsky *et al.* 1972. (30) Li, Sprott, and Clark 1976. (31) Jones *et al.* 1976. (32) Davison, Burnell, and Ives 1976. (33) Holt and Kaluzienski 1977. (34) Kellogg *et al.* 1971. (35) *Ariel 5* group 1976. (36) Eyles *et al.* 1975b. (37) Francey 1971.

five sources, three of which (Aql X-1, MX 0836−42, A1918+14) were originally reported by other experimenters (Friedman, Byram, and Chubb 1967; Markert *et al.* 1975b; Seward *et al.* 1976). No emission was detected from any of these sources during the *Uhuru* lifetime other than that present during the isolated outbursts which are described below.

a) *MX 0836−42 = 4U 0836−42*

First reported by Markert *et al.* (1975b) from *OSO 7* observations, MX 0836−42 is present in the *Uhuru* data during two week-long surveys of the galactic plane, in 1971 December and 1972 January, approximately 1 month apart. The peak intensity observed for this source is 60 counts s^{-1}; the minimum length of time the source appeared above background, τ_{min}, is 50 days. Figure 1a shows the X-ray light curve of MX 0836−42, which includes both *Uhuru* and *OSO 7* data (Markert *et al.* 1977).

Combining the *OSO 7* and *Uhuru* positional information yields the following improved 90% confidence location for the source:

Center:

$$\alpha = 129°130, \delta = -42°655;$$

Corners:

$$\alpha = 129°300, \delta = -42°520;$$

$$\alpha = 128°925, \delta = -42°760;$$

$$\alpha = 128°985, \delta = -42°780;$$

$$\alpha = 129°350, \delta = -42°540.$$

b) *4U 1730−22*

The source 4U 1730−22 (Forman *et al.* 1978) is typical of that subset of well-observed transient X-ray sources whose light curves are characterized by exponential decay from the peak with *e*-folding time constant τ_e. This decay may be interrupted by a plateau, secondary maxima, extreme variability, or another outburst. Although we do not scan the galactic plane during the initial rise in source intensity of 4U 1730−22, Figure 1b shows the observed part of the exponential decay, which has $\tau_e \approx 30$ days and a possible secondary maximum.

The energy spectrum of this source may be derived according to the procedure described by Jones (1977). For an exponential spectrum, $kT = 4 \pm 1$ keV and the upper limit on the low-energy absorption is 2.25 keV. For a power law spectrum, the energy spectral

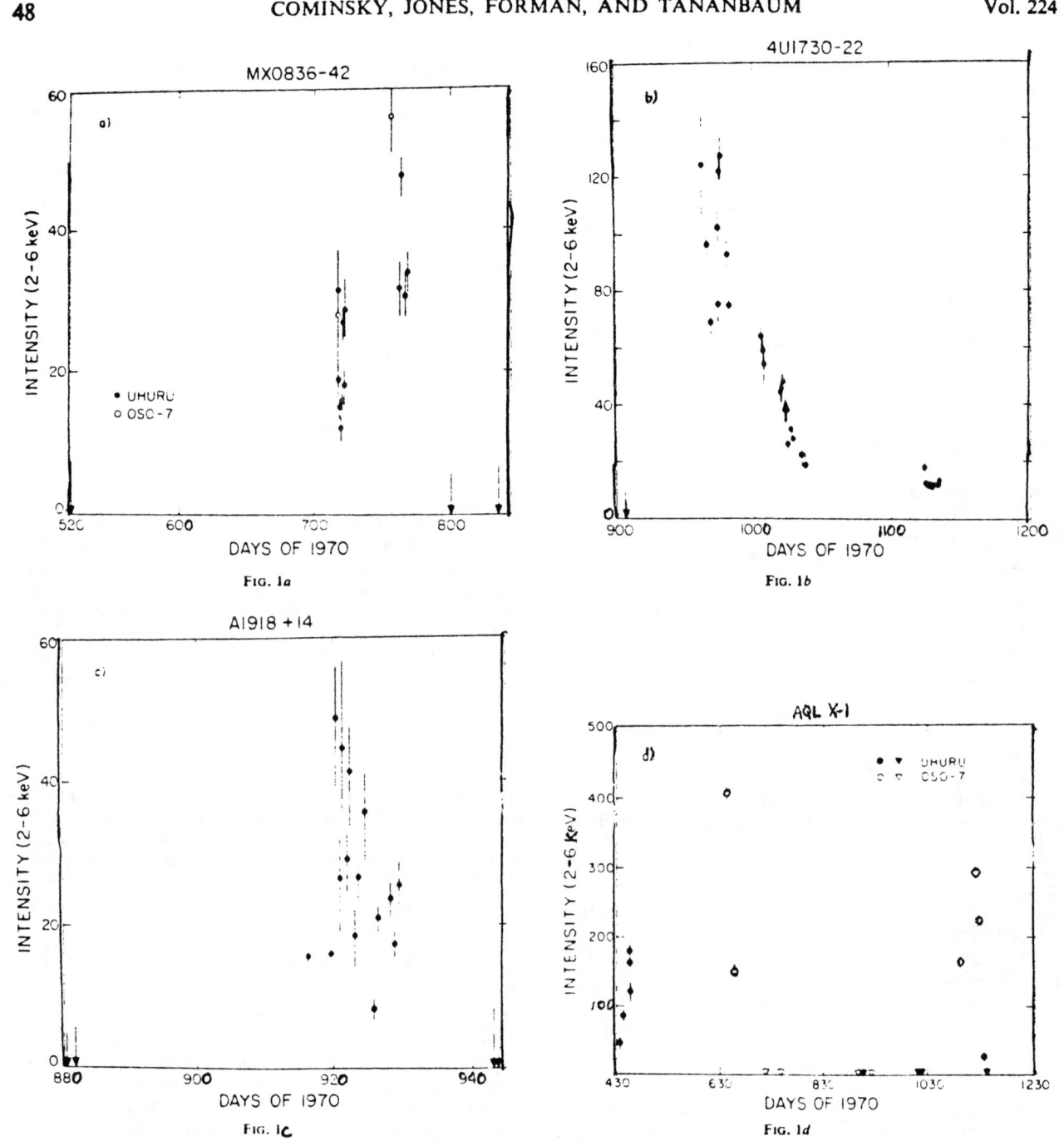

Fig. 1.—Light curves of (a) MX 0836−42, (b) 4U 1730−22, (c) A1918+14, and (d) Aql X-1 are shown which indicate the transient nature of these sources.

index $\alpha = 1.9 \pm 0.4$, with an upper limit on the low-energy cutoff $\bar{E}_a \gtrsim 3.00$ keV. The formalism used is that of Avni (1976) and of Lampton, Margon, and Bowyer (1976) for estimating two parameters simultaneously at the 90% confidence level, $\chi^2_{min} + 4.6$.

c) 4U 1807−10

Observed above the survey limit for only 1 day, at the end of the *Uhuru* data, 4U 1807−10 was discovered during the preparation of the 4U catalog (Forman *et al.* 1978). The observed intensity of 10 counts s^{-1} is a factor of 3 above adjacent upper limits but is too low to allow accurate spectra to be determined.

d) A1918+14

A1918+14 was initially reported by Seward *et al.* (1976) as a weak variable source with maximum intensity ~10 *Uhuru* counts. During 1972 July, an

X-ray source consistent in location with A1918+14 was detected above background for $\tau_{min} \approx 2$ weeks; assuming the source to be A1918+14, we find the peak intensity was ~50 counts s^{-1} (Fig. 1c). This is a factor of 10 increase above adjacent upper limits and a factor of 5 above the observed *Ariel* peak intensity. This outburst was the only emission from A1918+14 detected by *Uhuru*. Both exponential and power law spectral shapes adequately fit the *Uhuru* observations of A1918+14. For an exponential spectrum, $kT = 6 \pm 2$ keV with an upper limit on the low-energy absorption of 1.75 keV. When the observations are fitted to a power law spectrum, the upper limit on the low-energy cutoff is 2.75 keV and the energy spectral index is $\alpha = 1.25 \pm 0.25$.

e) Aql X-1 = 4U 1908+00

Kaluzienski et al. (1977a) have summarized the existing observations of Aql X-1, a recurrent transient X-ray source first detected by Friedman, Byram, and Chubb (1967). More recently, an additional outburst has been reported by Holt and Kaluzienski (1977). Figure 1d is the X-ray light curve of Aql X-1, from 1971 January to 1973 March, including observations from both *Uhuru* and *OSO 7* (Markert 1975). Note that persistent emission is not observed from Aql X-1 and that all of the positive detections occur during the repeated flares.

III. ANALYSIS

Using the *Uhuru* observations, it is possible to calculate the occurrence rate of outbursts at different luminosities from transient sources in the galactic plane. *Uhuru* surveyed the entire galactic plane, $|b^{II}| \leq 7°$, at irregular intervals for ~2½ years (~800 days) to a limit of 10 counts s^{-1}. For each non-plane scanning data segment lasting N days, a transient outburst of duration τ_{min} days would not have been detected during the first $N - \tau_{min}$ days of the interval. Sky coverage is therefore more complete for outbursts with larger values of τ_{min}, and the sky coverage correction factors C_i can be calculated by

$$C_i = \frac{1}{800}\left[800 - \left(\sum_{j=1}^{j=n} N_j - \tau_{min_i}\right)\right].$$

where the summation is taken over all the non-plane data segments of length $N_j > \tau_{min_i}$. The annual rate ρ at which transient outbursts occur may be calculated from the data in Table 2 as follows:

$$\rho = \sum_{i=1}^{n} \frac{N_{obs_i}}{2.5} \times \frac{1}{C_i}, \quad (1)$$

where N_{obs_i} is the number of outbursts observed in 2.5 years (col. [2]) in each interval in τ_{min}, C_i is the sky coverage correction factor (col. [3]) for each interval in τ_{min}, and n is the number of 10 day intervals in τ_{min}. By definition, τ_{min} is a lower limit to the actual duration of the outburst for those sources which were not continuously monitored; the sky coverage factors

TABLE 2
OBSERVED AND CORRECTED OCCURRENCE RATES FOR TRANSIENT X-RAY OUTBURSTS

τ_{min} (1)	N_{obs_i}* (2)	C_i† (3)	ρ_i (rate yr^{-1}) (4)
10–20	2	0.36	2.2
21–30	0	0.52	0
31–40	0	0.62	0
41–50	1	0.70	0.6
51–60	3	0.77	1.6
61–70	1	0.81	0.5
71–80	0	0.84	0
81–90	1	0.86	0.5
91–100	0	0.89	0
101–110	0	0.91	0
111–120	1	0.94	0.4
			$\rho = 5.8$ yr^{-1}

* In 2.5 yr.
† Sky coverage correction factor.

C_i were derived for the minimum τ_{min} in each 10 day interval. These factors combine to make ρ an upper limit. Since a source emitting 1.7×10^{37} ergs s^{-1} at the maximum galactic distance of ~25 kpc would be detectable above the *Uhuru* survey limit of 10 counts s^{-1} and since some of the outbursts may be from less luminous objects, the rate $\rho = 5.8$ yr^{-1} thus determined is an upper limit on the annual number of transient sources exceeding ~10^{37} ergs s^{-1}.

It is interesting to compare the outburst rate and luminosity limit thus calculated with those resulting from the assumption that all outbursts from transient sources occur at the same luminosity within a thin galactic disk of radius R and scale height $h \ll R$. In this case, the relationship between L and $\tau^{-1} = \rho$ can be derived by following Silk (1973) and Kaluzienski (1977), who define the number-flux relation for such transient sources as follows:

$$N(>S_0, t) = \int_0^{R_0} n_t(2\pi h r' dr'), \quad (2)$$

where $N(>S_0)$ is the number of outbursts observed above the survey limit S_0 in a time t; n_t is the number density of transients in the Galaxy; and the limiting radius $R_0 = (L/4\pi S_0)^{1/2}$, where the luminosity L is assumed constant for the entire group of sources. The diffuse flux due to unresolved sources in the disk as measured by an omnidirectional detector is

$$I_d = \frac{L \langle t_L \rangle}{\tau} \left(\frac{\langle \alpha \rangle}{4\pi^2 R^2}\right) \text{ergs cm}^{-2} \text{ s}^{-1} \text{ rad}^{-1}, \quad (3)$$

where

$$\langle t_L \rangle = \sum_{i=1}^{n} \frac{\rho_i \times \tau_{min_i}}{365 \times \rho}$$

gives the average duration of the outburst; and $\alpha = 1 + \ln(2R/h)$, where R is the radius and h is the scale height of the disk containing the transients. The mean α may be calculated from the limits on R/h

defined by the entire population of transient sources (Kaluzienski 1977). Substituting $\langle t_L \rangle = 0.10$ and $\langle \alpha \rangle = 6.25$ in equation (3) yields

$$I_d \geq \frac{L}{\tau}(1.4 \times 10^{-49}) \text{ ergs cm}^{-2}\text{ s}^{-1}\text{ deg}^{-1}. \quad (4)$$

The upper limit on the diffuse flux in the Galaxy I_d can be estimated from the difference in background levels observed by Uhuru between galactic and extragalactic scans. This flux does not exceed 2 Uhuru counts in the $\frac{1}{2}° \times 5°$ collimator (FWHM) or 6.8×10^{-11} ergs cm^{-2} s^{-1} deg^{-1} (2-6 keV). Thus the limiting value on L/τ can be found from equation (4):

$$4.9 \times 10^{38} \geq L/\tau.$$

This limit is similar to that found by Kaluzienski et al. (1977b) and implies that, if all transients were 10^{36} ergs s^{-1} at peak, then no more than 4.9 yr^{-1} could occur without violating the upper limit to a galactic "ridge," or, if all were 10^{37} ergs s^{-1}, 49 yr^{-1} could occur, and so on. Our observed limit, $\rho \leq 5.8$ yr^{-1}, on outbursts in excess of 10^{37} ergs s^{-1} therefore provides a significantly more sensitive limit on the number of 10^{37} ergs s^{-1} transient sources than can be obtained from the galactic ridge calculation.

Figure 2 is a plot of exponential temperature versus τ_{min} for all the well-observed transient sources listed in Table 1. The distribution of transients with respect to characteristic spectral temperature appears to be bimodal, while the distribution in τ_{min} seems continuous. The bimodal temperature distribution observed is similar to that found by Maraschi et al. (1976), who considered a subset of uniformly observed transients. There is an apparent deficiency of transient sources with temperatures between 7 and 15 keV relative to the nontransient galactic sources (Table 3). The data for the persistent sources were taken from Jones (1977), who analyzed Uhuru spectral data for all relatively bright galactic sources.

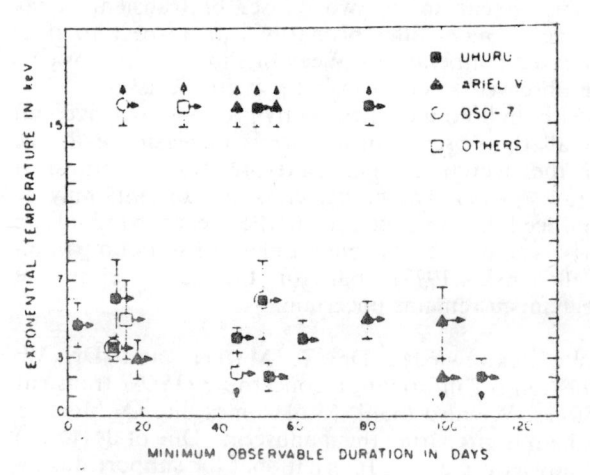

FIG. 2.—The minimum observable duration (τ_{min}, Table 1, col. [4]) is plotted versus the exponential spectral temperature (Table 1, col. [9]) for all well-observed transient sources in Table 1.

TABLE 3

COMPARISON OF SPECTRAL TEMPERATURE DISTRIBUTION FOR TRANSIENT AND PERSISTENT X-RAY SOURCES

Temp. (keV)	No. of Transients	No. of Nontransients
< 7	13	24
7-15	0	11
>15	6	6

IV. DISCUSSION

Many definitions of transient sources have been formulated (Kaluzienski 1977; Pounds 1976) in attempts to physically quantize what is basically an observational phenomenon. A source will appear transient if it is present above the survey limit of a detector for an interval which is short compared with the amount of time the region is observed. Highly variable sources may therefore appear transient, and there may not be a physical basis for considering the two as separate phenomena. Of the 24 sources listed in Table 1, two are identified with stars and exhibit sinusoidal modulations suggestive of orbital periods (A0620−00 = V616 Mon [Boley et al. 1976; Matilsky et al. 1976] and Aql X-1, a faint flaring star [Thorstensen, Charles, and Bowyer 1977; Watson 1976]); two are tentatively identified with OB stars (A1118−61 [Chevalier and Ilovaisky 1975] and A0535+26 [Liller 1975]); and one is identified with a late-type star (A1524−62 [Murdin et al. 1977]). One transient is associated with a globular cluster (MX 1746−20 = NGC 6440); two pulsate slowly (A0535+26 and A1118−61); three exhibit irregular intensity variations prior to the primary peak (A1524−62, A0535+26, and A1118−61); and four have apparent secondary maxima (4U 1543−47, 4U 1730−22, A0620−00, and A1524−62). (See Table 1 for references.)

Kaluzienski et al. (1977b) have suggested the existence of two classes of galactic plane transients, the first of longer duration, higher intrinsic luminosity, and lower X-ray temperature than the second. This suggestion was primarily based on data from the Ariel 5 All Sky Monitor, which has a survey limit a factor of 30 times higher than that of Uhuru and would therefore not be able to detect sources similar in intensity to 4U 1730−22. The division of transient sources into two classes based on the bimodal temperature distribution seems clear (Fig. 2). On the other hand, there is no convincing evidence for a difference in detectable duration between the two groups. The ranges in τ_{min} (the amount of time during which the outburst was detected) for the softer and harder transients are 2-110 and 15-80 days, respectively. A possible systematic effect to consider is that sources which have higher observed intensities or apparent luminosities can be observed above our detector threshold for longer durations and therefore may have larger values of τ_{min}. Since the percentage of sources with apparent peak luminosities greater than 100 counts s^{-1} is higher for the softer than for the harder transients, we might expect the softer transients to show values of τ_{min} systematically larger than those of the harder transients. The distribution of τ_{min} values is similar for both groups, however (see Fig. 2), and there is therefore no convincing evidence for a difference in detectable duration between the two groups.

There is also no convincing evidence for a difference in intrinsic luminosity between the two groups. Optical identifications and therefore reliable distance estimates exist for very few transient sources. The softer group has a higher percentage of sources with apparent luminosities in excess of 100 counts s^{-1} than does the harder group, but this effect may be due to a different spatial distribution and/or to an intrinsic luminosity difference. Furthermore, outbursts in which the maximum emission occurs above 15 keV may actually have intrinsic X-ray luminosities considerably greater than those estimated for the 2–6 keV energy band. Relatively weak outbursts of short duration are observed to occur in both spectral classes.

The quantity which is least affected by various observational biases is the intrinsic energy spectrum of a source. Jones (1977) has noted that regularly pulsating sources have temperatures greater than 10^8 K and has associated these characteristics with binary systems containing neutron stars. Two of the six transients with high temperatures exhibit pulsations. The other four sources with high temperatures were not observed in the same manner, and similar analyses were apparently either nonconclusive or not applicable to the type of observation. The similarity of this class of transient sources to persistent sources in binary systems containing neutron stars has been noted by Coe *et al.* (1975) and Avni, Fabian, and Pringle (1976) among others, as well as by Kaluzienski (1977), who suggests that this class of transient sources results from an increase in the stellar wind of a massive early-type optical companion. This interpretation seems reasonable and is supported by the observed pulsations and tentative identification of the two *Ariel 5* transients A0535+26 and A1118−61 with early-type stars (Ives, Sanford, and Bell-Burnell 1975; Rosenberg *et al.* 1975; Liller 1975; Chevalier and Ilovaisky 1975).

There are 13 transient sources listed in Table 1 that have exponential temperatures less than 7 keV. Their lack of observable pulsations or other short-timescale variability and their steeper characteristic spectra do not support the conventional neutron star binary system models for these transients. Three sources are confidently identified with stars, and one is associated with a globular cluster; but the positional uncertainties of the remaining nine sources are too large to permit identifications in the galactic plane.

Kaluzienski (1977) has associated this cooler class of transient sources with Roche lobe overflow of a low-mass, late-type optical star in a close binary system with luminosity near or at the Eddington limit ($\sim 10^{38}$ ergs $M_\odot^{-1}$). This mechanism for X-ray production is also believed to explain the observations of Her X-1 (cf. Pravdo 1976 and references therein), which has a flat, hard spectrum; is believed to be a neutron star; and emits two orders of magnitude below the Eddington limit (Schreier *et al.* 1972; Clark *et al.* 1972; Holt *et al.* 1974). Jones (1977) interprets hard spectra as indicating the presence of a neutron star in a close binary system, independent of the mechanism of mass transfer. If this interpretation is correct, then the difference in the spectrum between Her X-1 and the cooler class of transient sources may be accounted for by a different type of collapsed object, such as a white dwarf or black hole, or by an optically thick plasma surrounding the central hard source which degrades the initial photons by Compton scattering (Maraschi, Treves, and van den Heuvel 1977).

Another type of model has been proposed by Brecher, Ingham, and Morrison (1977) and by Gorenstein and Tucker (1976). In the Gorenstein and Tucker model, the transient outburst is the result of a hydrogen-burning flash on the surface of a white dwarf in a close binary system, which is embedded in an extended cloud of circumstellar material $\sim 10^{14}$ cm. The shell of matter ejected by the hydrogen flash would form a shock wave as it drove through the surrounding material which would heat the cloud and produce X-rays by thermal bremsstrahlung. The cloud would then expand and cool, which would account for the spectral softening often observed. This model also explains the lack of short-time-scale variability, but it predicts line emission from the hot circumstellar material and an absence of periodicities. These two additional constraints are not consistent with the observations of A0620−00, which has a reported 8 day period (Matilsky *et al.* 1976) and upper limits on line emission two orders of magnitude too low for an optically thin model (Griffiths, Ricketts, and Cooke 1976).

V. CONCLUSIONS

The number of transient outbursts with X-ray luminosities in excess of 10^{37} ergs s^{-1} is limited by the *Uhuru* observations to no more than 5.8 yr^{-1}. There appear to be two classes of transient X-ray sources, which differ primarily with respect to their observed temperatures. Weak or short-lived transients are observed to occur in both spectral classes.

Although no model currently proposed will account for all the observations, there is increasing evidence for the accreting binary interpretation of transient X-ray sources. The harder class of transients may be produced by an increase in the stellar wind of an early-type optical star onto a neutron star companion (Kaluzienski 1977), but for the softer class the mechanism remains uncertain.

We acknowledge Dr. T. Markert and Dr. W. Wheaton for information concerning *OSO 7* transient sources. We also thank M. Twomey and O. Messina for help in preparing the manuscript. One of us (L. C.) acknowledges Zonta International for support during part of the work.

This work was supported under NASA contract NAS5-20048.

REFERENCES

Ariel 5 group. 1976, *IAU Circ.*, No. 2934.
Avni, Y. 1976, preprint.
Avni, Y., Fabian, A., and Pringle, J. 1976, *M.N.R.A.S.*, **175**, 297.
Barnden, L. R., and Francey, R. J. 1969, *Proc. Astr. Soc. Australia*, **1**, 236.
Boley, F., Wolfson, R., Bradt, H., Doxsey, R., Jernigan, G., and Hiltner, W. A. 1976, *Ap. J. (Letters)*, **203**, L13.
Brecher, K., Ingham, W. H., and Morrison, P. 1977, *Ap. J.*, **213**, 492.
Carpenter, G. F., Eyles, C. J., Skinner, G. K., Willmore, A. P., and Wilson, A. M. 1975, *IAU Circ.*, No. 2852.
Chevalier, C., and Ilovaisky, S. A. 1975, *IAU Circ.*, No. 2778.
Chodil, G., Mark, H., Rodrigues, R., and Swift, C. D. 1968, *Ap. J. (Letters)*, **152**, L45.
Clark, G. W. 1975, *IAU Circ.*, No. 2843.
Clark, G. W., Bradt, H. V., Lewin, W. H. G., Markert, T. H., Schnopper, H. W., and Sprott, G. F. 1972, *Ap. J. (Letters)*, **177**, L109.
Clark, G. W., and Li, F. 1977, *IAU Circ.*, No. 3090.
Coe, M. J., Carpenter, G. F., Engel, A. R., and Quenby, J. J. 1975, *Nature*, **256**, 630.
Conner, J. P., Evans, W. D., and Belian, R. D. 1969, *Ap. J. (Letters)*, **157**, L157.
Cooke, B. A. 1976, *Nature*, **259**, 564.
Davison, P., Burnell, J., and Ives, J. 1976, *IAU Circ.*, No. 2925.
Elvis, M., Page, C. G., Pounds, K. A., Ricketts, M. J., and Turner, M. J. L. 1975, *Nature*, **257**, 656.
Evans, W. D., Belian, R. D., and Conner, J. P. 1970, *Ap. J. (Letters)*, **159**, L57.
Eyles, C. J., Skinner, G. K., Willmore, A. P., and Rosenberg, F. D. 1975a, *Nature*, **254**, 577.
———. 1975b, *Nature*, **257**, 291.
Forman, W., Jones, C., Cominsky, L., Julien, P., Murray, S., Peters, G., Tananbaum, H., and Giacconi, R. 1978, *Ap. J. Suppl.*, submitted. (Center for Astrophysics preprint, No. 763.)
Forman, W., Jones, C., and Tananbaum, H. 1976a, *Ap. J. (Letters)*, **206**, L29.
———. 1976b, *Ap. J. (Letters)*, **207**, L25.
Francey, R. J. 1971, *Nature Phys. Sci.*, **229**, 229.
Friedman, H., Byram, E. T., and Chubb, T. A. 1967, *Science*, **156**, 374.
Gorenstein, P., and Tucker, W. 1976, *Ann. Rev. Astr. Ap.*, **14**, 373.
Griffiths, R., Ricketts, M. J., and Cooke, B. A. 1976, *M.N.R.A.S.*, **177**, 429.
Harries, J. R., Tuohy, I. R., Broderick, A. J., Fenton, K. B., and Luyendyk, A. P. J. 1971, *Nature Phys. Sci.*, **234**, 149.
Holt, S. S., Boldt, E. A., Rothschild, R. E., Saba, J. L. R., and Serlemitsos, P. J. 1974, *Ap. J. (Letters)*, **190**, L109.
Holt, S. S., and Kaluzienski, L. J. 1977, *IAU Circ.*, No. 3031.
Ives, J. C., Sanford, P. W., and Bell-Burnell, S. J. 1975, *Nature*, **254**, 578.
Jernigan, G. 1976, *IAU Circ.*, No. 2957.
Jones, C. 1977, *Ap. J.*, **214**, 856.
Jones, C., Forman, W., Tananbaum, H., and Turner, M. J. L. 1976, *Ap. J. (Letters)*, **210**, L9.
Kaluzienski, L. J. 1977, GSFC X-661-77-107.
Kaluzienski, L. J., Holt, S. S., Boldt, E. A., and Serlemitsos, P. J. 1975a, *Nature*, **256**, 633.
———. 1976, *IAU Circ.*, No. 2935.
———. 1977a, *Nature*, **265**, 606.
———. 1977b, *Ap. J.*, **212**, 203.
Kaluzienski, L. J., Holt, S. S., Boldt, E. A., Serlemitsos, P. J., Eadie, G., Pounds, K. A., Ricketts, M. J., and Watson, M. 1975b, *Ap. J. (Letters)*, **201**, L121.
Kellogg, E., Gursky, H., Murray, S., Tananbaum, H., and Giacconi, R. 1971, *Ap. J. (Letters)*, **169**, L99.
Lampton, M., Margon, B., and Bowyer, S. 1976, *Ap. J.*, **208**, 177.
Li, F. K. 1976, *IAU Circ.*, No. 2936.
Li, F. K., Sprott, G. F., and Clark, G. W. 1976, *Ap. J.*, **203**, 187.
Liller, W. 1975, *IAU Circ.*, No. 2936.
Maraschi, L., Huckle, H., Ives, J., and Sanford, P. 1976, *Nature*, **263**, 34.
Maraschi, L., Treves, A., and van den Heuvel, E. P. J. 1977, *Ap. J.*, **216**, 819.
Markert, T. 1975, Ph.D. thesis, MIT.
Markert, T., Backman, D. E., Canizares, C. R., Clark, G. W., and Levine, A. M. 1975a, *Nature*, **257**, 32.
Markert, T., Bradt, H. V., Clark, G. W., Lewin, W. H. G., Li, F. K., Schnopper, H., Sprott, G. F., and Wargo, G. F. 1975b, *IAU Circ.*, No. 2765.
Markert, T. H., Canizares, C. R., Clark, G. W., Hearn, D. R., Li, F. K., Sprott, G. F., and Winkler, P. F. 1977, *Ap. J.*, **218**, 801.
Matilsky, T., et al. 1976, *Ap. J. (Letters)*, **210**, L127.
Matilsky, T. A., Giacconi, R., Gursky, H., Kellogg, E. M., and Tananbaum, H. D. 1972, *Ap. J. (Letters)*, **174**, L53.
Murdin, P., Griffiths, R. E., Pounds, K. A., Watson, M. G., and Longmore, A. J. 1977, *M.N.R.A.S.*, **178**, 27P.
Pounds, K. 1976, *Comments Ap.*, **6**, No. 5, 145.
Pravdo, S. 1976, GSFC X-661-76-280.
Rappaport, S., Buff, J., Clark, G., Lewin, W. H. G., Matilsky, T., and McClintock, J. 1976, *Ap. J. (Letters)*, **206**, 139.
Ricker, G., and Primini, F. 1977, *IAU Circ.*, No. 3078.
Ricketts, M. J., Cooke, B. A., and Pounds, K. A. 1976, *Nature*, **259**, 546.
Rosenberg, F. D., Eyles, C. J., Skinner, G. K., and Willmore, A. P. 1975, *Nature*, **256**, 628.
Schreier, E., Levinson, R., Gursky, H., Kellogg, E., Tananbaum, H., and Giacconi, R. 1972, *Ap. J. (Letters)*, **172**, L79.
Seward, F. D., Page, C. G., Turner, M. J. L., and Pounds, K. A. 1976, *M.N.R.A.S.*, **175**, 39P.
Shukla, P. G., and Wilson, B. G. 1971, *Ap. J.*, **164**, 265.
Silk, J. 1973, *Ap. J.*, **181**, 747.
Tananbaum, H., Chaisson, L. J., Forman, W., Jones, C., and Matilsky, T. A. 1976, *Ap. J. (Letters)*, **209**, L125.
Thorstensen, J. R., Charles, P. A., and Bowyer, S. 1977, *IAU Circ.*, No. 3088.
Ulmer, M. P., Baity, W. A., Wheaton, W. A., and Peterson, L. E. 1973, *Ap. J. (Letters)*, **178**, L121.
Watson, M. G. 1976, *M.N.R.A.S.*, **176**, 19P.
Wheaton, W. A., Baity, W. A., and Peterson, L. E. 1975, *IAU Circ.*, No. 2761.

L. COMINSKY: Department of Physics and Center for Space Research, Massachusetts Institute of Technology, Cambridge, MA 02139

W. FORMAN, C. JONES, and H. TANANBAUM: Center for Astrophysics, Smithsonian Astrophysical Observatory/Harvard College Observatory, 60 Garden Street, Cambridge, MA 02138

The Astrophysical Journal, 205:L127–L130, 1976 May 1
© 1976. The American Astronomical Society. All rights reserved. Printed in U.S.A.

Reprinted with permission from *The Astrophysical Journal*, 205 pp. L127-L130, J. Grindlay, *et al.*, "Discovery of Intense X-Ray Bursts from the Globular Cluster NGC 6624." © 1976 The American Astronomical Society.

DISCOVERY OF INTENSE X-RAY BURSTS FROM THE GLOBULAR CLUSTER NGC 6624

J. Grindlay, H. Gursky, and H. Schnopper
Center for Astrophysics, Harvard College Observatory and Smithsonian Astrophysical Observatory, Cambridge, MA 02138

D. R. Parsignault
American Science and Engineering, Cambridge, MA 02139

AND

J. Heise, A. C. Brinkman, and J. Schrijver
The Astronomical Institute, Space Research Laboratory, Beneluxlaan 21, Utrecht, The Netherlands
Received 1976 January 14; revised 1976 February 3

ABSTRACT

A new type of time variation of cosmic X-ray sources has been found from the Astronomical Netherlands Satellite (ANS) observations of the source 3U 1820−30 associated with the globular cluster NGC 6624. Two bursts in the ∼1–30 keV X-ray intensity of this source are reported. Each displayed a rapid rise in flux (≤ 1 s) by a factor of 20–30 followed by a ∼8 s exponential decay. These bursts appear to be qualitatively different from short time variations previously reported from X-ray sources. Analysis for further source variability, energy spectra, and position is presented. The characteristics of these events may imply the existence of a collapsed core in the globular cluster.

Subject headings: clusters: globular — stars: black holes — X-rays: sources — X-rays: transient sources

I. INTRODUCTION

The X-ray source 3U 1820−30 is the brightest of five X-ray sources associated with globular clusters, and is located within ∼1′ of the nucleus of NGC 6624 (Giacconi et al. 1974; Jernigan et al. 1975). It has been observed to be highly variable (factor of ≤ 5) on time scales of minutes to days and months (Canizares and Neighbours 1975; Forman, Jones, and Tananbaum 1976) but no regular variations or periodicities have been reported. Recently N. Bahcall (1976) has provided much-needed optical data and stellar density distributions for NGC 6624 and has found that the nucleus contains a bright unresolved core. We shall describe the results of observations of 3U 1820−30 by the X-ray detectors on ANS in 1975 March and September. During the second observing period, two very fast (∼0.5 s rise, ∼8 s decay) and intense (factor of 20–30 increase) bursts of ∼1–30 keV X-rays were detected from this source. The source intensity and spectrum were otherwise rather constant during the two observing periods, although the source flux (1.3–7 keV) in September was a factor of ∼4 lower than in March. This information was first reported by us in an IAU Telegram (Grindlay and Heise 1975).

The bursts appear to be qualitatively different from previously reported intensity variations in X-ray sources. In an accompanying *Letter*, Grindlay and Gursky (1976) explain certain characteristics of the bursts by Compton scattering of primary X-rays in a hot, surrounding cloud. This model implies the existence of a massive, collapsed object within the cloud.

II. TIME STRUCTURE AND LOCATION OF THE BURSTS

The soft X-ray (SXX) detector on ANS used in these observations was the medium-energy (1–7 keV) proportional counter with a 34′ × 82′ (FWHM) field of view (Brinkman, Heise, and de Jager 1974). The hard X-ray (HXX) experiment includes two large-area detectors (LAD1 and LAD2) each of 64 cm² effective area and sensitive from ∼1 to 30 keV (Gursky, Schnopper, and Parsignault 1975). The two LADs have 10′ × 3° (FWHM) fields of view separated by 3.7 so that source positions within a few arc minutes (in one dimension) may be obtained from the LAD1/LAD2 ratio of the source detection. The sensitivities of the two experiments are comparable and in the range 1–7 keV, 1 ANS count s^{-1} ≈ 15 *Uhuru* counts s^{-1} for a Crab-type spectrum.

About 3 hours of pointed observations of 3U 1820−30 were conducted in 1975 between March 24.4 and 28.0 and for a comparable exposure between September 27.4 and 30.7. A typical observation lasted ∼10 min and was either a continuous pointing on the source or a series of alternating pointings (16 or 64 s durations) on and off source for determination of background and source spectra. The 1.3–7 keV X-ray fluxes observed by the sum of the LADs for these two observing periods is shown in Figure 1. The mean intensity in the March observations was 13.5 ± 0.2 counts s^{-1} (∼200 *Uhuru* counts, or near the 3U catalog value) and only 3.2 ± 0.2 counts s^{-1} in September. Apart from ∼30 percent changes over time scales ∼1 hr observed in March, no factor of ≥ 2 variations over 5–10 minutes, such as

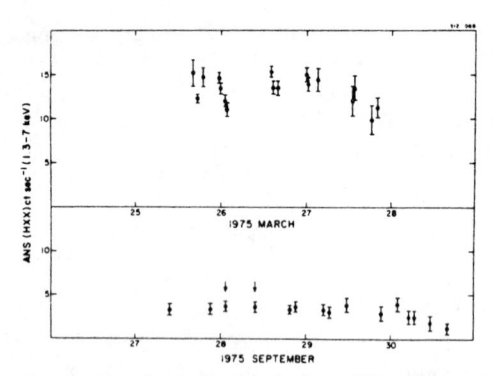

FIG. 1.—Average intensity (typically ~300 s integrations) of 3U 1820−30 as seen by the HXX experiment during 1975 March and September. The arrows mark the two observations in September during which the X-ray bursts were detected.

reported by Canizares and Neighbours (1975), were detected. The source showed no evidence of eclipses with periods in the range ~3 hours to 2 days. However, very large increases (bursts) in the apparent flux over much shorter times (~10 s) were found in the September data.

The first such burst to be identified in the data occurred 1975 September 28 in the middle of a 252 s continuous pointing on source in a region of the satellite orbit (longitude 122°, latitude +3°, height 280 km), where the the particle background is both the lowest and most constant. In Figure 2a, we plot the sum of LAD1 and LAD2 counts accumulated each second in the range 1.3–7 keV. Beginning at 09:49:40 UT, the total count rate increased from 11.2 ± 0.3 counts s^{-1} (approximately 8.1 counts s^{-1} background +3.1 counts s^{-1} source) to ~100 counts s^{-1} within 2 s and then decayed approximately exponentially over the next 10 s. In Figure 2b, we show the count rate observed in the SXX detector, which recorded total counts in the range 1–6 keV each 0.125 s. These data resolve the burst rise time to be ~0.5 s to 70 percent of maximum intensity. The burst profile was quite smooth with fluctuations consistent with statistics. The total increase in the burst observed by SXX is a factor of ~25, or very close to the HXX value.

A second burst was found after carefully examining all remaining data taken at nearly the same time. The burst occurred at 01:31:51 UT September 28 during a 64 s pointing on 3U 1820−30. The observation was conducted in the offset pointing mode, alternating between 64 s on source and 16 s on background. The count rate observed by the SXX detector around the time of the burst is plotted in Figure 3b, where the time on source, slew and background is also marked. The satellite was in a region of higher background (longitude 262°, latitude −61°, height 550 km) at the time of the event, and the rate increased from 10.1 ± 0.2 counts s^{-1} (approximately 7.1 counts s^{-1} background +2.9 counts s^{-1} source) to ~66 counts s^{-1} within 2 s followed by an approximately exponential decay over the next 10 s. The HXX experiment, recording also in the 1 s mode, recorded a nearly identical relative increase, as shown in Figure 3a.

One other candidate event was found in March (13^{h}14^{m}08^s, 27 March) in the SXX data. This was a factor of 25 increase in apparent source flux during three adjacent 125 ms bins during a pointing on 3U 1820−30. The event was not seen in the HXX, and no background counter, spectral data, or location information is available. Thus, this event could not be confirmed as being due to X-rays.

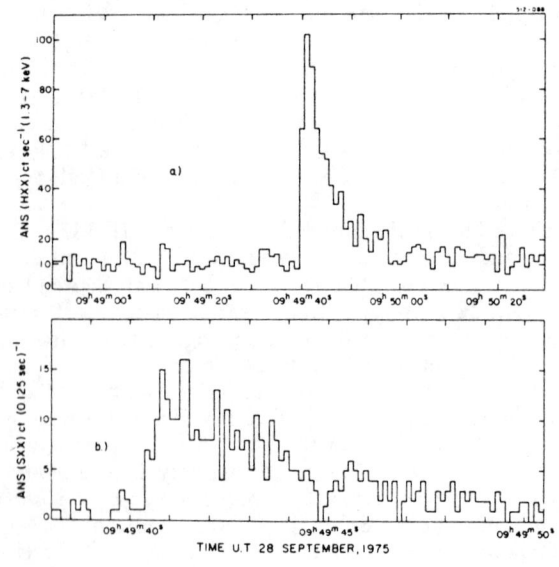

FIG. 2.—Burst of 09^{h}49^m September 28. Upper histogram is intensity profile as seen by HXX with 1 s integration intervals. Lower histogram is the event as seen by SXX with 0.125 s integration intervals.

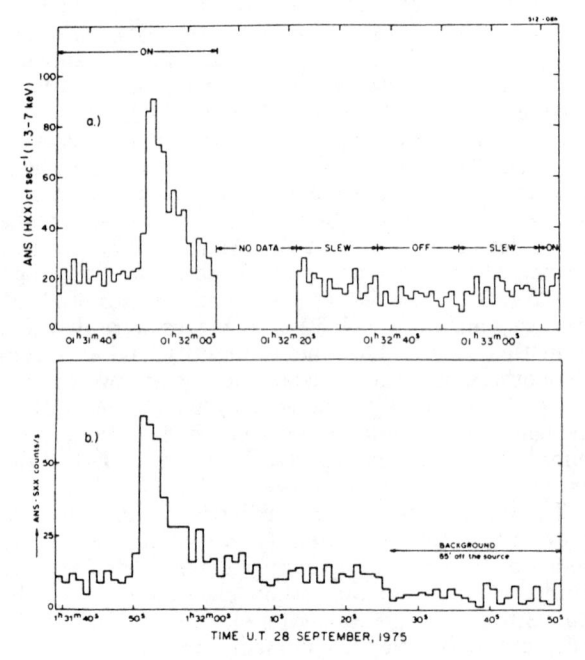

FIG. 3.—Burst of 01^{h}31^m September 28. *Upper histogram*, intensity profile as seen by HXX. *Lower histogram*, the event as seen by SXX. Integration time is 1 s for both.

We have also examined the data from both observing periods for possible periodic variations and excess fluctuations of the quiescent source intensity. No significant deviations in the summed power spectra were found for either set of data. The upper limits (95% confidence level) for the source power in pulsations with periods between 2 and ~100 s are ≤10 percent and ≤30 percent for the March and September data, respectively. Neither is there evidence (at the level of 2.5 σ) for excess 1 s variability corresponding to rate increases of a factor ~2 in March and ~4 in September.

The position of the bursts is determined from the X-ray count-rate data independent of any other information. The ratio of counts LAD1/LAD2 is proportional to the offset between the source position and the satellite pointing direction in one dimension. For the bursts these ratios were 1.04 ± 0.13 and 1.08 ± 0.17, respectively. The quiescent ratio source was 1.10 ± 0.30, where the uncertainty is due to the 1' uncertainty in the satellite pointing system. Thus, the location of the burst is determined to be within 1' of 3U 1820−30 in declination. In the other dimension the ratio of counts between HXX and the SXX limits the right ascension of the bursts to within 30' of 3U 1820−30. No other known X-ray sources are included in this combined field of view.

III. THE POSSIBILITY OF A PARTICLE ORIGIN

There are three distinct kinds of particle events which may mimic the observed bursts, namely, particles which penetrate the spacecraft, particles which can only penetrate the counter windows, and those at very low energy which produce fluorescence X-rays on the collimators which are then detected in the counters. We have data which eliminates each of these possibilities.

The HXX background counter records each 64 sec the number of particle events rejected by the X-ray counter PSD system as well as anti-coincidence and PHA overflow events (Gursky, Schnopper, and Parsignault 1975), and showed no increase during the 64 s block containing the burst. During the second burst the SXX background counter, which recorded similar events every 4 s, also recorded no increase. A significant increase in these background counters would have been recorded if the bursts were due to penetrating particles. Also, such particle events are seen in the HXX detectors as a much higher increase in the LAD1 rate than LAD2, since LAD1 is located exterior to LAD2 in the spacecraft. As noted above, however, the LAD1/LAD2 ratio is as expected for X-rays from 3U 1820−30.

Regarding particles which can only penetrate the counter windows, the very different detector window mass densities (9.3×10^{-3} g cm^{-2} for HXX versus 8.4×10^{-4} g cm^{-2} for SXX) would lead to a very different apparent increase in the two instruments if the event were due to particles, whereas both HXX and SXX record essentially the same total rates for both bursts.

Finally, bursts due to fluorescence X-rays produced in the collimators of the two experiments by soft electrons would also result in different rates because of the different collimator materials (copper for HXX versus aluminum for SXX). The thick-target bremsstrahlung spectrum of ~20 keV electrons (the maximum energy which do not penetrate the SXX window) on the HXX and SXX collimators would yield 1–7 keV continuum intensities in a ratio of 10:1 between the HXX and SXX rates, whereas they are observed to be nearly equal.

IV. SPECTRAL DISTRIBUTIONS

We have fitted both thermal and power-law spectra to the HXX data (15 logarithmically spaced channels in the range 1–30 keV) and SXX data (6 channels in the range 1–7 keV) obtained on 3U 1820−30 from both the March and September observations. The results of the 3-parameter fits are given in Table 1; the ±1 σ errors are χ^2 min + 3.5. In general, the thermal fits gave lower

TABLE 1

ANS SPECTRAL FIT RESULTS FOR 3U 1820−30

Observation	Experiment	Thermal χ^2 min	n_H (10^{21} cm^{-2})	kT(keV)	Power Law χ^2 min	n_H (10^{21} cm^{-2})	α
1975 March Avg.........	HXX	21	1(+1, −0.5)	9±2	28	5(+5, −4)	0.9±0.1
	SXX	2.2	5(+2, −4)	7.5±1	0.7	3(+3, −1)	0.6±0.1
1975 September Avg.....	HXX	22	9(+11, −8)	10(+19, −5)	24	10(+15, −9)	0.9±0.5
	SXX	1.2	5(+2, −4)	10(+?, −3)	0.5	7(+7, −6)	0.35(+0.35, −0.25)
Burst 1:							
Total..............	HXX	27	12(+15, −10)	16±8	32	20(+15, −12)	0.8±0.3
Peak...............	HXX	24	20(+15, −12)	5(+3, −2)	25	40(+40, −30)	2.1±0.3
Decay..............	HXX	43	20(+15, −12)	20±8	47	20(+15, −12)	0.6±0.3
Burst 2:							
Total..............	HXX	12	10(+20, −8)	17(+13, −10)	13	10(+20, −8)	0.4(+0.6, −0.4)
	SXX	1.2	15(+5, −9)	≥8	0.6	6(+14, −6)	−0.3(+0.9, −0.5)

NOTE.—Minimum χ^2 values are for 3-parameter fits with 12 degrees (9 for burst 2) of freedom (HXX) and 3 degrees of freedom (SXX). The errors quoted are ±1 σ and are χ^2 min + 3.5. For burst 1 (HXX spectra only) at 09^{h}49^{m}40^s, the peak includes seconds 40–41 and the decay is seconds 42–49. Burst 2 is the event at 01^{h}31^{m}51^s, also on UT 1975 September 28. The thermal spectra include a Gaunt factor; and for the power-law spectra, α is the energy spectral index.

$\chi^2_{\min}$ values with temperatures and cutoffs in agreement with the *Uhuru* value (Forman *et al.* 1976). We note that our absorption column densities given in Table 1 are generally consistent with that derived (2.2×10^{21} cm^{-2}) from the optical extinction measurements of Burstein and McDonald (1975) using the relation between extinction and column density as given by Gorenstein (1975).

During the low background burst of Figure 2, only HXX spectra were recorded (every 4 s). The fit to the spectrum of 10 s during the burst yielded a temperature and cutoff consistent with the average (September) values. We have separately fitted the spectra for the 2 s at the burst peak and the subsequent 8 s of decay. The low-energy absorption was found to remain constant during the burst, whereas the temperature increased significantly during the decay. The results given in Table 1 suggest that the spectrum at the burst peak is similar to the average spectrum in the March "high state" and is softer than the average in the September "low state."

Spectra were recorded by both HXX (every 4 s) and SXX (every 1 s) for the second burst identified (Fig. 3). The high background environment precluded fitting the HXX spectra above 10 keV as the fluctuations are then significant. The best fit spectral parameters (Table 1) are for the total burst and include 11 s after the burst rise. Since both fits were necessarily below 10 keV and because of the adverse background, it was not possible to search for the spectral hardening in the decay. We note that both bursts may show an intrinsic low-energy cutoff relative to the quiescent source.

V. DISCUSSION

We have detected X-ray bursts while viewing the source 3U 1820−30, which is within 1′ of the nucleus of the globular cluster NGC 6624. The available data allow us to exclude magnetospheric particle events as their origin, and to localize the events within a region $2' \times 1°$ which includes 3U 1820−30, but no other reported X-ray sources. Thus, we conclude that 3U 1820−30 (NGC 6624) is the likely origin of these events.

The two observed bursts were similar in rise time ($\lesssim 1$ s), source intensity increase (20–30), and decay time (~ 8 s). At least one burst spectrum was definitely softer at the peak than during the subsequent decay. Adopting a distance of 10 kpc to NGC 6624 (N. Bahcall 1976) the X-ray luminosity in March was $\sim 10^{38}$ ergs s^{-1} and increased from $\sim 3 \times 10^{37}$ ergs s^{-1} to $\sim 10^{39}$ ergs s^{-1} during the bursts in September. The detected energy in both bursts was $\sim 10^{-7}$ ergs cm^{-2}, implying a total burst energy of $\sim 2 \times 10^{39}$ ergs.

We cannot exclude a possible relation between these events and the γ-ray bursts discovered by Klebesadel, Strong and Olson (1973), which have yet to be identified with known cosmic objects. The fact that these bursts are so much faster and larger than "flares" previously reported for cosmic X-ray sources precludes obvious comparison with other galactic sources. They are very different from the X-ray flares detected (Heise *et al.* 1975) from flare stars (~ 20 s rise time, maximum luminosity $\sim 10^{31}$ ergs s^{-1}) or very large increases from known binary sources (e.g., the giant flare of 3U 0900−40: an intensity increase by a factor of ~ 30 in about 12 min [Forman *et al.* 1973]). This lack of similarity suggests to us that perhaps the X-ray sources in globular clusters are themselves different from the usual binary X-ray stars. This hypothesis requires, of course, confirmation by detection of additional large bursts from globular cluster sources and possibly similar (e.g., Canizares 1975) "galactic bulge" sources and not from known binaries.

J. Bahcall and Ostriker (1975) and Silk and Arons (1975) have suggested that the globular X-ray sources, including NGC 6624, may result from accretion onto a massive black hole which was formed by the collapse of the cluster core. N. Bahcall (1976) has shown that the star distribution in the core of NGC 6624 is consistent with the presence there of a massive black hole. Grindlay and Gursky (1976) in an accompanying *Letter* show that the spectral hardening and the exponential intensity falloff seen during the decaying portion of the burst are what is expected from Compton scattering in a hot gas cloud surrounding the X-ray source. Their model implies the existence of a collapsed core of at least several hundred solar masses within the cloud. In fact, the finite rise time of 0.5 s is the expected time scale for rapid changes in the X-ray emission from the vicinity of a $10^3 M_\odot$ black hole.

We are pleased to acknowledge discussions with Ethan Schreier which helped clarify our interpretation of these events. This work was partially supported by NASA contract NAS5-23282 and NAS5-11350.

REFERENCES

Bahcall, J. N., and Ostriker, J. P. 1975, *Nature*, **256**, 23.
Bahcall, N. 1976, *Ap. J. (Letters)*, in press.
Brinkman, A., Heise, J., and de Jager, C. 1974, *Philips Tech. Rev.*, **34**, 43.
Burstein, D., and McDonald, L. 1975, *A.J.*, **80**, 17.
Canizares, C. 1975, *Ap. J.*, **201**, 589.
Canizares, C., and Neighbours, J. 1975, *Ap. J. (Letters)*, **199**, L97.
Forman, W., Jones, C., and Tananbaum, H. 1976, submitted to *Ap. J.*
Forman, W., Jones, C., Tananbaum, H., Gursky, H., Kellogg, E., and Giacconi, R. 1973, *Ap. J. (Letters)*, **187**, L103.
Giacconi, R., Murray, S., Gursky, H., Kellogg, E., Schreier, E., Matilsky, T., Koch, D., and Tananbaum, H. 1974, *Ap. J. Suppl.*, No. 237, **27**, 37.
Gorenstein, P. 1975, *Ap. J.*, **198**, 95.
Grindlay, J., and Gursky, H. 1976, *Ap. J. (Letters)*, **205**, L131.
Grindlay, J., and Heise, J. 1975, IAU Circ. No. 2879.
Gursky, H., Schnopper, H., and Parsignault, D. R. 1975, *Ap. J. (Letters)*, **201**, L127.
Heise, J., Brinkman, A. C., Schrijver, J., Mewe, R., Gronenschild, E., den Boggende, A., and Grindlay, J. 1975, *Ap. J. (Letters)*, **202**, L73.
Jernigan, G., Canizares, C., Clark, G., Doxsey, R., Epstein, A., Matilsky, T., Mayer, W., and McClintock, J. 1975, *Bull. AAS*. **7**, 443.
Klebesadel, R. W., Strong, I. B., and Olson, R. A. 1973, *Ap. J. (Letters)*, **182**, L85.
Silk, J., and Arons, J. 1975, *Ap. J. (Letters)*, **200**, L131.

THE DISCOVERY OF RAPIDLY REPETITIVE X-RAY BURSTS FROM A NEW SOURCE IN SCORPIUS

W. H. G. Lewin, J. Doty, G. W. Clark, S. A. Rappaport,* H. V. D. Bradt, R. Doxsey, D. R. Hearn, J. A. Hoffman, J. G. Jernigan, F. K. Li, W. Mayer, J. McClintock, F. Primini, and J. Richardson

Department of Physics and Center for Space Research, Massachusetts Institute of Technology

Received 1976 March 10; revised 1976 April 26

ABSTRACT

Rapidly repetitive X-ray bursts have been observed from a new X-ray source in Scorpius. More than 2000 bursts were observed during the ∼4 day continual SAS-3 observations of this source which we designated MXB 1730−335. The time interval between bursts varied from a minimum of ∼6 s to a maximum of ∼5 minutes. The energy in a given burst is approximately linearly proportional to the time interval to the next burst. The largest bursts observed last for ∼60 s and represent an energy release of ∼10^{40} ergs for an assumed distance to the source of 10 kpc. The smallest bursts observed last only for a few seconds. We suggest that the bursts are caused by sporadic precipitations of plasma from a reservoir in the magnetosphere of a neutron star. The reservoir is replenished at a nearly constant rate by mass transferred from a binary companion.

Subject headings: X-rays: sources — X-rays: transient sources

I. INTRODUCTION

Several groups have now reported X-ray burst sources which were detected with satellite-borne experiments (Babushkina *et al.* 1975; Grindlay *et al.* 1976; Clark *et al.* 1976; Lewin 1976*a*; Belian, Connor, and Evans 1976). At present there are known to be at least nine X-ray burst sources which typically exhibit a rapid rise time ($\lesssim 1$ s), peak brightness of the order of magnitude of the Crab Nebula, exponential decay time of a few seconds, and repetition rates of a fraction of a day to several days. With one exception these sources are located near the galactic plane with $|b^{II}| \lesssim 14°$ and with $263° \lesssim l^{II} \lesssim 45°$ (Clark *et al.* 1976; Grindlay *et al.* 1976; Jones and Forman 1976; Belian, Connor, and Evans 1976; Clark 1976; Doty 1976; Lewin 1976*a,b*; Hoffman 1976).

While investigating a region which we knew from previous observations to contain at least one of these sources, we pointed the Y-axis detectors of the SAS-3 observatory (see next section) in the vicinity of 3U 1727−33. A source of rapid repetitive X-ray bursts was immediately apparent from the first data we received, and from the aspect solutions it was clear that the source was not 3U 1727−33 (Lewin 1976*b*). This new and unique source was designated MXB 1730−335.

During the ∼4 day continual observations of MXB 1730−335 more than 2000 X-ray bursts were observed from this source. Here we discuss some of the salient features of this source and, in particular, a relation between burst-energy and burst-spacings which implies a relaxation-oscillator model.

II. OBSERVATIONS

The new source in Scorpius (MXB 1730−335) was observed by the SAS-3 X-ray observatory simultaneously in independent detector systems with four different fields of view. The counting rate data from the four systems are shown in Figure 1 for an ∼40 min period of pointed observations near 1976 March 2.3 (UT). Forty-six bursts can clearly be distinguished. The smallest and largest time intervals between the bursts during this portion of the observations are ∼13 s and ∼6.5 min, respectively.

The detection systems are labeled, in Figure 1: (*a*) left slat, (*b*) center slat, (*c*) right slat, and (*d*) horizontal tube. The center slat has a $1° \times \sim40°$ FWHM field of view; its direction is perpendicular to the equatorial plane of the satellite. The left and right slat have $0°.5 \times \sim40°$ FWHM fields of view which are inclined $\pm 30°$ with respect to the center slat. The horizontal tube has a circular field of view of $1°.7$ FWHM. These four detection systems and a star camera are coaligned to within 6′ (the alignment is known to better than ∼$0'.5$) on the equatorial plane of the satellite. The four different fields of view have an area of ∼1.1 square degrees in the sky in common. The effective area of each of the four systems is ∼80 cm². The energy range of the left and right slat detectors is ∼1.3–15 keV, while the energy range of the center slat and the horizontal tube is ∼1.3–50 keV.

The position of the source was first determined by a careful comparison of the burst counting rates in the various systems using some of the data from seven orbits. The position of the source obtained in this was is in an ∼50 arcmin² area in the sky (Hearn 1976). It is centered on $l^{II} = 354°.80$, $b^{II} = -0°.22$; $\alpha(1950) = 17^h 30^m 2$, $\delta(1950) = -33°25'$.

Counting-rate plots for five energy channels of the horizontal tube detection system, for an ∼170 s period near 2 March $7^h 39^m$ (UT), are shown in Figure 2. During this period (also shown in Fig. 1) six bursts

---

* Alfred P. Sloan Research Fellow.

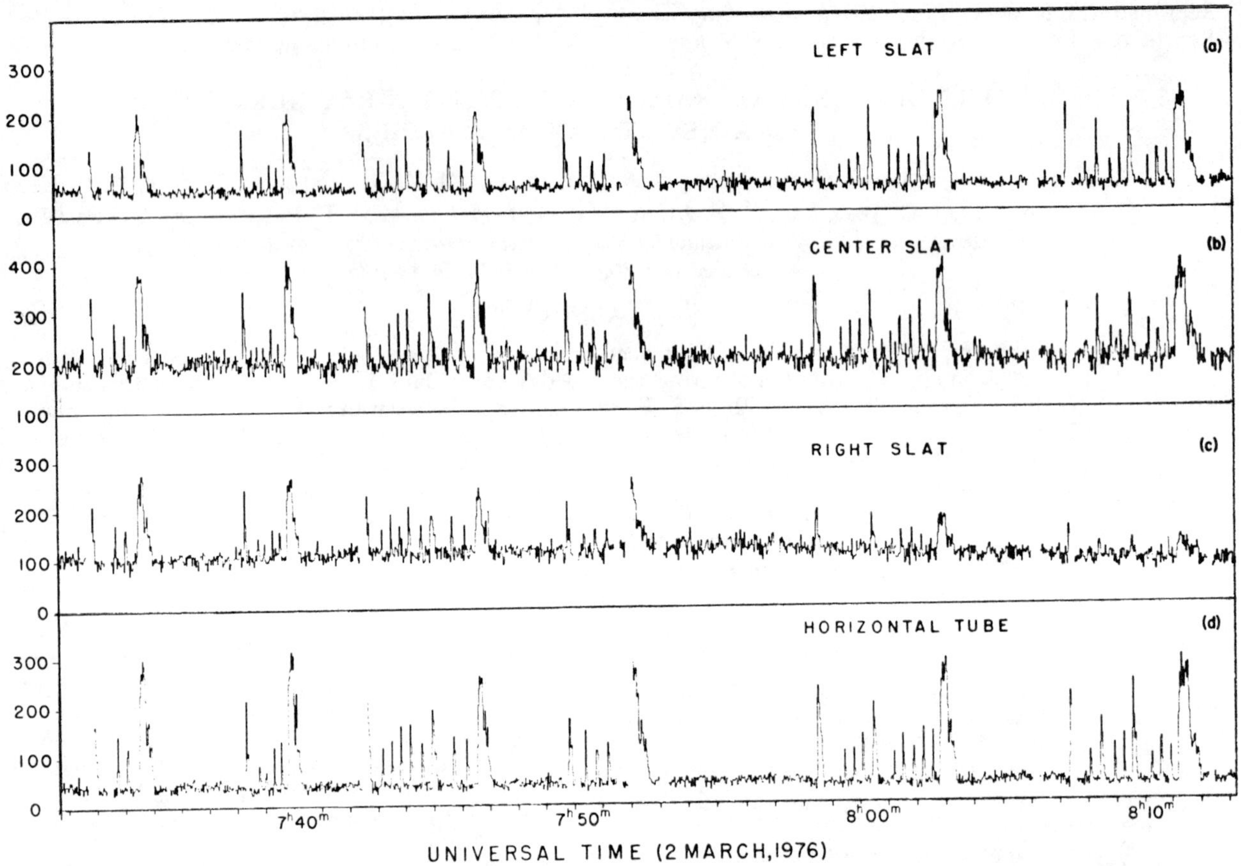

FIG. 1.—Counting rate data (∼1.5–10 keV) from MXB 1730−335 which illustrates the burst activity typical of the observations near 2.3 March 1976 (UT). The source was within the fields of view of four of the collimated proportional counter systems of SAS-3. Toward the end of the data segment in (c), small motions of the satellite (∼10′) caused the source to drift toward the edge of the field of view of the right slat collimator. Occasional small gaps in the data are due to losses in the transmission link via telephone line. The non-burst counting rate in (a), (b), and (c) is due largely to (i) other X-ray sources which are simultaneously in the fields of view of the extended slat-type collimators, (ii) isotropic X-ray background, and (iii) non-X-ray background. In (d) the non-burst counting rate could be due to (i) the known X-ray source 3U 1727−33 which was in the circular field of view of the horizontal tube detection system, (ii) a non-X-ray background, and (iii) possibly, a more or less steady component of MXB 1730−335.

of widely varying magnitude were observed. It is apparent from this figure that the spectra of individual bursts are about the same, although the energy in the bursts varies by about a factor of 30. There seems to be one burst near $7^h39^m30^s$ which appears only in the 1.3–3 keV energy channel. We have no way of knowing, however, whether this burst comes from MXB 1730−335. In general, the largest bursts (one is shown in Fig. 2) show a multiple-peak structure.

The burst patterns sometimes change significantly in less than half an hour; we give two examples only. On March 3 between 02^h10^m and 02^h40^m (UT) the time separations between bursts varied very little from a minimum of ∼40 s to a maximum of ∼80 s; very small bursts and very large ones were not observed. About half an hour later an entirely different pattern, similar to the one shown in Figure 1, was observed. On March 4, from 10^h00^m to 10^h20^m (UT) the burst pattern was again similar to the one in Figure 1; however, about half an hour later from 10^h55^m to 11^h40^m the bursts were almost equal in size and their time separations were typically ∼35 s.

An area of the sky which includes MXB 1730−335 was also observed by SAS-3 (*right slat*) between 1976 February 1.3 and 6.8 (UT), and no repetitive bursts were observed at that time.

A striking feature is evident from Figure 1. The larger the energy in a burst, the longer it takes for the onset of the *next* burst. In Figure 3 we show the approximate energy (∼1.5–10 keV) in ∼250 bursts versus the time separation to the following burst. The bursts were taken from horizontal tube data of six orbits which showed significantly different burst patterns as discussed above. If the energy in a given burst were linearly proportional to the time separation between it and the following burst, then the points would fall on a straight line which is inclined at 45° (one such line is shown). Near the low-energy end of the scale the time intervals are uncertain by ∼40 percent. The data in Figure 3 have not yet been corrected for effective exposure to the source, which varied during the six orbits by about

X-RAY BURSTS FROM SCORPIUS SOURCE

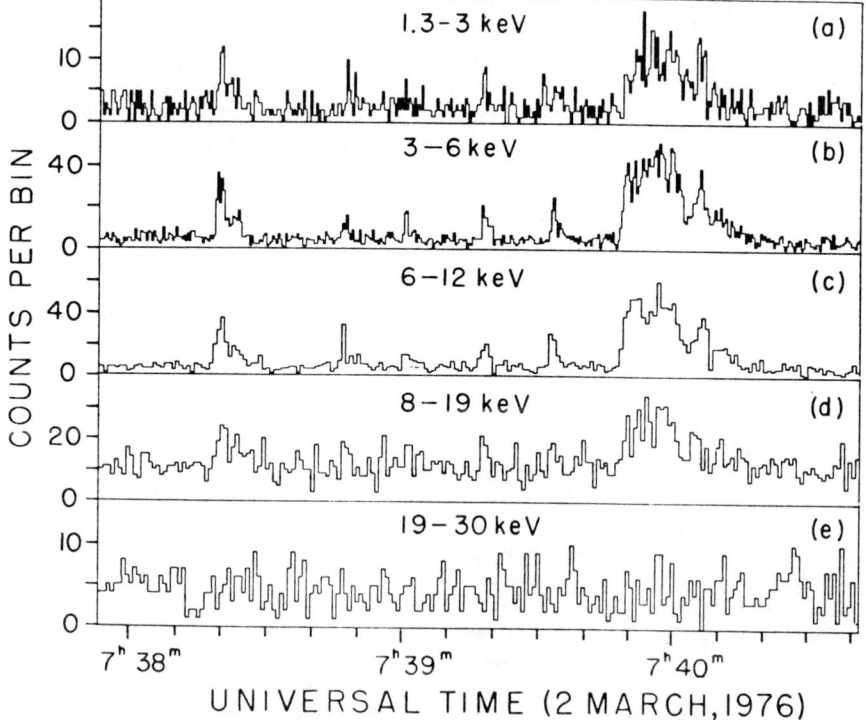

FIG. 2.—Counting rate data from the burst source MXB 1730−335 in five energy intervals. The time-bin size in (a) and (b) is 0.416 s while for (c), (d), and (e) the bin size is 0.832 s. The spectra for the bursts are similar. Note the multiple peak structure in the largest burst which occurs near 7^h40^m (UT).

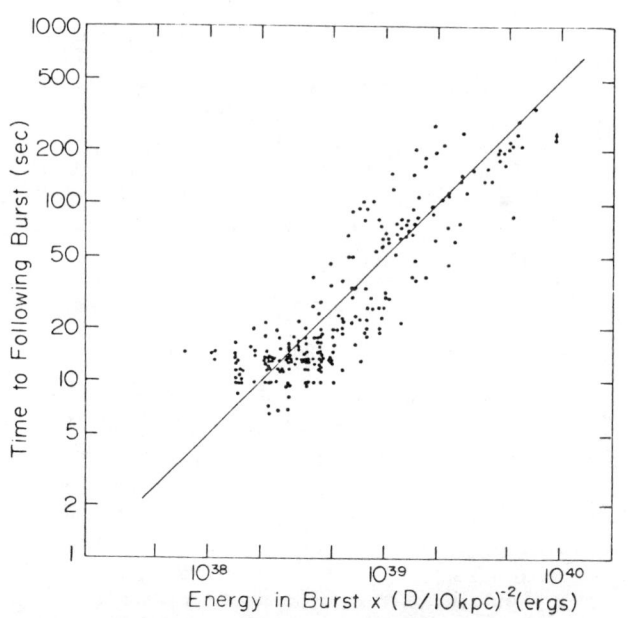

FIG. 3.—The relation between the X-ray energy in a burst and the time interval to the following burst. The burst energy (∼1.5–10 keV) is normalized to an assumed source distance, D, of 10 kpc. If the energy in the first burst were linearly proportional to the time separation between this and the following burst, then the points would fall on a straight line which is inclined at 45° (one such line is drawn). Uncertainties in the time intervals could be ∼40% (only at the low end of the scale). Uncertainties in the relative burst energies are ±20% (see text).

±20 percent from a mean value. As a result, there is a ±20 percent jitter in the relative energies in the data points.

As can be seen from Figure 3, the approximate linear relation between the energy in a burst and the time interval to the next burst holds quite well for bursts in the range $\sim 2 \times 10^{38}$ ergs to $\sim 6 \times 10^{39}$ ergs (for an assumed distance of 10 kpc). One can see from Figure 3 that no bursts were observed with energies either larger than $\sim 10^{40}$ ergs or with energies less than $\sim 10^{38}$ ergs. Also, time intervals between bursts of less than ∼6 s were not observed.

There is a maximum peak luminosity for the X-ray bursts (regardless of whether they last 5 s or 60 s). This luminosity is equivalent to about 150 counts s^{-1} (horizontal tubes) and amounts to $\sim 2 \times 10^{38}$ ergs s^{-1} for an assumed source distance of 10 kpc. All of the six giant bursts shown in Figure 1 have reached this maximum level.

When MXB 1730−335 is Earth occulted, the measured X-ray intensity in the horizontal tube system is about half the non–X-ray burst intensity when the source is not occulted. The difference between these two levels of X-ray intensity can be due to (i) a steady luminosity from 3U 1727−33 (only ∼0°.7 from MXB 1730−335); (ii) a non–X-ray background; (iii) a possible steady luminosity (not in X-ray bursts) from MXB 1730−335. We have no knowledge of the contributions of (i) and (ii); therefore, we can only set an upper limit to (iii). This upper limit is ∼18 counts s^{-1}

in the horizontal tube which represents about 4×10^{37} ergs s^{-1}. The time averaged X-ray burst luminosity of the source is about 2×10^{37} ergs s^{-1} (for an assumed distance to the source of 10 kpc).

III. DISCUSSION

The following appear to be the salient properties of these rapidly repetitive bursts: (1) Their rise times are almost always $\lesssim 1$ s. (2) The repetition rate is highly and irregularly variable. (3) The burst patterns can change significantly in about half an hour. (4) The interval between a given burst and the next burst is approximately proportional to the total energy of X-rays in the given burst. (5) The energy release in the bursts varies by ~ 2 orders of magnitude from $\sim 10^{38}$ to $\sim 10^{40}$ ergs (for an assumed distance to the source of 10 kpc). (6) The X-ray luminosity during a burst does not exceed a certain level ($\sim 2 \times 10^{38}$ ergs s^{-1} for the same assumed distance). (7) The average X-ray luminosity in bursts amounts to $\sim 2 \times 10^{37}$ ergs s^{-1} for the same assumed distance. (8) The average non-burst luminosity is less than 4×10^{37} ergs s^{-1} for the same assumed distance. (9) The burst spectrum does not exhibit a substantial low-energy cutoff. (10) Rapid repetitive bursts were not detected during extended observations of the same region several weeks earlier.

The short rise times suggest that the energy in a burst is released in a region small compared to the diameter of a normal nondegenerate star. The large and irregular variability of the repetition rate implies that the bursts are not controlled by a periodic mechanism such as an orbiting or rotating system, or by a mechanical oscillator with one or more characteristic frequencies. Rather, the approximately proportional relation between the burst energy and the time interval to the following burst is the behavior that one expects of a relaxation oscillator powered by a steady flow of energy.

It seems to us that the most likely source of this energy is accretion of matter onto a compact star. The high value of the average luminosity in the bursts suggests that the energy released in bursts, per unit of accreted mass, is large compared to nuclear binding energies, i.e., that it is a substantial fraction of mc^2 as is the case for accretion onto a neutron star or black hole. In particular, the bursts probably cannot be the result of nuclear burning (hydrogen flashes) on the surface of either a neutron star or a white dwarf. In the case of a neutron star, the steady X-ray emission would dominate the time-averaged burst emission by more than an order of magnitude because of the higher energy yield in the accretion process. Nuclear burning on a white dwarf would lead to a (recurrent) nova phenomenon with known temporal properties which do not resemble those of the present source.

With these considerations in mind, we suggest that the bursts are caused by plasma instabilities in an orderly magnetosphere such as may exist in the neighborhood of a magnetized neutron star. (This mechanism was suggested to us by F. K. Lamb [1976] in connection with previously detected burst phenomena.) Matter falling toward the star is arrested in a reservoir at a distance from the surface that is large compared to the radius of the neutron star. Only a small fraction of the gravitational energy is released up to that point. The matter piles up in the reservoir until its total mass exceeds a critical value, whereupon the reservoir springs a leak and some matter falls toward the surface, releasing most of its gravitational potential energy as a burst of thermal X-rays. The leak heals, and the source is relatively quiet until the reservoir fills up again to the point of instability, whereupon another burst occurs. If such a process emptied the reservoir during each burst, then the size of a burst would be correlated with the time interval to the previous burst. In contrast, the observed correlation of burst size with the succeeding interval suggests that the reservoir is partially drained whenever its level reaches a critical upper limit.

The absence of rapid repetitive bursts only a few weeks earlier indicates that the material for accretion is likely to be drawn not from a large cloud of interstellar material but rather from a binary companion. The latter source of accretion, as we know from other binary systems, can be highly variable or even intermittent.

The conditions for the stability of plasma in the magnetosphere of a neutron star have been examined by Lamb (1974; 1975), Elsner and Lamb (1976), and Arons and Lea (1976). They showed that a plasma cloud supported by magnetic pressure is subject to a Rayleigh-Taylor instability if its temperature falls below a critical value. They discuss only the onset of the instability (which is expected to allow plasma to penetrate over most of the magnetosphere boundary except near the poles) and steady mass flows. We note that our observations indicate that the critical parameter for stability is not temperature but total mass. Furthermore, the data show that stability is frequently restored before the reservoir is emptied.

We thank M. Scholtes for assistance in the data analysis. We are grateful to the many persons who have contributed to the successful fabrication, launch, and operations of SAS-3. We thank the staffs of the Laboratory for Space Experiments and the Center for Space Research at M.I.T., the Applied Physics Laboratory of Johns Hopkins University, the Goddard Space Flight Center, and the Centro Ricerche Aerospaziali. This work was supported in part by NASA under contract NAS5-11450.

REFERENCES

Arons, J., and Lea, S. M. 1976, *Ap. J.*, in press.
Babushkina, O. P., Bratolyubova-Tsulukidze, L. S., Kudryavtsev, M. I., Melioranskii, A. S., Savenko, I. A., and Yushkov, B. Yu. 1975, *Soviet Astr. Letters*, 1, 32.
Belian, R. D., Conner, J. P., and Evans, W. D. 1976, Los Alamos preprint UR-76-309.
Clark, G. W. 1976, *IAU Circ.*, No. 2915.
Clark, G. W., Jernigan, J. G., Bradt, H., Canizares, C., Lewin,

W. H. G., Li, F. K., Mayer, W., McClintock, J. 1976, *Ap. J. (Letters)*, **206**, L107.
Doty, J. 1976, *IAU Circ.*, No. 2922.
Elsner, R. F., and Lamb, F. K. 1976, preprint.
Grindlay, J., Gursky, H., Schnopper, H., Parsignault, D. R., Heise, J., Brinkman, A. C., and Schrijver, J. 1976, *Ap. J. (Letters)*, **205**, L127.
Hearn, D. 1976, *IAU Circ.*, No. 2925.
Hoffman, J. 1976, *IAU Circ.*, No. 2946.
Jones, C., and Forman, W. 1976, *IAU Circ.*, No. 2913.
Lamb, F. K. 1975, *Seventh Texas Symposium on Relativistic Astrophysics*, pp. 262, 331.
Lewin, W. H. G. 1976a, *IAU. Circ.*, No. 2918.
———. 1976b, *ibid.*, No. 2922.

H. V. D. BRADT, G. W. CLARK, J. DOTY, R. DOXSEY, D. R. HEARN, J. A. HOFFMAN, J. G. JERNIGAN, W. H. G. LEWIN, F. K. LI, W. MAYER, J. MCCLINTOCK, F. PRIMINI, S. A. RAPPAPORT, and J. RICHARDSON: Center for Space Research, Building 37-627, Massachusetts Institute of Technology, Cambridge, MA 02139

The Astrophysical Journal, 234:L69–L72, 1979 November 15
© 1979. The American Astronomical Society. All rights reserved. Printed in U.S.A.

Reprinted with permission from *The Astrophysical Journal*, 243 pp. L69-L72, S.S. Murray, G. Fabbiano, A.C. Fabian, A. Epstein, and R. Giacconi, "High-Resolution X-Ray Observations of the Cassiopeia A Supernova Remnant with the *Einstein* Observatory."
© 1979 The American Astronomical Society.

HIGH-RESOLUTION X-RAY OBSERVATIONS OF THE CASSIOPEIA A SUPERNOVA REMNANT WITH THE *EINSTEIN* OBSERVATORY

S. S. Murray, G. Fabbiano, A. C. Fabian,[1] A. Epstein, and R. Giacconi

Harvard-Smithsonian Center for Astrophysics, Cambridge, Massachusetts

Received 1979 June 29; accepted 1979 July 31

ABSTRACT

Cas A was observed with the imaging detectors on the *Einstein* X-ray Observatory. We determine the mass of the expanding shell due to the interaction of the blast wave and the ambient medium. This indicates that Cas A is probably not in the free expansion phase, but has not yet reached the adiabatic phase of expansion. Detailed comparison of X-ray, optical, and radio images shows various degrees of correlation, indicating that several processes for emission are present. An upper limit of 1.5×10^6 K is established for any stellar remnant that is below the expected temperature of a neutron star by a factor of about 5. The total mass of the SNR is found to be 10–30 $M_\odot$.

Subject headings: nebulae: supernova remnants — X-rays: general

I. INTRODUCTION

The Cassiopeia supernova (Cas A) is the youngest known supernova in our Galaxy, and is of great interest in our understanding of these explosive processes. Such processes as the interaction of the blast wave from the explosion with the surrounding medium, the formation of collapsed remnants, and the mixing of enriched material with the interstellar medium can best be studied in relatively recent SNRs such as Cas A. Previous observations of young SNRs of small angular size have been limited by the resolution of the X-ray detectors available. The spatial resolution of previous observations of Cas A (e.g., Charles, Culhane, and Fabian 1977) has been too coarse ($\sim 3'$) to permit comparison with optical and radio features, which are typically of arcsec angular scale.

We have observed Cas A with both of the imaging detectors on the *Einstein* X-ray Observatory, and present results of a preliminary analysis of these data. The high-resolution picture illustrates the complex structure of this source in the 0.1–4 keV band, which is in general similar to both radio (5 GHz, Bell 1977) and optical (Kamper and van den Bergh 1976) images of similar resolution. We observe a ringlike structure with X-ray peaks which has a diameter of about $4'$–$5'$. Surrounding this is an emission plateau corresponding to one observed in radio with an outer diameter of $\sim 5\rlap{.}'5$ and thickness of at least $0\rlap{.}'5$. We interpret this as the region associated with the expanding shock front of the SNR blast wave. The complex interior structure is probably due to several phenomena, including shock heating of previously ejected material from the presupernova (see Woltjer 1972; McKee 1974; Gull 1973) evaporation of material into the relatively hot ambient gas (Gull 1975), conductive heating of high-velocity knots (Chevalier 1975), and perhaps nonthermal processes such as synchrotron emission from high-energy electrons produced in the ring by tangled magnetic fields (Scott and Chevalier 1975).

We have searched for X-ray emission from a stellar remnant and find an upper limit corresponding to a neutron star blackbody with temperature less than 1.5×10^6 K. This is about a factor of 5 less than the theoretically lowest temperature for a neutron star using standard cooling models (Tsuruta 1978).

We have developed a model for Cas A based on several simplifying assumptions which allow us to interpret the high-resolution image in terms of a temperature profile. This results in an estimate of the total SNR mass of between 10 and 30 $M_\odot$.

II. DATA/OBSERVATIONS

Cas A has been observed with both of the imaging detectors on the *Einstein* Observatory. The imaging proportional counter (IPC) picture is shown in Figure 1a (Plate L20). The angular resolution is a few arcmin and the detector also provides moderate energy resolution, which allows spectral variation to be measured over the image. The high-resolution imaging detector (HRI) picture is shown in Figure 1b (Plate L21). The angular resolution is about $4''$ FWHM, and there is no spectral resolution over the 0.1–4 keV bandwidth of the telescope. These detectors and various features of the Observatory are described in more detail by Giacconi *et al.* (1979). In both pictures it is clear that the image contains complex structure, but that the general appearance of the remnant is roughly circular. From the higher resolution picture we see that the brightest regions lie in a ringlike structure and that there is an emission plateau extending beyond this ring surrounding the source. This is similar to the emission plateau observed at 5 GHz by Bell (1977). The measured radius is $170''$ with a thickness of $\sim 30''$. Taking a distance to Cas A of 2.8 kpc (van den Bergh and Dodd 1970), the corresponding linear sizes are:

$$r = 2.3 \text{ pc}, \quad \Delta r = 0.4 \text{ pc}.$$

---
[1] Also Institute of Astronomy, Cambridge, England.

PLATE L20

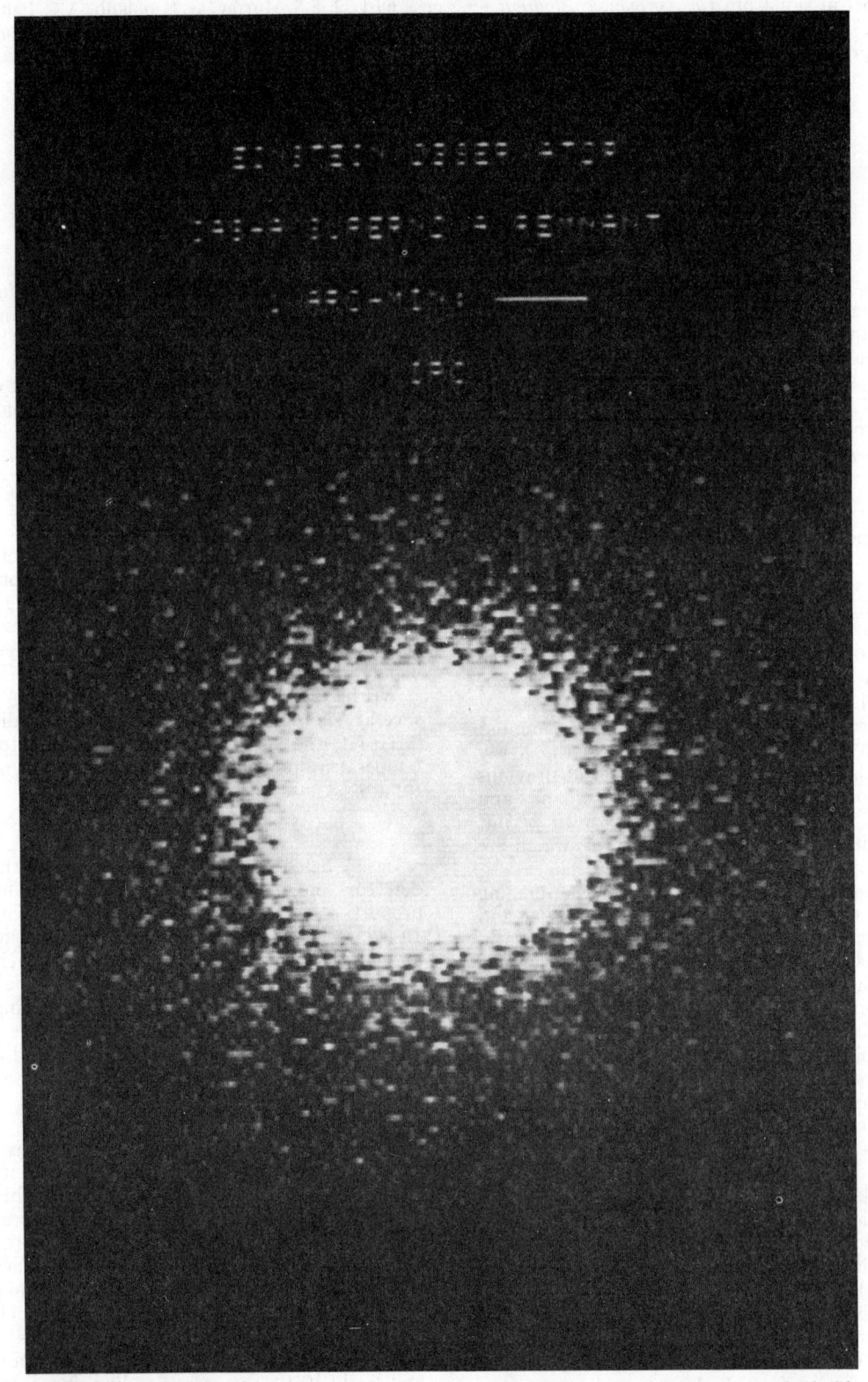

Fig. 1a.—IPC image of Cas A. Each pixel is 8″ × 8″ and the exposure time is 1660 s. The brightness at each pixel is proportional to the logarithm of the counts in that pixel.

Murray *et al.* (see page L69)

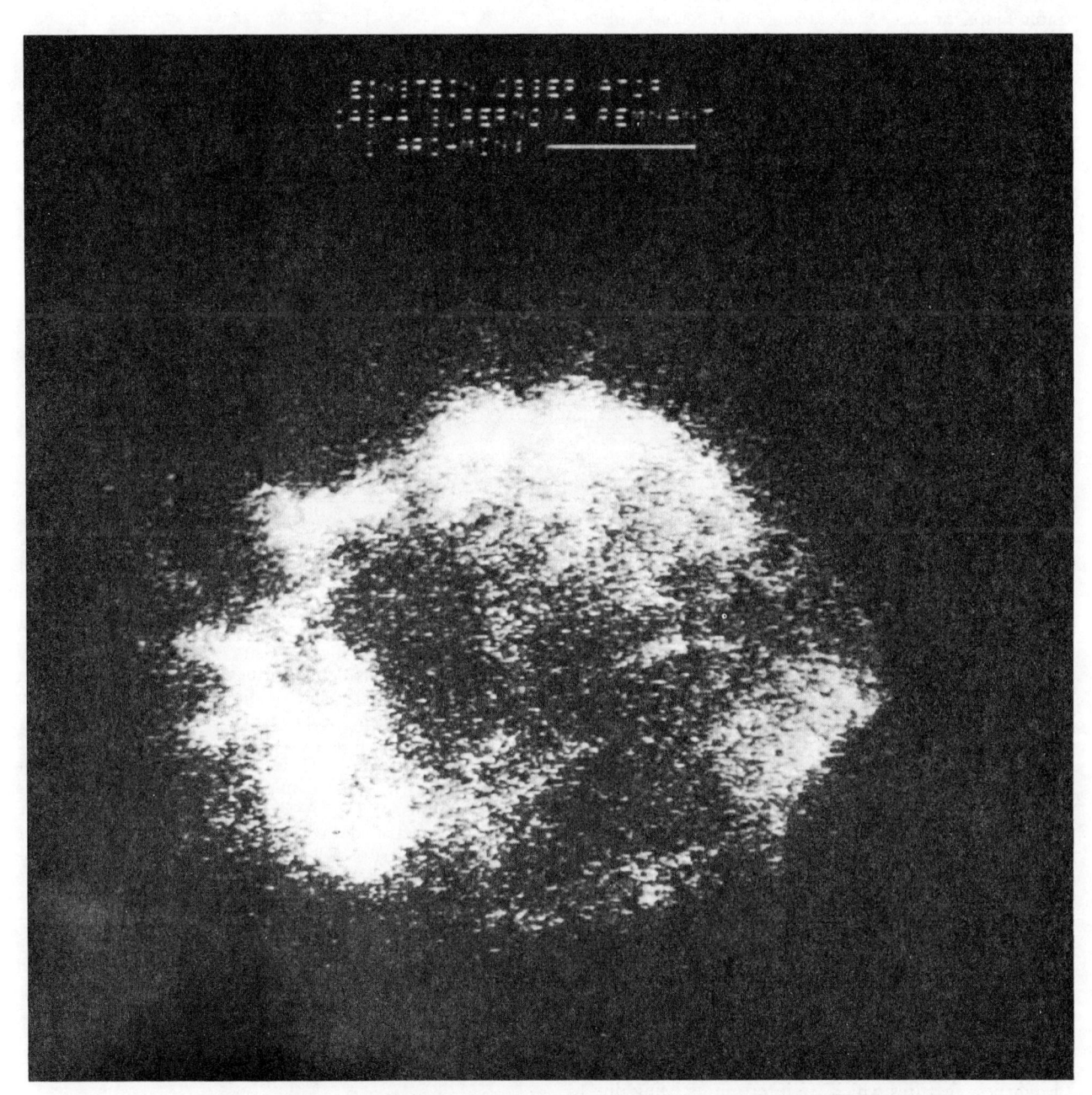

FIG. 1b.—HRI image of Cas A. Each pixel is 2" × 2", and the exposure time is 32,519 s

MURRAY et al. (see page L69)

The ring of brightest X-ray emission corresponds in general with the radio ring with a radius of about 2′ (1.7 pc). Detailed comparison of the X-ray and 5 GHz radio maps for Cas A indicate a general correspondence between features such as the emission plateau and the ring structure. However, this correlation is not always maintained on the 2″ scale where we find several isolated individual X-ray or radio peaks. Similarly, with regard to optical features we find a general correlation between the optical ring of knots and flocculi and the X-ray emission ring. There appears to be a stronger correlation between the X-ray emission and optical knots with strong [S II] emission relative to [O III] than with other features. The IPC data indicate that most of the X-ray emission in the optical nebulosity region is in the 1–2.5 keV band. This is the energy range where S and Si lines are expected. The solid-state spectrometer experiment has reported a very strong contribution from these elements as well as Ca and Ar in the Cas A spectrum (Becker et al. 1979). However, there is also at least one example of an X-ray feature coincident with a quasi-stationary flocculus (2Q2; Kirshner and Chevalier 1977) that has primarily Hα and [N II] optical emission.

The IPC image for Cas A also shows the features discussed above, but at significantly lower resolution. We use the IPC data to study spectral variations over the remnant, which are shown in Figure 2 (Plate 22). Here we have constructed images which show the ratio of either hard-to-total or soft-to-total flux as a function of position. The soft band is from 1 to 2 keV (Fig. 2a), and the hard band is from 2.4 to 3.6 keV (Fig. 2b). The structure shown in Figure 2 shows an east/west asymmetry which corresponds to differences in the radio and optical images. The radio emission feature in the western section of Cas A does not have a corresponding optical feature. Figure 3 shows the radial dependence of these ratios and indicates quantitatively the variation of the relative intensity in the 1–2 keV band seen in Figure 2. In Figure 4 we have plotted the spectral data taken from the northeast and southwest sections of the SNR. We observe that the spectral distributions are significantly different. The southwest region has more flux at high energies (≥ 1.5 keV) and is also more cutoff at lower energies.

III. DISCUSSION

a) Shell

As described above, there is an emission plateau surrounding the complex structure of maximum surface brightness features which we interpret as being due to the blast wave shock front interacting with the ambient ISM. From the measurements of Bell (1977), the mean expansion velocity of the radio shell is about 1700 km s^{-1}, while the optical knots have an expansion rate of ~ 6000 km s^{-1} (Kamper and van den Bergh 1976). Bell interprets this as evidence for deceleration of the shell, which requires that the mass swept up by the shock be greater than the initial mass during the shock. In this view, the fast-moving knots are explained as a slower component of the original ejecta

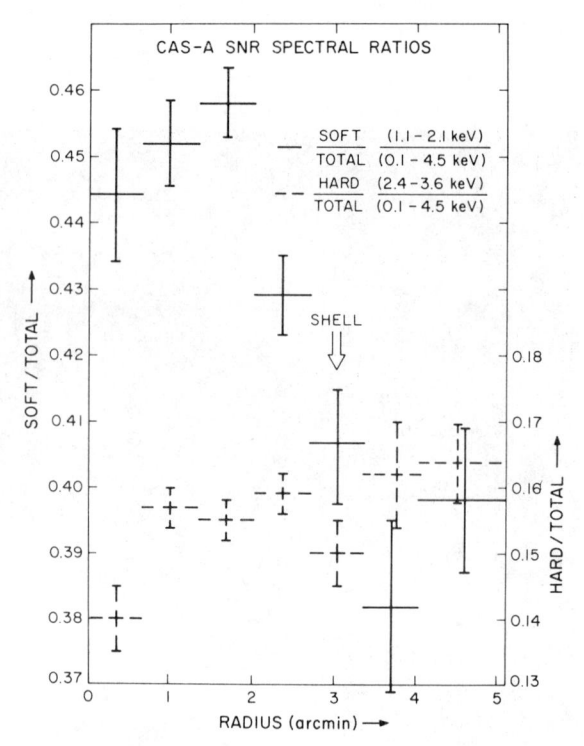

Fig. 3.—Plot of the spectral ratios versus radial annulus for Cas A.

which have not undergone appreciable deceleration because of their high density. The quasi-stationary flocculi could instead be due to presupernova ejecta and contribute to the medium in which the main shock has been propagating. We cannot directly measure the temperature of the shell because the resolution of the IPC is too poor to separate out features with sub-arcmin extent. However, various observations of Cas A (which are nonimaging) have yielded spectral data best fitted by a two-component thermal spectrum (Charles et al. 1975; Davison, Culhane, and Mitchell 1976; Pravdo et al. 1976) with the high-temperature component ranging from about 3.5 to 5 keV and the low-temperature component from ~ 0.7 to 1.0 keV. It is likely that the higher temperature component is due to emission from a shock-heated shell because, first, the expected shell temperature as given by

$$T_s \sim \frac{3M_p}{16k} \mu V_s^2$$

$$\sim 14 V_s^2 (\text{km s}^{-1}) \text{ K}$$

implies a shock velocity of about 2000 km s^{-1} consistent with the radio expansion; and second, the spectral plots of Figure 3 indicate that the low-temperature, line-dominated emission is from the interior ring, while the X-ray emission plateau appears to have a flatter spectrum. The X-ray flux for this shell is obtained from the HRI image by integrating the observed counts in the annulus of radius 170″ and thickness 30″.

PLATE L22

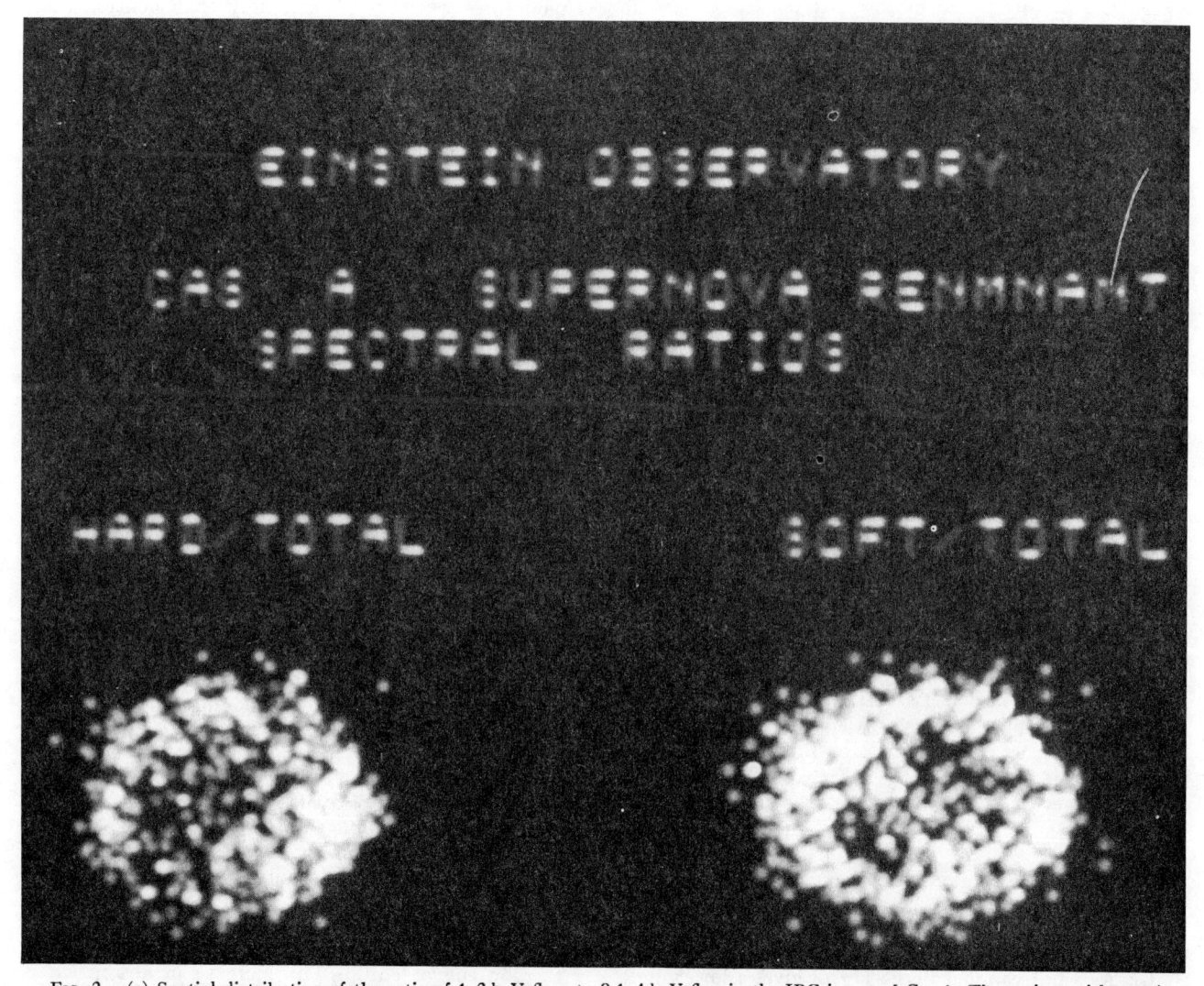

FIG. 2.—(a) Spatial distribution of the ratio of 1–2 keV flux, to 0.1–4 keV flux in the IPC image of Cas A. The regions with a ratio greater than 0.25 are shown. (b) Spatial distribution of the ratio of 2.4–3.6 keV flux, to 0.1–4 keV flux in the IPC image of Cas A. Regions with a ratio greater than 0.25 are shown.

MURRAY et al. (see page L70)

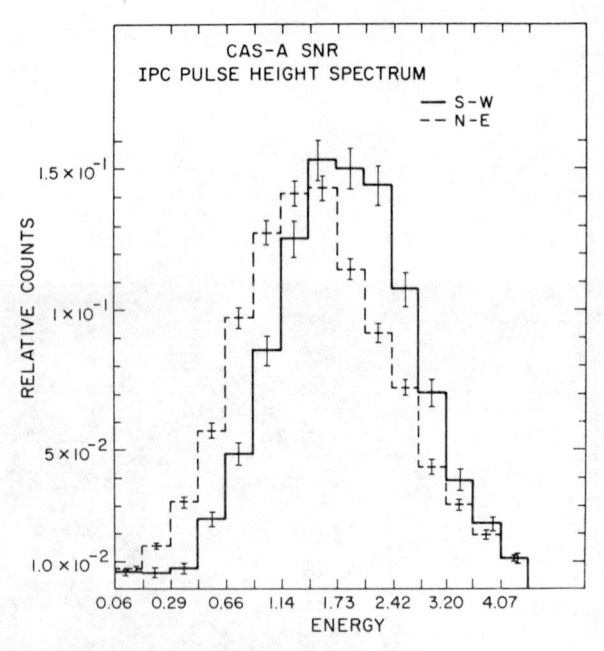

FIG. 4.—Spectral data for features in the northeast and southwest of the Cas A image.

We convert this to flux assuming a temperature of 4 keV and a hydrogen column density of $\sim 1 \times 10^{22}$ cm^{-2} (see below). We obtain $F_{\text{shell}}(0.1-4) \approx 2.3 \times 10^{-10}$ ergs cm^{-2} s^{-1} with an estimated uncertainty of 25% due mainly to the corrections in counting rate for background subtraction and spillover from bright regions. The shell luminosity is then $L_x(0.1-4) \approx 2.2 \times 10^{35}$ ergs s^{-1} (distance = 2.8 kpc). This implies a density of X-ray emitting material of about 6 cm^{-3} and a total mass (shell plus swept-up matter) of ~ 3.5 $M_\odot$. The ambient density is estimated to be $\gtrsim 1.5$ cm^{-3} (Bell 1977), and the mass of swept-up material is then $\gtrsim 2.2\ M_\odot$. These values indicate that Cas A is in a deceleration phase but not yet adiabatically expanding. (cf. Woltjer 1972).

b) Neutron Star

Supernova explosions are thought to play an important role in producing pulsars, with the Crab Nebula being a classic example. In the case of Cas A, no such object is detected, and an optical search by Kamper and van den Bergh (1976) gives a lower limit of $M_r \geq 8$ mag for any stellar remnant near the explosion center. We have searched the X-ray image for a possible point source and determine an upper limit of 7.5×10^{-3} counts s^{-1} (3 σ) for the HRI counting rate. The conversion of this count rate to flux (and luminosity) depends on the absorption and spectrum assumed. We take $N_H \sim 1 \times 10^{22}$ cm^{-2} as determined from our IPC data and assume that a collapsed remnant will be typically 10 km in radius and radiate as a blackbody of temperature T_{BB}. We then compare our upper limit to the corresponding maximum blackbody temperature which predicts a flux less than that limit. We find $T_{\text{BB}} \lesssim 1.5 \times 10^6$ K is required. Helfand and Novick (1979) discuss the expected temperature of neutron stars produced in supernova explosions and give a minimum theoretically expected temperature of $\gtrsim 7.5 \times 10^6$ K for neutron stars which cool by the "standard" neutrino process (also Tsuruta 1978). Our measurement excludes the possibility for such a neutron star having been formed. However, Helfand and Novick point out that collapsed objects with pion cores will cool more rapidly than neutron cores and a temperature as low as 8×10^5 K could be expected. In that case, our upper limit is consistent with such a model. Both the X-ray and optical data provide only upper limits for a stellar remnant; it may be more likely that no such remnant is produced in the supernova. Should this be true for many supernovae, we must look toward other processes to produce the large number of pulsars (neutron stars) observed in the Galaxy.

c) Interior Region

The high-resolution picture of Cas A and the observation of spectral variations within the interior indicate that the emission processes are both complex and varied. With our current data we cannot uniquely determine the emission mechanism for each region of the SNR. However, the degree of correlation among X-ray, radio, and optical features indicates that there are underlying relations present that must be addressed in models for the object. In general, thermal processes dominate the X-ray emission, as indicated by the high-resolution spectrum by Becker et al. (1979) and from previous observations at higher energies (e.g., Davison, Culhane, and Mitchell 1976). The deceleration of the expanding shell as discussed earlier is consistent with the presence of a reverse shock that will heat the interior material to X-ray temperatures (McKee 1974; Gull 1975). Chevalier (1975) has also suggested a heating mechanism for fast-moving knots which is basically conductive heating as these knots plow through the ambient medium. Examples where these processes are likely to be occurring are given by the X-ray filament associated with the quasi-stationary flocculus 2Q2 (Kirshner and Chevalier 1978) and the X-ray "jet" in the northeast region which corresponds to the jet of high-speed knots (Kamper and van den Bergh 1977). We have also noted that the variation in spectrum within the SNR is characterized both by an increase in the "hardness" of the spectrum and in the low-energy cutoff. The increase in absorption is similar to that observed by Greisen (1973) of about a factor of 2 in hydrogen column density from the east to the west ($N_H(0.5-1) \times 10^{22}$ cm^{-2}). We note that the western structure which corresponds to the bright radio peaks does not have an optical counterpart. If this is not due to obscuration, it may indicate a contribution from nonthermal processes that would require recent acceleration of electrons to high energies. A proposed mechanism for such cosmic acceleration has been given by Scott and Chevalier (1975). The required electron energy and estimated magnetic field strength limit the synchrotron lifetime for these electrons to about 10 years, thus requiring an ongoing acceleration process.

In order to estimate the amount of material associated with the X-ray emission from Cas A we have developed a simple model by assuming basically only thermal processes (with line emission) and pressure equilibrium. This analysis is still in progress and will be described in detail in a later paper (Fabian et al. 1979). The general conclusions are that a self-consistent description for the X-ray emission can be found with variations in temperature and density throughout the SNR and constant abundances of iron, sulfur, and silicon. Best fits are for cosmic abundance of Fe and twice cosmic abundances of S and Si. This is consistent with the spectrum of Becker et al. (1979) and suggests that emission is largely connected with the surfur knots which, as noted earlier, correlate well with the X-ray image. There is, in fact, a range of solutions consistent with present observations which results in a range of total mass inside the shell from about 10 to 30 $M_\odot$. This range of mass is consistent with a mass estimate by Chevalier and Kirshner (1978) for Cas A that assumes complete chemical mixing. A more recent analysis by Chevalier and Kirshner (1979) points out that incomplete chemical mixing, which is likely to be the case for Cas A, results in large uncertainties in mass estimates based on the relative abundance of heavy atoms.

IV. CONCLUSIONS

Our observations of Cas A with the imaging detectors of the *Einstein* X-ray Observatory provide detailed data which enable us to establish that the supernova is in the adiabatic expansive phase and is decelerating as the shock wave sweeps up interstellar material. We have placed an upper limit of 1.5×10^6 K on the temperature of a collapsed stellar remnant, a limit that is below the temperature for normal neutron stars created in SNRs. This may indicate that a pion core was formed that can cool more rapidly than a neutron star, or that no collapsed object was formed. We find that there is general correlation of X-ray, radio, and optical features, but that in detail the level of correlation is lessened. This suggests that several emission processes are involved in the production of X-rays from Cas A, some of which are shock heating, conductive heating, and nonthermal emission. Finally, taking a simple model with pressure equilibrium and thermal emission (bremsstrahlung and line emission) as the dominant mechanism we estimate the total X-ray emitting mass of the SNR to be between 10 and 30 $M_\odot$.

We thank Dr. J. Ostriker for his stimulating comments and interesting discussions. We also appreciate the helpful comments of Drs. P. Gorenstein, R. Griffiths, M. Elvis, and J. P. Henry. One of us (A. C. F.) acknowledges the financial support from the Radcliffe Trust, and thanks Dr. R. Giacconi and his colleagues for generous hospitality. The authors also wish to thank the referee for his timely and thoughtful suggestions.

This research was sponsored by NASA contract NAS8-30751.

REFERENCES

Becker, R. H., Holt, S. S., Smith, B. W., White, N. E., Boldt, E. A., Mushotzky, R. F., and Serlemitsos, P. J. 1979, *Ap. J. (Letters)*, **234**, L73.
Bell, A. R. 1977, *M.N.R.A.S.*, **179**, 573.
Charles, P. A., Culhane, J. L., and Fabian, A. C. 1977, *M.N.R.A.S.*, **178**, 307.
Charles, P. A., Culhane, J. L., Zarnecki, J. C., and Fabian, A. C. 1975, *Ap. J. (Letters)*, **197**, L61.
Chevalier, R. A. 1975, *Ap. J.*, **200**, 698.
Chevalier, R. A., and Kirshner, R. P. 1978, *Ap. J.*, **218**, 142.
———. 1979, *Ap. J.*, submitted.
Davison, P. J. N., Culhane, J. L., and Mitchell, R. J. 1976, *Ap. J. (Letters)*, **206**, L37.
Fabian, A. C., Murray, S. S., Fabbiano, G., and Epstein, A. 1979, in preparation.
Giacconi, R., et al. 1979, *Ap. J.*, **230**, 540.
Greisen, E. W. 1973, *Ap. J.*, **184**, 363.
Gull, S. F. 1973, *M.N.R.A.S.*, **162**, 135.
———. 1975, *M.N.R.A.S.*, **171**, 263.
Helfand, D., and Novick, R. 1979, private communication.
Kamper, K., and van den Bergh, S. 1976, *Ap. J. (Suppl.)*, **32**, 351.
Kirshner, R. P., and Chevalier, R. A. 1977, *Ap. J.*, **218**, 142.
McKee, C. F. 1974, *Ap. J.*, **188**, 335.
Pravdo, S. H., Becker, R. H., Boldt, E. A., Holt, S. S., Rothschild, R. E., Serlemitsos, P. J., and Swank, J. H. 1976, *Ap. J. (Letters)*, **206**, L41.
Scott, J. S., and Chevalier, R. A. 1975, *Ap. J. (Letters)*, **197**, L5.
Tsuruta, S. 1978, Research Institute for Fundamental Physics, preprint RIFP-344, Kyoto University, Japan.
van den Bergh, S., and Dodd, W. W. 1970, *Ap. J.*, **162**, 485.
Woltjer, L. 1972, *Ann. Rev. Astr. AP.*, **10**, 129.

A. EPSTEIN, G. FABBIANO, R. GIACCONI, and S. S. MURRAY: Harvard-Smithsonian Center for Astrophysics, 60 Garden Street, Cambridge, MA 02138

A. C. FABIAN: Institute of Astronomy, University of Cambridge, Madingley Road, Cambridge CB3 0HA, England

THE ASTROPHYSICAL JOURNAL, 234:L73–L76, 1979 November 15
© 1979. The American Astronomical Society. All rights reserved. Printed in U.S.A.

Reprinted with permission from *The Astrophysical Journal*, 234 pp. L73-L76, R.H. Becker, et al., "X-Ray Spectrum of Cassiopeia A Measured with the *Einstein* SSS." © 1979 The American Astronomical Society.

X-RAY SPECTRUM OF CASSIOPEIA A MEASURED WITH THE *EINSTEIN* SSS

R. H. BECKER,[1] S. S. HOLT, B. W. SMITH,[1] N. E. WHITE,[1] E. A. BOLDT,
R. F. MUSHOTZKY,[2] AND P. J. SERLEMITSOS

Laboratory for High Energy Astrophysics, NASA Goddard Space Flight Center, Greenbelt, Maryland

Received 1979 June 25; accepted 1979 August 8

ABSTRACT

The solid state spectrometer (SSS) on the *Einstein* Observatory observed the X-ray spectrum of Cas A between 0.8 and 4.5 keV with a FWHM energy resolution of 160 eV. Line emission consistent with transitions of helium-like ions of Si, S, and Ar was well defined. Comparison between the data and the emission expected from a gas containing two distinct thermal components requires additional line emission from Mg, Al, Ca, and Fe. These results are discussed in the context of both equilibrium and nonequilibrium situations.

Subject headings: nebulae: abundances — nebulae: supernova remnants — X-rays: spectra

I. INTRODUCTION

Approximately 15 supernova remnants (SNRs) have been detected as X-ray sources. The X-ray spectra of these objects, except for the Crab nebula, are consistent with thermal emission from hot plasma. In particular, several SNRs exhibit line emission at 6.7 keV from ionized Fe, as would be expected for a hot plasma of $\sim 10^7$ K temperature (Serlemitsos *et al.* 1973; Pravdo *et al.* 1976; Davison, Culhane, and Mitchell 1976). These sources should also produce line emission from lighter elements, and previous experiments have reported evidence for unresolved spectral enhancements consistent with Si line emission from SNRs (Hill, Burginyon, and Seward 1975*a*, 1975*b*; Mason, Charles, and Bowyer 1978). The solid state spectrometer (SSS) onboard the *Einstein* Observatory (*HEAO 2*) was specifically designed to detect such line emission between 1 and 4 keV. This *Letter* will present the evidence for line emission obtained with the SSS in the spectrum of Cas A, and discuss some of its model implications.

II. EXPERIMENT AND ANALYSIS

The SSS instrument parameters are described in detail by Joyce *et al.* (1978), and a brief summary of most of its characteristics relevant to this analysis appears in an accompanying *Letter* (Holt *et al.* 1979). However, an additional complication in the case of emission from an extended source arises from the fact that the detector gain varies slightly across its surface. At the extremes, the measured energy of a line can vary up to 30 eV depending on whether the focussed X-rays fall on the center or the edge of the detector. Although this uncertainty is small relative to the energy resolution of 160 eV FWHM, it can make a significant systematic contribution to χ^2 when models are compared to data with strong line features. In addition, there is some probability that the measured energy of any photon can be significantly less than its true energy. The percentage of events that fall into this "low energy plateau" increases from about 4% at the detector's center to 8% at its edges, further complicating our fitting procedures. For an extended source such as Cas A, all parts of the detector's surface are sampled. Cas A approximately fills the whole 6' diameter SSS field of view, and comparison with data from *HEAO* A-2 (Pravdo 1979) indicates that we are capturing >80% of the total emission.

Spectral data from the SSS are analyzed by comparing them to model spectra which have been folded through the response of the SSS to incident X-rays. A final calibration of the SSS was achieved by comparing the observed spectrum of the Crab SNR to a power-law model of number index 2.1 (see Fig. 1). The calculated effective area of the SSS was altered to force an agreement between the data and the assumed model. The best-fit column density of 5×10^{21} cm^{-2} is higher than previous measurements (e.g., Charles *et al.* [1979] measured 3×10^{21} cm^{-2}) and possibly reflects the uncertainty in our estimate of the instantaneous amount of ice cryopumped onto the detector surface from ambient outgassing material in the spacecraft (see Holt *et al.* 1979).

Since the X-ray emission from Cas A is believed to originate from a hot plasma, we have compared the Cas A spectrum to models for emission from a hot plasma in collisional equilibrium based on the work of Raymond and Smith (1977). These models have the inherent weakness that they are based on atomic physics data which have uncertainties sufficient to allow factor-of-2 errors in elemental abundances (for discussion, see Smith, Mushotzky, and Serlemitsos 1979). Furthermore, it is not clear that the material in Cas A is in collisional equilibrium (Itoh 1977; Gorenstein, Harnden, and Tucker 1974). Nonetheless, an extension of this analytic work by Raymond and Smith (1979) is the closest approximation to the conditions in Cas A currently available to us, and we

---
[1] Also Department Physics and Astronomy, University of Maryland.
[2] NAS/NRC Research Associate.

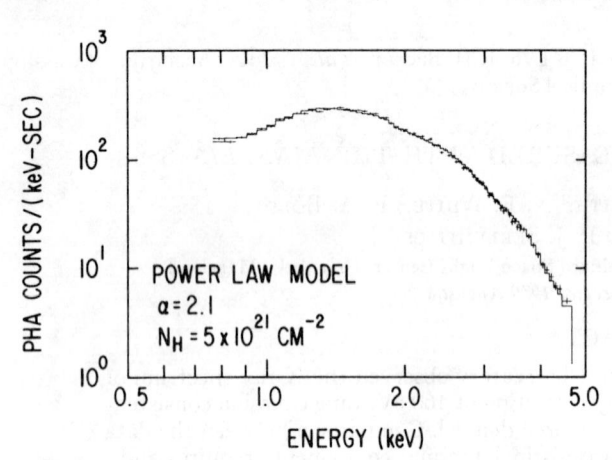

FIG. 1.—X-ray spectrum of the Crab nebula as observed by the SSS on *HEAO 2*. Superposed upon the data is the expected response of the SSS to a power law with a 2.1 number index.

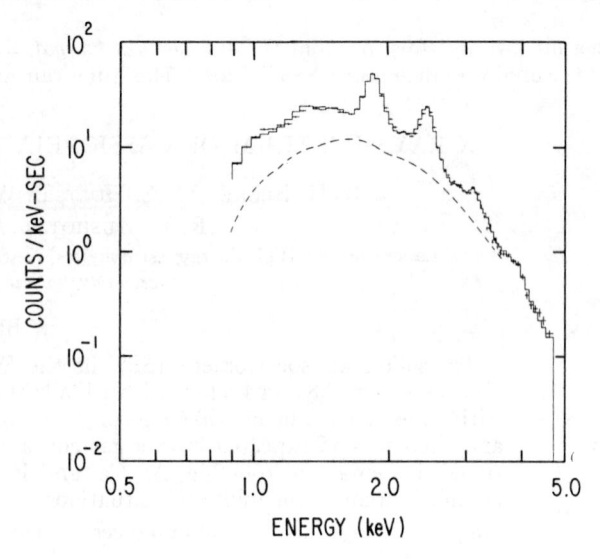

FIG. 2.—X-ray spectrum of the Cas A supernova remnant as observed by the SSS on *HEAO 2*. Superposed upon the data is the expected response of the SSS to a two-component isothermal model. The lower second trace is the contribution to the spectrum from H, He, C, N, O, and Ne; this represents the continuum above which the line constituents tabulated in Table 1 are observed.

have therefore, attempted a detailed comparison of the Cas A spectrum to that expected for collisional equilibrium.

Previous observations of the Cas A X-ray spectrum indicate that it is composed of at least two thermal components (Davison, Culhane, and Mitchell 1976; Charles et al. 1975; Pravdo et al. 1976). Hence we constructed models of a similar nature, fixing the higher temperature component at 4 keV, while the temperature of the low-energy component was a free parameter. Our final results should be insensitive to the precise value of the high-temperature component, as its continuum is relatively flat over the SSS dynamic range and it contributes very little to the lines. In addition, the relative abundances of Mg, Al, Si, S, Ar, Ca, and Fe, which were constrained to be the same for both thermal components, were all free parameters, as were the overall normalizations of the two thermal components. In this procedure, the relative abundances determine the normalization for the lines from each element, but the temperature determines which ionic species of each element are present. Lastly, the equivalent column density of N_H to Cas A using the interstellar abundances of Brown and Gould (1970) with the addition of iron (Fireman 1974) was a free parameter. The thickness of the ice layer on the detector was assumed to increase through the observation at a predetermined rate and was included in the analysis as a time-dependent column density of water.

III. RESULTS

The SSS pulse-height spectrum of Cas A is shown in Figure 2. The dominant features at 1.86 and 2.46 keV are consistent with line emission from helium-like Si and S, respectively. In attempting to fit collisional-equilibrium thermal models to the data, two limitations became apparent. First, the column density N_H and the abundance of Fe are strongly coupled, higher values of N_H requiring higher Fe abundance. Similarly, the abundance of the low-Z elements He, C, N, O, and Ne are strongly coupled to the abundance of all higher-Z elements such that increasing the assumed abundances of the low-Z elements increases the inferred abundances of all the higher-Z elements (see Table 1). We have dealt with these ambiguities by fixing the value of $N_H = 1.5 \times 10^{22}$ cm^{-2} (3 times the value determined by the 21 cm radio measurements [Greisen 1973], the same factor found for the Crab nebula) and the abundance for He, C, N, O, and Ne at their

TABLE 1

ELEMENTAL ABUNDANCES INFERRED FOR CASSIOPEIA A

Element	Abundance[a]	Ratio to Solar Abundance[b]	Primary Line[c] Energy (keV)	Equivalent Width (eV)	Photons (cm^{-2} s^{-1})
Mg	2.0×10^{-5}	0.49 ± 0.05	1.35	150	0.017
Al	4.0×10^{-6}	1.2 ± 0.25	1.61	20	0.003
Si	6.1×10^{-5}	1.7 ± 0.1	1.86	500	0.050
S	5.5×10^{-5}	3.3 ± 0.2	2.46	500	0.030
Ar	1.8×10^{-5}	6.9 ± 0.6	3.1	150	0.005
Ca	7.0×10^{-6}	3.0 ± 0.6	3.9	100	0.003
Fe	2.6×10^{-5}	0.8 ± 0.1	0.8–1.2	...	...

[a] Strongly dependent on assumed abundances of C, N, O, and Ne. If abundances of C, N, O, and Ne increased by a factor of 20 over solar abundance, the inferred abundances for high-Z elements in Table 1 increase by factors of 6–9. Quoted values are by number relative to hydrogen.

[b] Based on estimates of solar abundances by Meyer 1978. Error estimates are 1 σ based on scatter from six independent data sets.

[c] Energy corresponds to centroid of line complex.

nominal solar values of 0.11, 4.8×10^{-4}, 8.5×10^{-5}, 8.5×10^{-4}, and 1.3×10^{-4}, respectively (Meyer 1978).

With these constraints, we find a best-fit temperature of 0.63 keV for the low-temperature component of the Cas A spectrum. The best-fit values for the elemental abundances of Mg, Al, Si, S, Ar, Ca, and Fe are tabulated in Table 1. The emission measure of the 0.63 keV component is approximately 7.5 times that of the 4.0 keV component. The model has been folded through the instrument response and superposed upon the data in Figure 2. As a further aid, we have drawn in Figure 2 the estimated contribution from continuum emission. Since line emission dominates the X-ray spectrum below 2.5 keV, the lines and not the underlying continuum determined the temperature of the best-fit model. The overall χ^2 for this model is 700 for 55 degrees of freedom. We can reduce this χ^2 drastically by "tweaking" the response function to better accommodate the model (note, in particular, the apparent offset by $\sim$10 eV of the helium-like S feature), but we have decided to use an objectively determined response function until such time as we have accumulated enough data to infer the true differential corrections to the detector and telescope responses. The conclusions discussed here should be insensitive to such corrections, as we are convinced that the large χ^2 is caused predominantly by the combination of few-percent systematic uncertainties with the high statistical quality of the data, rather than by a poor fit to the shape of the spectrum.

We emphasize that the abundances in Table 1 depend strongly on the initial assumptions concerning the low-Z elements. Therefore, we have also given the equivalent widths of the primary features. These values should be much less sensitive to those elements that are responsible for the underlying continuum. Absolute photon fluxes are also listed in Table 1. Apart from other errors, these may be underestimates of the true flux to the extent that some of the Cas A emission may have been outside the field of view of the SSS.

IV. DISCUSSION

The abundances in Table 1 can be compared to previous estimates of the elemental abundances in Cas A. Optical line intensities have been observed for both the fast-moving and the slower-moving, quasi-stationary flocculi (Kirshner and Chevalier 1977; Searle 1971). Chevalier (1977) concluded from these data that the fast-moving knots are highly enriched in O, S, Ar, and Ca and significantly deficient in H and N, while in the slower-moving material He and N are overabundant by an order of magnitude. These observations are consistent with the conjecture that the fast-moving knots are ejecta from the SN explosion while the slower flocculi are composed of material shed by the star prior to its demise.

The plasma responsible for the X-rays is an unknown combination of the two components just discussed and the swept-up interstellar material. Any abundance anomalies present in the knots and flocculi will thus be moderated. Nonetheless, the present data do indicate some enrichment of S and Ar relative to the other elements which contribute to the X-ray spectra. As we have discussed, the Fe abundance depends strongly on the value of N_H, but it is encouraging to note the excellent agreement between the present estimate (0.79) based on L-shell emission and that (0.64) based on the 6.7 keV K emission (Pravdo et al. 1976).

The clear predominance of helium-like Si and S over the hydrogen-like species sets a strong upper limit of 10^7 K for the ionization temperature of the lower temperature plasma, where we define ionization temperature as the temperature which for collisional equilibrium would result in the same ionic distribution. However, we cannot place a similar constraint on the electron temperature because the low-temperature continuum is buried beneath the line emission. The high χ^2 for the best-fit collisional equilibrium model does not rule out a collisional equilibrium model for Cas A. As discussed earlier, there are significant contributions to χ^2 from systematic uncertainties in the response of the SSS to X-rays. Additional contributions to χ^2 may arise from the uncertainty of various transition rates used by Raymond and Smith (1979). However, it is unlikely on theoretical grounds that the hot plasmas in Cas A are in collisional equilibrium. In a shock-heated plasma, the ionization temperature of the gas will increase more slowly than the electron temperature so that lower ionization states will be overpopulated relative to their expected equilibrium population (Itoh 1977; Gorenstein et al. 1974). We emphasize that the quoted abundances are derived from the hypothesis of collisional equilibrium with solar abundances assumed for $Z \leq 10$.

We have been careful to qualify the abundance estimates in Table 1 with the reasons why the details of our spectral-fitting procedures may be oversimplifications of the physical conditions in Cas A. Nevertheless, it is important to note that the most likely lowest-order "correction" that would be applied to all of the abundances is a common multiplicative factor, and that this factor is almost certainly close to unity if the abundances are taken relative to oxygen. Since the masses of the ejecta and the swept-up interstellar medium for Cas A should be roughly comparable, our results would indicate that the relative abundances of the high-Z to low-Z elements are not strikingly different from solar. The low Fe abundance we observe is especially noteworthy, as it suggests that explosive nucleosynthesis has not substantially transformed Si-group to Fe-group material (Arnett 1978).

These results are illustrative of the most line-rich celestial X-ray spectra we expect to observe with the SSS. Below about 1 keV we are limited to testing for consistency between the data and a complex model, but at higher energies the SSS is well suited for studying line emission from ionized Mg, Si, S, Ar, and Ca. The SSS can easily distinguish the line emission of hydrogen-like species from that of lower ionization states for these elements. In particular, for Cas A the helium-like lines of Si and S are at least an order of magnitude stronger than their hydrogen-like counterparts, placing severe constraints on the ionization structure of the emitting gas. Observations of SNR at later stages of evolution will allow the time development of this ionization structure to be studied.

We wish to acknowledge Howard Eiserike and Nancy Laubenthal for programming an efficient and accessible data retrieval system.

REFERENCES

Arnett, W. D. 1978, *Ap. J.*, **219**, 1008.
Brown, R. L., and Gould, R. J. 1970, *Phys. Rev. D*, **1**, 2252.
Charles, P. A., Culhane, J. L., Zarnecki, J. C., and Fabian, A. C. 1975, *Ap. J. (Letters)*, **197**, L61.
Charles, P. A., Kahn, S. M., Bowyer, S., Blissett, R. J., Culhane, J. L., Cruise, A. M., and Garmire, G. 1979, preprint.
Chevalier, R. A. 1977, Proc. 8th Texas Symposium Relativistic Astrophysics (*Ann. NY Acad. Sci.*, **302**, 106).
Davison, P. J. N., Culhane, J. L., and Mitchell, R. J. 1976, *Ap. J. (Letters)*, **206**, L37.
Fireman, E. L. 1974, *Ap. J.*, **187**, 57.
Gorenstein, P., Harnden, F. J., and Tucker, W. H. 1974, *Ap. J.*, **192**, 661.
Greisen, E. W. 1973, *Ap. J.*, **184**, 363.
Hill, R. W., Burginyon, G. A., and Seward, F. D. 1975a, *Ap. J.*, **200**, 158.
———. 1975b, *Ap. J.*, **200**, 163.
Holt, S. S., White, N. E., Becker, R. H., Boldt, E. A., Mushotzky, R. E., Serlemitsos, P. J., and Smith, B. W. 1979, *Ap. J. (Letters)*, **234**, L65.
Itoh, H. 1977, *Pub. Astr. Soc. Japan*, **29**, 813.
Joyce, R. M., Becker, R. H., Birsa, F. B., Holt, S. S., and Noordzy, M. P. 1978, *IEEE Trans. Nucl. Sci.*, **25**, 453.
Kirshner, R. P., and Chevalier, R. A. 1977, *Ap. J.*, **218**, 142.
Mason, K. O., Charles, P. A., and Bowyer, S. 1978. *Bull. AAS*, **10**, 508.
Meyer, J.-P. 1978, private communication.
Pravdo, S. H. 1979, private communication.
Pravdo, S. H., Becker, R. H., Boldt, E. A., Holt, S. S., Rothschild, R. E., Serlemitsos, P. J., and Swank J. H. 1976, *Ap. J. (Letters)*, **206**, L41.
Raymond, J. C., and Smith, B. W. 1977, *Ap. J. Suppl.*, **35**, 419.
———. 1979, in preparation.
Searle, L. 1971, *Ap. J.*, **168**, 41.
Serlemitsos, P. J., Boldt, E. A., Holt, S. S., Ramaty, R., and Brisken, A. F. 1973, *Ap. J. (Letters)*, **184**, L1.
Smith, B. W., Mushotzky, R. F., and Serlemitsos, P. J. 1979, *Ap. J.*, **227**, 37.

R. H. BECKER, E. A. BOLDT, S. S. HOLT, R. F. MUSHOTZKY, P. J. SERLEMITSOS, B. W. SMITH, and N. E. WHITE: Code 661, Laboratory for High Energy Astrophysics, NASA Goddard Space Flight Center, Greenbelt, MD 20771

The Astrophysical Journal, 246:L33–L36, 1981 May 15
© 1981. The American Astronomical Society. All rights reserved. Printed in U.S.A.

Reprinted with permission from *The Astrophysical Journal*, 246 pp. L33-L36, C.R. Canizares and P.F. Winkler, "Evidence for Elemental Enrichment of Puppis A by a Type II Supernova." © 1981 The American Astronomical Society.

EVIDENCE FOR ELEMENTAL ENRICHMENT OF PUPPIS A BY A TYPE II SUPERNOVA

C. R. CANIZARES[1]
Department of Physics and Center for Space Research, Massachusetts Institute of Technology

AND

P. F. WINKLER
Department of Physics, Middlebury College
Received 1981 January 12; accepted 1981 February 11

ABSTRACT

We use the measurements of X-ray emission lines from Puppis A obtained with the Focal Plane Crystal Spectrometer on the *Einstein* Observatory to deduce the relative abundances of O, Ne, and Fe in the remnant. We find that O:Fe and Ne:Fe abundance ratios are 3–5 times their cosmic values. Our conclusion requires no strong assumptions about the condition of the plasma. We argue that the abundance ratios result from the enrichment of the emitting plasma by at least 3 $M_\odot$ of oxygen and neon ejected by the supernova and mixed with $\gtrsim 100$ $M_\odot$ of interstellar material. A Type II supernova in a star of ≥ 25 $M_\odot$ can provide the necessary oxygen and neon.

Subject headings: abundances — nebulae: supernova remnants

I. INTRODUCTION

In a companion *Letter* (Winkler *et al.* 1981*a*, hereafter Paper I) we present the initial results from a spectral survey of the supernova remnant Puppis A carried out with the Focal Plane Crystal Spectrometer on the *Einstein* Observatory. The survey covers the range 500–1100 eV for an approximately $3 \times 30'$ region centered on the brightest portion of the remnant. It reveals numerous strong lines from O VII, O VIII, Ne IX, Ne X, and Fe XVII, all of which are also prominent in the spectra of solar active regions. The strengths of these lines are used as diagnostics of the conditions in the emitting plasma.

In the present *Letter* we explore the consequences of our earlier findings. We use our deductions about the plasma conditions to draw conclusions about the elemental abundances in Puppis A and find that oxygen and neon are overabundant relative to iron by factors of 3–5. The qualitative aspects of this conclusion are based on very general deductions about the dominant ionization states of the emitting elements in Puppis A and in solar active regions. They do not depend on any assumptions about the equilibrium of the plasma or about its detailed temperature distribution, nor do they rely on calculations of atomic collision strengths and line emissivities. Quantitative estimates of the degree of overabundance do require calculated emissivities but still do not rely on the assumption of exact plasma equilibrium. We find it plausible to attribute the overabundance to an enrichment of the plasma by at least 3 $M_\odot$ of oxygen and neon ejected by a Type II supernova explosion in a star of $\gtrsim 25$ $M_\odot$.

II. RELATIVE ABUNDANCES

a) Qualitative Analysis

Paper I notes that in Puppis A the Fe XVII line at 826 eV is unusually weak relative to the neighboring lines of O and Ne, when compared to spectra of solar active regions. The relevant line ratios are given in Table 1 for Puppis A (Paper I) and the Sun (Walker, Rugge, and Weiss 1974*a*). We have chosen lines which are strong, well resolved, and close enough in energy that the ratios are relatively insensitive to uncertainties in the column density of absorbing material along the line of sight (which we take to be $4 \pm 2 \times 10^{21}$ cm^{-2} as in Paper I). The nearly equal line energies also give nearly equal Boltzmann factors in the expressions for the line emissivities (e.g., see Winkler *et al.* 1981*b*). Thus the temperature dependence of the line ratios is primarily in the fractional population of the relevant ionization stage of the emitting atoms (the ionization fraction).

The O/Fe and Ne/Fe line ratios in Table 1 for Puppis A are larger than those for the solar active regions by a factor of ~4. In principle this could be the result of true abundance differences or of differences in the ionization fraction. We argue here that the ionization fractions are in fact similar in Puppis A and in solar active regions, so the abundance differences are real.

In spectra of both Puppis A and solar active regions, lines from iron ionization stages other than neon-like Fe XVII are absent or very weak, implying an ionization

---

[1] Alfred P. Sloan Foundation Fellow.

fraction near unity for this stage (see Paper I). This is in contrast to hotter solar flare spectra, for example, which contain numerous strong lines of higher-ionization stages of iron within the energy range we observed (McKenzie et al. 1980). Predominance of the stable, neon-like ionization stage in Puppis A is theoretically plausible as well. Equilibrium models (both with and without autoionization effects) show large Fe XVII ionization fractions over the relevant temperature range 2–4×10^6 K (Jacobs et al. 1977). Similarly, in ionizing nonequilibrium plasma, iron will rapidly ionize to its neon-like form and then proceed much more slowly to higher stages. Following Mewe and Gronenschild (1980, and references therein) we calculate that the ionization time scale for Fe XVI → Fe XVII is 50–125 times shorter than that for Fe XVII → Fe XVIII (for electron temperatures of 10^7 to 5×10^6 K). Furthermore, the Fe XVI → Fe XVII time scale is nearly 50 times smaller than that for O VII → O VIII, whereas the time scale for Fe XVII → Fe XVIII is a factor of 1.2 (at 10^7 K) to 2 (at 5×10^6 K) times longer. Thus, in any ionizing plasma containing both O VII and O VIII, the dominant species of iron is Fe XVII. Finally, the Fe XVI → Fe XVII ionization time scale is extremely short ($\lesssim 10$ yr) compared to the estimated age of Puppis A (~ 5000 yr; Culhane 1977), so less-ionized iron should not be present.

The relative population of the He-like and H-like stages of oxygen are nearly the same in Puppis A and the Sun as indicated by the nearly equal line ratio for O VII ($3p \to 1s$)/O VIII Lyα of Table 1. So comparison of O VIII Lyβ and Fe XVII lines should give a direct measure of the O:Fe abundance ratio. The data of Table 1 indicate that this ratio is higher in Puppis A by a factor of 4.

The Ne IX/Ne X line ratio is about twice as large in Puppis A as in the Sun. However, an analysis similar to that represented above for iron indicates that nearly all the neon will be in the H-like or He-like form, in both equilibrium and nonequilibrium plasmas. Thus the Ne IX/Ne X line ratio implies an ~ 35% higher Ne IX ionization fraction in Puppis A than in the Sun. This is insufficient to explain the factor of 4 larger Ne IX/Fe XVII line ratio, which again indicates a significant overabundance of neon relative to iron.

b) Quantitative Analysis

Quantitative estimates of the relative abundances in Puppis A can be made by comparing the measured line ratios of Table 1 with calculated emissivities. We use the calculations of Raymond and Smith (1979) and Mewe and Gronenschild (1980) for plasmas in statistical equilibrium. We argue in Paper I that the plasma in Puppis A is probably not too far from equilibrium, but the abundance estimates do not depend on that conclusion. This is because of our deductions concerning the dominant ionization stage of iron and our choice of line ratios to reduce temperature dependences other than in the ionization fractions. Since we find that iron is predominantly in the form of Fe XVII, even ionizing, nonequilibrium plasmas will have line fluxes similar to those given by equilibrium models at the appropriate ionization temperatures.

In Figure 1 we show the abundance ratios derived for various temperatures. We have assumed an absorbing column of $4 \pm 2 \times 10^{21}$ cm^{-2}, and the error bars in Figure 1 reflect the uncertainty in this number. In Paper I we derive ionization temperatures of 2.2×10^6 K for oxygen and 4×10^6 K for neon, and we argue that the remnant probably contains material distributed more or less uniformly through the range 2–5×10^6 K. The derivation of precise abundance ratios requires a detailed model for the distribution to permit an appropriately weighted average of the values in Figure 1. However, it is clear that the uncertainties in the emissivities indicated by the discrepancy between the two models would limit the accuracy.

Figure 1 gives an O:Fe abundance ratio by number of 50 to 100 compared to "cosmic" values of 17 (Allen 1973) or 27 (Cameron 1974; Meyer 1979; Walker, Rugge, and Weiss 1974b). The O:Ne ratio of 4–6 is marginally smaller than the cosmic values 7.9 (Allen 1973), 8.5 (Meyer 1979), and 6.3 (Cameron 1974). However, the cosmic neon abundance is not well determined and

TABLE 1
LINE FLUX RATIOS

Source	O VII ($3p \to 1s$)/O VIII Lyα	Ne IX[a]/Ne X Lyα[b]	O VIII Lyβ/Fe XVII[c]	Ne IX[a]/Fe XVII[c]
Solar active region[d] ...	0.16	4.2	0.16	0.23
Puppis A[e]	0.12 ± 0.03	9.3 ± 1.4	0.62 ± 0.06	0.76 ± 0.06

[a] Includes resonance, forbidden, and intercombination lines (905–921 eV).
[b] Corrected for an assumed 25% contamination from an Fe XVII line: see Paper I.
[c] Blend of Fe XVII and O VIII lines (812–837 eV).
[d] From Walker, Rugge, and Weiss 1974a.
[e] From Paper I, Table 1 corrected for interstellar absorption with $N_H = 4 \pm 2 \times 10^{21}$ cm^{-2}.

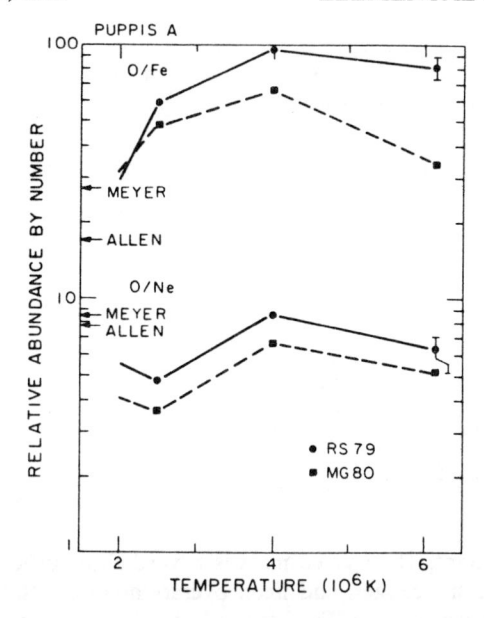

FIG. 1.— The relative abundances in Puppis A of O:Fe (*top*) and O:Ne (*bottom*) by number of atoms vs. assumed plasma temperature. The abundances are deduced from the line ratios of Table 1 for plasmas at equilibrium using the emissivities of RS79 and MG80 and for a column density of 4×10^{21} atoms cm^{-2} along the line of sight. The error bars show the effect of varying the column density by $\pm 2 \times 10^{21}$ cm^{-2}; they apply equally to all points on the solid or neighboring dashed curves. The arrows mark the cosmic abundance ratios given by Allen (1973) or Meyer (1979).

O:Ne ratios from solar coronal measurements range from 4 (Acton, Catura, and Joki 1975) to 13 (Walker, Rugge, and Weiss 1974*b*).

III. DISCUSSION

Our data show that the composition of the dominant, low-temperature component of the X-ray-emitting plasma in Puppis A differs substantially from that of normal cosmic material. From our observations alone, we are so far unable to distinguish between the alternatives of iron depletion versus oxygen and neon enhancement, but the latter seems much more acceptable in the overall context of this source.

The only plausible cause of widespread depletion of gaseous iron is the trapping of iron in interstellar grains (Savage and Bohlin 1979). Observationally, if gaseous iron were depleted due to grain trapping, one would expect silicon to be similarly depleted (Duley 1980). But spectra of Puppis A obtained with the *Einstein* Observatory Solid State Spectrometer show that if anything the silicon abundance is enhanced (Szymkowiak 1980). Furthermore, depletion by this mechanism seems untenable for Puppis A even though the remnant is so large that most of its material is probably of interstellar origin. This is because grains originally present in the material would most likely be destroyed during the heating process. If the dominant process is shock heating, then the observed postshock temperature of $\gtrsim 2 \times 10^6$ K implies a shock speed $V_s \gtrsim 400$ km s^{-1}, whereas Draine and Salpeter (1979*b*) show that iron grains are destroyed by even a modest shock with $V_s > 100$ km s^{-1}. If instead most of the material evaporates from dense clouds through which the shock has propagated more slowly, then the grains could initially survive. However, they would be subject to destruction by sputtering (Draine and Salpeter 1979*a*) in the hot, high-density region surrounding the evaporating cloud (Tsunemi and Inoue 1980).

The second alternative, an enhancement of oxygen and neon, can result from enrichment of predominantly interstellar material by supernova ejecta. Baade and Minkowski (1954) and Dopita, Mathewson, and Ford (1977) found high oxygen and nitrogen abundances in several optical filaments of Puppis A, suggesting that supernova ejecta are present in at least a few locations. Our observations of a much larger volume of hot diffuse plasma indicate that material with anomalous abundances is widespread.

Enrichment of the entire remnant is plausible despite its large size and advanced age. The quantity of oxygen in Puppis A implied by our observations can be estimated using our measured O VIII Lyα line flux, corrected for interstellar absorption, together with the corresponding emissivity of RS79 (which varies by less than 20% over the range 1.6 to 4×10^6 K). Extrapolation to the entire remnant is justified by the qualitative similarity between spectra we measured at various locations. For a plasma with oxygen abundance X_O relative to solar (Allen 1973), we obtain for the mass in oxygen $m_O = 0.08 \, d^{5/2} X_O^{1/2} f^{-1} \, M_\odot$, where d is the distance to Puppis A in kpc, and f is the fraction of the emission included in our aperture. For $f = 0.1$ (see Paper I), $d = 1.2$–2.5 (Milne 1971; Clark and Caswell 1976) and $X_O = 3$–5 (from § II), we obtain $m_O = 2$–18 $M_\odot$. (The prodigious mass requirement at large distances favors values of $d \lesssim 2$ kpc.)

A Type II supernova could have ejected the large quantity of oxygen which our observations imply. In Table 2 we list the mass of the ejecta for models of Type II nucleosynthesis by Weaver, Zimmerman, and Woosley (1978), Weaver and Woosley (1980), and Arnett (1978). The models for stars of $\geq 25 \, M_\odot$ yield 3–4 $M_\odot$ of oxygen and O:Ne ratios of 3–5, which are consistent with our data. Most of this O and Ne is synthesized through carbon burning during the presupernova evolution of the star, rather than during the explosion itself. The oxygen-neon mantle is merely ejected by the supernova, and so its quantity is relatively independent of possible uncertainties in explosive synthesis calculations (Johnston and Joss 1980). A 15 $M_\odot$ star ejects between $\sim 0.3 \, M_\odot$ and 0.8 $M_\odot$ of oxygen, which is too little, and

TABLE 2
MODEL SUPERNOVA EJECTA

Stellar Mass ($M_\odot$)	Oxygen Mass ($M_\odot$)	Neon Mass ($M_\odot$)	O:Ne (by number)
15[a] (pre-SN)	0.1	0.1	...
(explosive)	0.2	0.1	...
Total	0.3	0.2	2.6
25[a] (pre-SN)	2.0	1.0	...
(explosive)	0.6	0.1	...
Total	2.6	1.1	3.8
16[b]	0.8	0.4	3.3
22[b]	1.7	0.8	3.5
28[b]	3.8	1.0	5.6

[a] From Weaver, Zimmerman, and Woosley 1978 and Weaver and Woosley 1980.
[b] From Arnett 1978. Total stellar masses listed correspond to helium core masses of 6, 8, and 12 $M_\odot$, respectively.

it gives an O:Ne ratio which is considerably smaller than we observe.

If the supernova ejecta contain negligible iron, then they must have mixed with interstellar material of mass $m_{ISM} \approx 50\ m_O X_{Fe}$ to give our observed O:Fe ratio of ~ 75. Here X_{Fe} is the iron abundance in the interstellar matter relative to solar. Thus $M_{ISM} \sim 100-200\ M_\odot$ for the case of a 25 $M_\odot$ supernova. This value implies a reasonable average interstellar density of $n \sim 2-5\ d^{-3}$ cm^{-3} in the vicinity of Puppis A, assuming the mass is swept up from a spherical region of diameter 50'.

We conclude that Puppis A is probably the remnant of a Type II supernova explosion in a $\gtrsim 25\ M_\odot$ star. The oxygen- and neon-rich ejecta from this event have been mixed with the much larger quantity of interstellar material swept up over the several thousand year life of the remnant. The ejected mass is nevertheless sufficient to cause the oxygen and neon overabundance (relative to iron) we observe. Further analysis of data from the FPCS and from other *Einstein* Observatory instruments should enable us to refine these conclusions and to study further heavy-element enrichment through the supernova process.

We thank J. C. Raymond, B. W. Smith, R. Mewe, and E. Gronenschild for their emissivity calculations and A. Szymkowiak for prepublication results. We are grateful to R. Becker, M. Johnston, J. Ostriker, and the referee for their comments, and to Brenda Parsons for her help in preparing the manuscript. We acknowledge the support of NASA under contract NAS-8-30752 and grant NAG-8389.

REFERENCES

Acton, L. W., Catura, R. C., and Joki, E. G. 1975 *Ap. J. (Letters)*, **195**, L93.
Allen, C. W. 1973, *Astrophysical Quantities* (3d ed.; London: Athlone).
Arnett, W. D. 1978, *Ap. J.*, **219**, 1278.
Baade, W., and Minkowski, R. 1954, *Ap. J.*, **119**, 206.
Cameron, A. G. W. 1974, *Space Sci. Rev.*, **15**, 121.
Clark, D. H., and Caswell, J. L. 1976, *M.N.R.A.S.*, **174**, 267.
Culhane, J. L. 1977, in *Supernovae*, ed. D. N. Schramm (Dordrecht: Reidel), pp. 29–51.
Dopita, M. A., Mathewson, D. S., and Ford, V. L. 1977, *Ap. J.*, **214**, 179.
Draine, B. T., and Salpeter, E. E. 1979a, *Ap. J.*, **231**, 77.
———. 1979b, *Ap. J.*, **231**, 438.
Duley, W. W. 1980, *Ap. J. (Letters)*, **240**, L47.
Jacobs, V. L., Davis, J., Kepple, P. C., and Blaha, M. 1977, *Ap. J.*, **211**, 605.
Johnston, M. D., and Joss, P. C. 1980, *Ap. J.*, **242**, 1124.
McKenzie, D. L., Landecker, P. B., Broussard, R. M., Rugge, H. R., Young, R. M., Feldman, U., and Doschek, G. A. 1980, *Ap. J.*, **241**, 409.
Mewe, R., and Gronenschild, E. H. B. M. 1980, *Astr. Ap.*, submitted (MG 80).
Meyer, J. P. 1979, *Proc. 16th Internat. Cosmic Ray Conf.—Kyoto*, **2**, 115.
Milne, D. K. 1971, *Australian J. Phys.*, **24**, 757.
Raymond, J. C., and Smith, B. W. 1979, private communication (RS 79).
Savage, B. D., and Bohlin, R. C. 1979, *Ap. J.*, **229**, 136.
Szymkowiak, A. E. 1980 in *Proc. Conf. Supernova Remnants, Austin, Texas*, ed. J. C. Wheeler, in preparation.
Tsunemi, H., and Inoue, H. 1980, *Pub. Astr. Soc. Japan*, **32**, 247.
Walker, A. B. C., Jr., Rugge, H. R., and Weiss, K. 1974a, *Ap. J.*, **192**, 169.
———. 1974b, *Ap. J.*, **194**, 471.
Weaver, T. A., and Woosley, S. E. 1980, *Annals NY Acad. Sci.*, **336**, 335.
Weaver, T. A., Zimmerman, G. B., and Woosley, S. E. 1978, *Ap. J.*, **225**, 1021.
Winkler, P. F., Canizares, C. R., Clark, G. W., Markert, T. H., Kalata, K., and Schnopper, H. W. 1981a, *Ap. J. (Letters)*, **246**, L27 (Paper I).
Winkler, P. F., Canizares, C. R., Clark, G. W., Markert, T. H., and Petre, R. 1981b, *Ap. J.*, **245**, 574.

C. R. CANIZARES: Massachusetts Institute of Technology, Room 37-501, Cambridge, MA 02139

P. F. WINKLER: Department of Physics, Middlebury College, Middlebury, VT 05753

THE ASTROPHYSICAL JOURNAL, 234:L45–L49, 1979 November 15
© 1979. The American Astronomical Society. All rights reserved. Printed in U.S.A.

Reprinted with permission from *The Astrophysical Journal*, 234 pp. L45-L49, L. Van Speybroeck, *et al.*, "Observations of X-Ray Sources in M31." © 1979 The American Astronomical Society.

OBSERVATIONS OF X-RAY SOURCES IN M31

L. VAN SPEYBROECK, A. EPSTEIN, W. FORMAN, R. GIACCONI, C. JONES, W. LILLER, AND L. SMARR[1]
Harvard-Smithsonian Center for Astrophysics, Cambridge, Massachusetts
Received 1979 July 3; accepted 1979 July 31

ABSTRACT

We have observed 69 unresolved X-ray sources and seven diffuse or confused source regions in M31 with the *Einstein* Observatory. The typical limiting sensitivities in the 0.5–4.5 keV band were 9×10^{36} ergs s^{-1}. There are 21 sources in a compact inner bulge, seven globular clusters, 40 sources we categorize as Population I, and 8 sources near globular clusters but with position errors large enough so that four chance coincidences would be expected. The Population I sources are associated with bright visible objects, and with neutral hydrogen, dust, and other Population I tracers. Luminosity distributions of these classes are given. The nucleus of M31 is coincident with a 10^{38} ergs s^{-1} X-ray source at approximately the 90% confidence level.

Subject headings: clusters: globular — galaxies: stellar content — X-rays: sources

I. INTRODUCTION

The nearby galaxy M31 provides certain advantages for the study of typical galactic phenomena. The distances are known and absolute luminosities can be determined. The morphology is more apparent than that of our own Galaxy, and consequently the significance of source positions can be more easily understood. Perhaps more important, M31 is not exactly like our own Galaxy, and thus permits us to begin to study the correlation between galactic type and the characteristics of X-ray sources. These of course are familiar advantages, and in fact M31 was the first external galaxy to be studied with an optical telescope (Marius, 1612 December 15), and also the first to be studied in radio (inconclusively, by Reber, 1939 October 11; Reber 1940; however, see also van der Kruit and Allen 1976). It has been detected previously as an aggregate source in X-rays; for example, it is listed in the 4th *Uhuru* catalog by Forman *et al.* (1978) and in the *Ariel* catalog by Cooke *et al.* (1978), and was detected in soft X-rays by Margon *et al.* (1974). The *Einstein* Observatory provides for the first time adequate sensitivity and angular resolution to study individual X-ray sources in this galaxy.

We used the *Einstein* Observatory (Giacconi *et al.* 1979) on 1979 January 11–14 to study three fields in M31 with the imaging proportional counter detector (IPC), and one field with the high-resolution imaging detector (HRI). Preliminary results from two of the IPC fields and the HRI field are reported here.

II. OBSERVATIONS

The field centers, observation times, and typical limiting sensitivities are given in Table 1. The central IPC field is illustrated in Figure 1 (Plate L11). The field consists of approximately 20 identifiable sources in addition to a large, confused region near the center of the field. This region is resolved into about 45 individual sources in the HRI data. The center of the HRI field is shown in Figure 2 (Plate L12). The source near the nucleus and a source near our detection threshold are indicated. The northeast field, which is not shown, contains four obvious sources.

In almost all cases, the individual sources were identified using the standard *Einstein* data reduction programs, which are based upon a sliding acceptance box algorithm. The photons associated with each source were then used to estimate its position and luminosity. The typical systematic position errors are about 2″ and 1′ for the HRI and IPC, respectively. Statistical fluctuations also contribute to the position errors of the faint sources. All values of luminosities given in Table 1 and throughout this *Letter* refer to the 0.5–4.5 keV interval, which is the band normally observed with the *Einstein* IPC. Unfortunately, the IPC was operated at an anomalously high gain during these observations, which resulted in the unusually narrow energy acceptance bands given in Table 1. We choose to quote luminosities in the wider interval of 0.5–4.5 keV because most X-ray sources emit over a wide energy band, and we believe that stating luminosities for the anomalously narrow energy intervals observed is misleading, and that the use of the wider interval will facilitate a comparison of these data with other *Einstein* observations. The data associated with each IPC source were fitted to power-law and thermal bremsstrahlung models to determine spectral parameters. We estimated luminosities based upon the best-fit spectral parameters in those cases which were sensitive to these quantities. Typically, however, the models were not sensitive to the spectral parameters because of the low counting rates and the narrow acceptance bands; in these cases we estimated luminosities assuming a 5 keV thermal bremsstrahlung, or an energy index of -2, the results of the two models differing by only a few percent. All luminosities were corrected for absorption in our Galaxy corresponding to $N_H = 8 \times 10^{20}$ cm^{-2}; Margon *et al.* (1974) observed values of N_H between 6.1×10^{20} and

[1] Jr. Fellow, Harvard Society of Fellows.

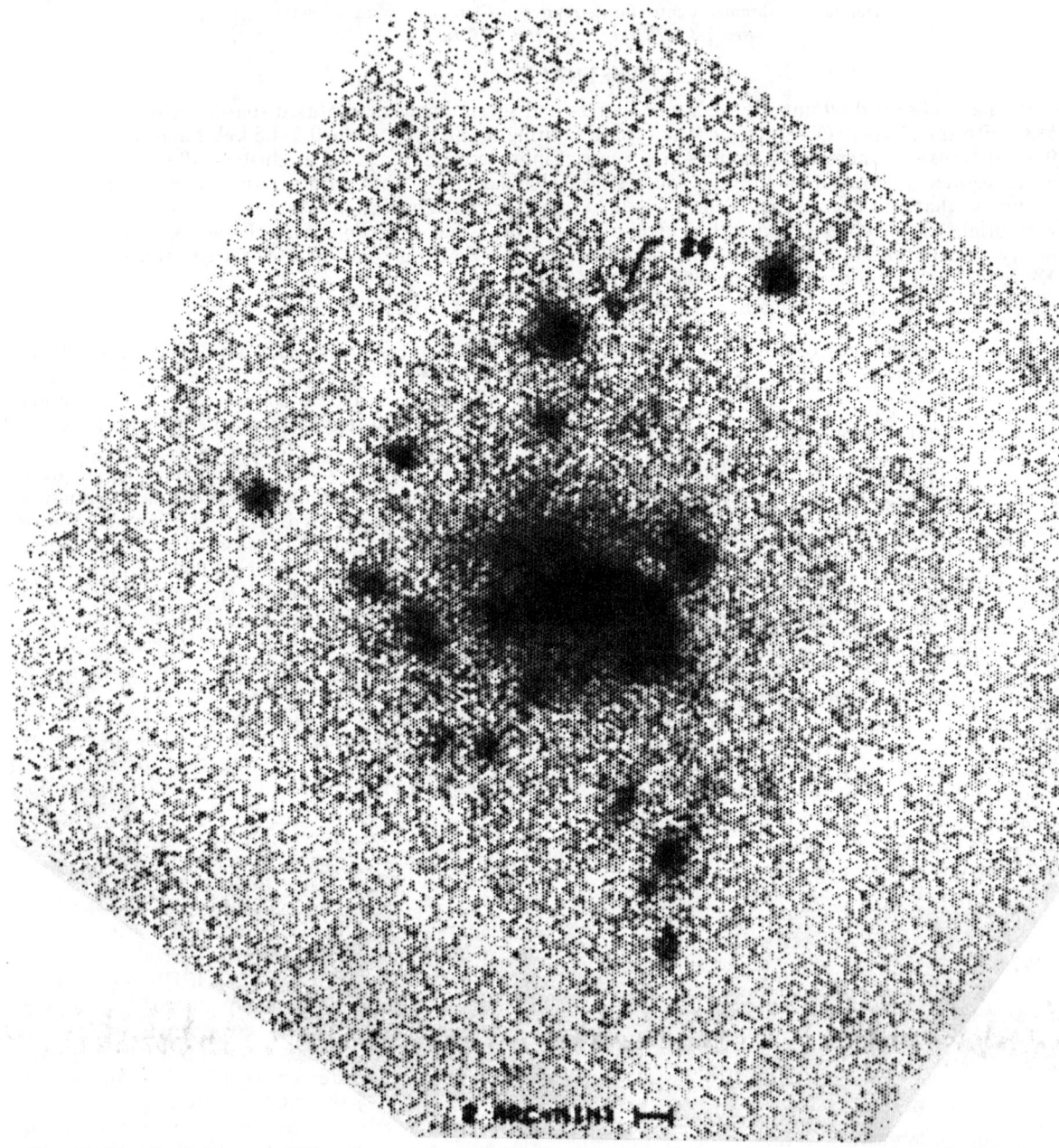

Fig. 1.—*Einstein* Observatory IPC exposure of the central field in M31. North is at the top and east is at the left. Source 26, which is indicated, is near the detection threshold. The nuclear region is unresolved.

Van Speybroeck *et al.* (*see* page L45)

PLATE L12

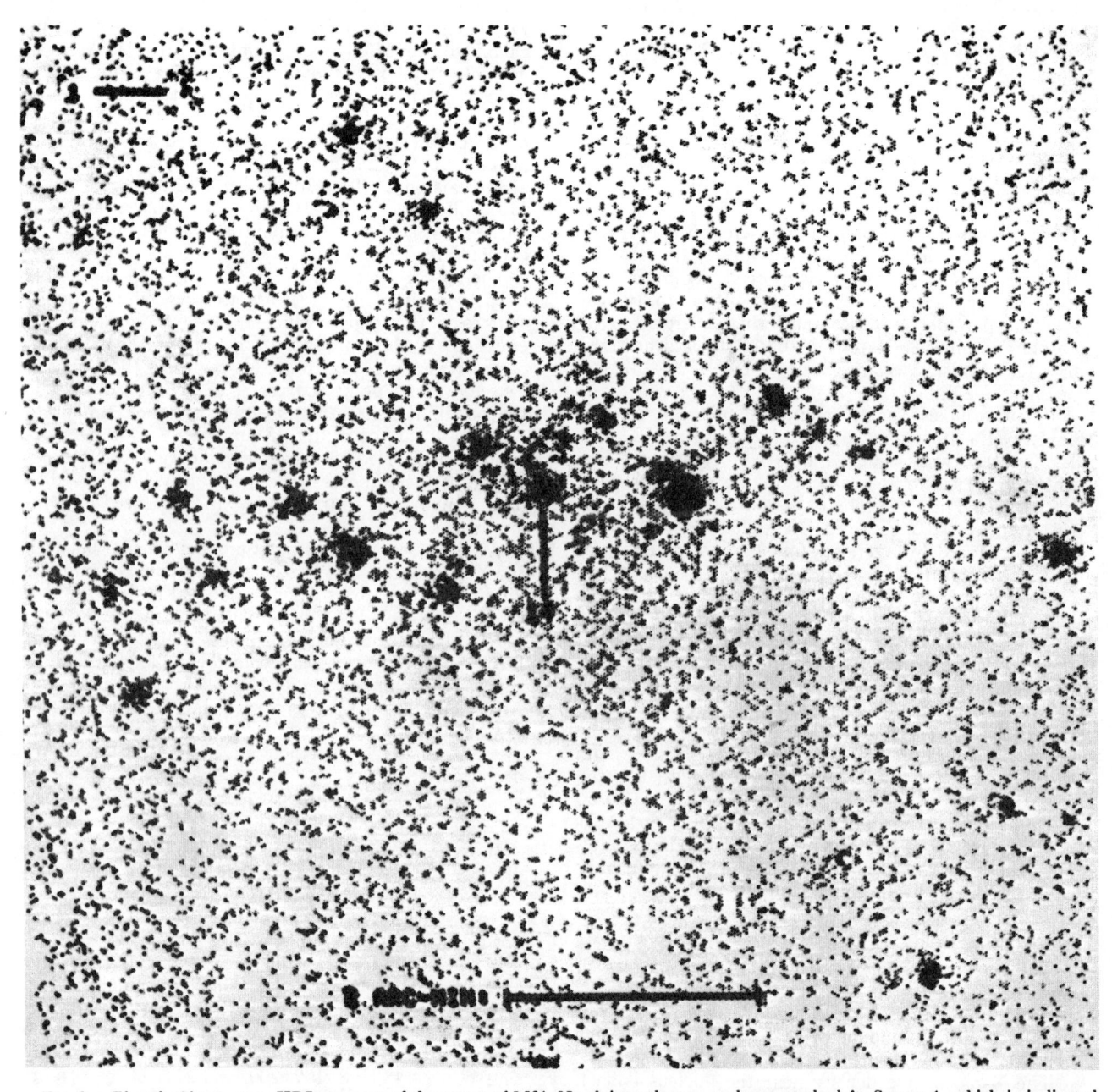

FIG. 2.—*Einstein* Observatory HRI exposure of the center of M31. North is at the top, and east at the left. Source 1, which is indicated, is near the detection threshold. The nucleus of M31 is coincident with source 14 within our measurement accuracy. The sources classified as the inner-bulge group are contained within 2' of the nucleus.

VAN SPEYBROECK *et al.* (*see* page L45)

TABLE 1
Einstein Observations of M31

Sequence	574	575	579
Instrument	IPC	IPC	HRI
Field	center	NE	center
Exposure (s)	36,060	31,785	27,328
Effective energy band (keV)	0.35–2.9	0.30–2.4	0.5–4.5
Threshold luminosity[a] (ergs s^{-1})	4.4×10^{36}	4.4×10^{36}	9.1×10^{36}

[a] 0.5–4.5 keV, near field center; this is spectrum and position dependent.

1.1×10^{21} cm^{-2}. Most spectral fits were consistent with this value, thus indicating little intrinsic source absorption. The HRI does not have spectral resolution, but its counting rate is relatively insensitive to typical spectral parameters other than the absorption. All absolute luminosity statements based upon either IPC or HRI in this *Letter* are uncertain by ±50% because of the above difficulties.

Relative fluxes can be determined more accurately. Seven of the 10 sources detected by both the HRI and IPC are too weak or exhibit variability and are not suited for verifying the calibration of the HRI relative to the IPC. The remaining three sources are consistent within the stated uncertainties. The effects of variability can be reduced by averaging many sources. The sum of the luminosities of HRI sources within 5' of the nucleus is 1.31×10^{39} ergs s^{-1}, whereas the IPC estimate of this region is 1.17×10^{39} ergs s^{-1}, or about 11% less. This indicates reasonable consistency between the two data sets and also argues against a large contribution by many sources which happen to be slightly below our threshold. The total luminosity of all detected sources is 2.3×10^{39} ergs s^{-1}; if we obtain the same flux from the unanalyzed southwest field as from the northeast, then the total luminosity due to detected sources in the 0.5–4.5 keV band will be about 2.7×10^{39} ergs s^{-1}, which can be compared to the *Uhuru* M31 estimate of 2.3×10^{39} ergs s^{-1} in the 2–6 keV band. We believe that most of the X-ray emission from M31 comes from the strong sources we are observing.

III. INDIVIDUAL SOURCE CHARACTERISTICS

The most interesting single object is source 14, which nominally is 2".1 from the nucleus of M31. Globular clusters, to be discussed later, provide a measure of our location accuracies and indicate that 5" is a conservative coincidence requirement.

The X-ray source density near the nucleus is such that the probability of an ordinary X-ray source being located less than 5" from the nucleus by chance is about 0.1. The association with the nucleus also is supported by the lack of detected variability in this source, such as is typical of bright X-ray sources in our Galaxy and which we also have found for the luminous sources in M31. Assuming that this 9.6×10^{37} ergs s^{-1} source is the nucleus, we discern a marked contrast with the center of our own Galaxy. The X-ray luminosity of the nucleus of M31 would be approximately 10^3 times that of any steady point source within 0°.5 of our own galactic nucleus (Epstein *et al.* 1979), although comparable to strong transient sources which have been observed near the galactic center, and still much weaker than the emission from the nuclei of active galaxies. The M31 nucleus emission at 1415 MHz is only about 0.06 that of Sgr A (van der Kruit 1972). This suggests a thermal source of the M31 X-ray radiation—possibly accretion of gas clouds onto a massive object, as has been suggested to be present near the center of our own Galaxy (see, e.g., Rodriguez and Chaisson 1979). Future *Einstein* observations of M31 will allow us to determine if the nuclear emission varies on a time scale of months, which would not be expected for such models.

The region northeast of the nucleus (near source 14) is particularly complex. There are three obvious point sources in this region which account for most of the emission. We interpret the data to indicate that much of the remaining flux from this region is the result of two additional point sources, but we cannot exclude the possibility that this is diffuse emission similar to that seen near the center of our Galaxy. Our interpretation is supported by marginal evidence for periodicity in one of these sources, the probability being ~0.055 of a nonvarying source giving the observed Fourier amplitude.

We have searched for position coincidences between the X-ray sources and two lists of supernovae remnants kindly provided to us prior to publication by Kirshner (1979) and by D'Odorico, Dopita, and Benvenuti (1979). These combined lists include 13 remnants in our fields of view; one clear coincidence was found—a supernova remnant associated with the Baade and Arp (1964) region 521 was observed with both HRI and IPC. The X-ray luminosity is about 2.2×10^{37} ergs s^{-1}. We do not detect emission at the position of S. Andromedae 1885.

IV. GROUP PROPERTIES

It is useful to divide the X-ray sources into three groups—inner bulge, Population I, and globular clusters—although further subclassifications eventually will be necessary. We have compared the HRI source positions with the globular cluster list of Sargent *et al.* (1977), and find five coincidences. Four of these five are in the inner part of the HRI field, where our position errors are less; these have an average separation of 3".5, the worst being 4.9. The globular cluster numbers from the list of Sargent *et al.* (1977) thus identified as X-ray sources are 148, 198, 199, 200, and 213. There also are eight IPC sources within 1' of listed globular clusters for which no HRI positions are available; approximately four would be expected by chance. We find five X-ray sources among the 60 globular clusters Sargent *et al.* (1977) list in the HRI field, and estimate approximately nine X-ray sources in the 237 globular clusters listed in our total observed fields. This is similar to the results obtained in our Galaxy, where seven of 150 clusters searched have such luminosities (Grindlay 1979). Two additional HRI sources which are coincident with bright objects have been observed optically, and a very preliminary inspection of these spectra suggests that these sources also are globular clusters.

The source positions and detector fields are plotted on an optical photograph of M31 in Figure 3 (Plate L13). The association of the outer X-ray sources with spiral features is quite apparent, although certain regions, such as the northeast, seem relatively overpopulated. The X-ray source positions are plotted on the H I map of Emerson (1976) in Figure 4. The sources possibly associated with globular clusters are indicated; the correlation of the remaining X-ray sources with the neutral hydrogen is striking. This, however, should not be surprising because we expect X-ray sources to be associated with young massive stars, and in M31 the indicators of Population I activity such as OB associations, H I, H II, CO, and dust are concentrated in a ring (see Fig. 4) centered at about 9 kpc from the nucleus (Emerson 1974, 1976, and 1978; Baade and Arp 1964; Berkhuijsen 1977; Combes *et al.* 1977*a, b*).

The X-ray positions in the inner part of M31 are

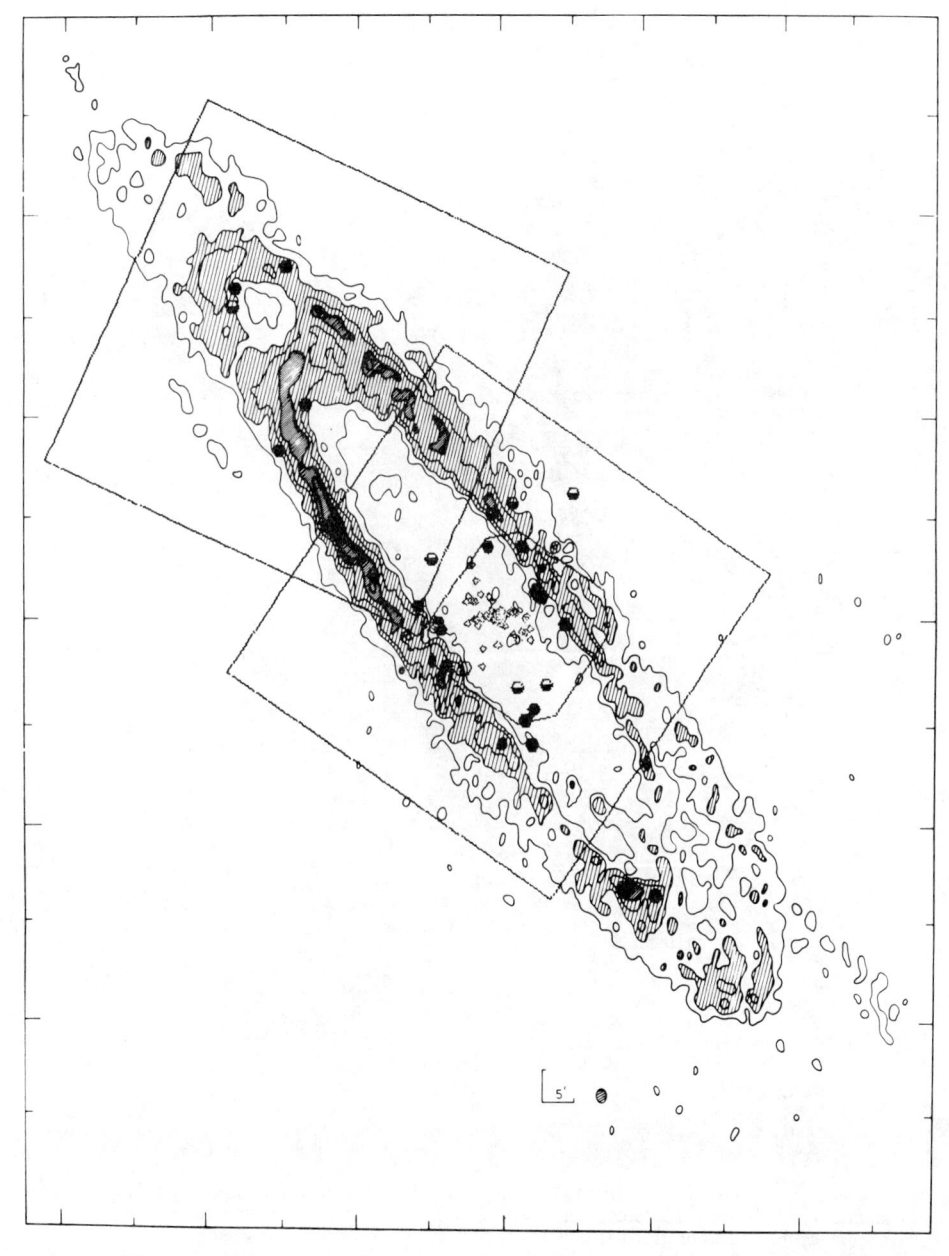

FIG. 4.—X-ray source positions superposed on the H I map of Emerson (1976). The larger symbols are IPC detections. Filled IPC symbols probably are Population I objects. Half-filled symbols are IPC sources located near globular clusters, but with position errors such that about one-half are expected to be spurious associations. The open IPC symbol is identified with a globular cluster through HRI observations. The correlation between H I emission and Population I X-ray sources is apparent.

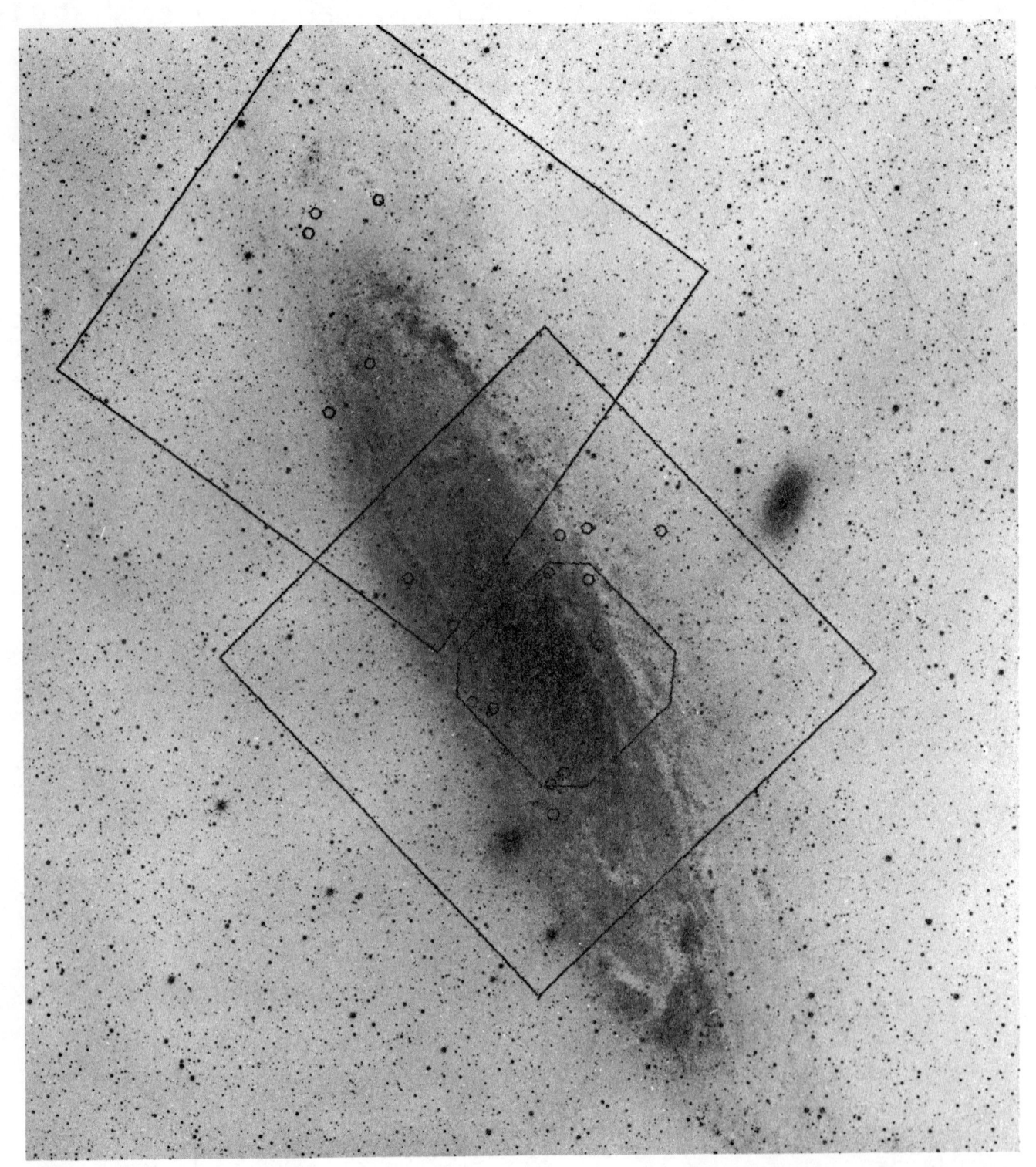

FIG. 3.—X-ray source positions superposed on an optical photograph of M31. The larger symbols are IPC detections with typical position errors of 1'. The smaller symbols are HRI detections; most of these are contained in the overexposed central region. The outer sources are associated with spiral-arm features or globular clusters. The detector fields are indicated.

VAN SPEYBROECK et al. (see page L47)

superposed on a Hale Observatory photograph in Figure 5 (Plate L14). The X-ray sources include a group concentrated around the nucleus and outlying sources which tend to be associated with inner spiral-arm features. We have divided these sources into Population I and inner bulge, placing a source into Population I if any dust or spiral-arm features are observed at the source position. This separation results in 18 inner bulge point sources including the nucleus candidate plus three diffuse or confused regions located within $2'$ of the nucleus, or 400 pc, assuming a spherical distribution so that projection effects can be ignored.

The average M31 inner bulge source luminosity is 4.5×10^{37} ergs s^{-1}, whereas the average Population I source luminosity is 2.8×10^{37}; both of these results are biased toward higher values by our threshold of approximately 9×10^{36} ergs s^{-1}, but, as discussed previously, there is evidence that most of the X-ray luminosity is provided by the observed sources in the nuclear regions, and so a lower threshold, if anything, would tend to increase the separation of the two source groups. The observed difference is not a result of a lower threshold in the IPC which contributes only to Population I sources; the average luminosity of the 29 HRI Population I point sources is 2.1×10^{37} ergs s^{-1}. The optical bulge of M31 is larger than our effective radius of 400 pc; Morton, Andereck, and Bernard (1977), for example, quote 1000 pc. There are 16 sources we call Population I inside of 1000 pc; this is higher than the typical density of sources in the disk, but the average luminosity of this group is 2.2×10^{37} ergs s^{-1}, which is slightly less than the average Population I luminosity found outside the optical bulge and therefore supports our present assignment of these to Population I.

The luminosity distributions for these groups of sources and the total source set are plotted in Figure 6. We have binned the luminosities to facilitate comparisons with the distributions found by Clark et al. (1978) for the luminosities of sources in our galaxy and the Magellanic Clouds. The M31 results are quoted for the 0.5–4.5 keV band, whereas Clark et al. (1978) chose the 2–11 keV interval, and so some care must be taken in comparing the data sets. The absolute values of the bulge source luminosities observed in our Galaxy and M31 are quite similar; Clark et al. (1978) find six bulge sources in our Galaxy between our M31 threshold and 3.2×10^{37}, and 10 between 3.2×10^{37} and 3.2×10^{38}. We find 11 point sources in the lower interval and seven in the upper interval. The spatial distribution of these sources, however, is very different. The sources we consider to be inner bulge objects in M31 are contained in a 400 pc radius, which would correspond to $2°.3$ at the position of our galactic nucleus, whereas the sources considered bulge objects in our Galaxy span 40° in longitude. There are only three sources listed in the fourth *Uhuru* catalog (Forman et al. 1978) within $2°.3$ of the nucleus, and one of these is transient, compared to 18 point sources and three diffuse or confused sources in this region of M31. The difference is unlikely to be due to the bulge sources in

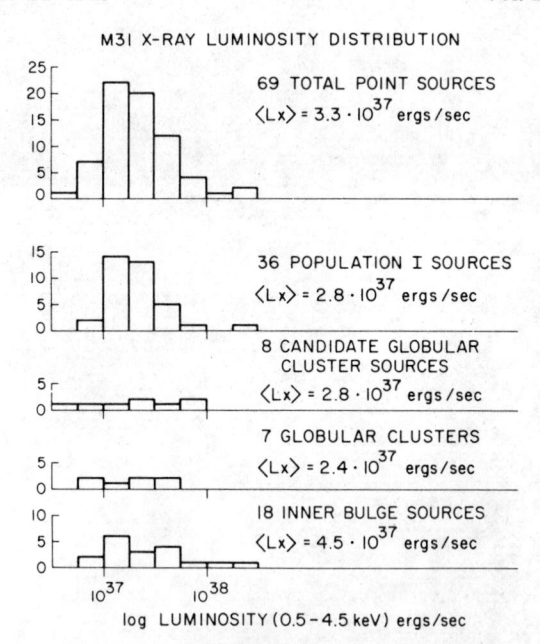

FIG. 6.—X-ray luminosity (0.5–4.5 keV) distributions for various categories of X-ray sources in M31.

M31 being so soft that they would not be detectable near our own galactic center; the best-fit temperature to the aggregate emission from this region detected with the *Einstein* IPC is 5 keV, which is high enough so that the ~ 3 keV absorption cutoff to our own galactic center would not have precluded observation of such sources in our Galaxy. We therefore conclude that the X-ray source density near the center of M31 is far greater than that of our own Galaxy. Clark et al. (1978) find 26 sources in our Galaxy with luminosities above their completeness threshold of about 5×10^{36} ergs s^{-1}; we find 76 in M31 and can expect about 80 when the southwest field is added, or a factor of 3.1. This is only slightly higher than the mass ratio of 2.4 found by van den Bergh (1975). Our ratio of the number of sources called bulge and Population I is about (1:2), which is quite different from the ratio (2.7:1) found by Clark et al. (1978), and there is a question of whether this results from the different criteria used or a difference between the two galaxies. Markert et al. (1977) and Jones (1977) argue that the bulge sources in our Galaxy can be distinguished from the disk population by softer spectra as well as higher luminosities. Our division is based upon morphology, but we also find higher luminosity sources in our inner bulge group. Clearly, some sources could be misclassified in either survey, but it is likely that the ratio of bulge to disk X-ray sources is lower in M31 than in the Galaxy.

The 400 pc radius defining the X-ray inner bulge group is near that where Rubin, Ford, and Kumar (1973) find a local minimum in the emission-line velocities (see, however, Peterson 1978). Rubin and Ford (1971) also find that the gas motions within this region show a complex velocity field superposed on the rapid rotation, with evidence for expansion at velocities up to

PLATE L14

FIG. 5.—X-ray source positions superposed on a Hale Observatory photograph of the central region of M31. The dense group of sources categorized as inner bulge are located within 2' of the nucleus; the detector field opposite boundaries are approximately 25' apart and can be used as a scale.

VAN SPEYBROECK *et al.* (*see* page L48)

100 km s^{-1}. This ~400 pc radius region thus may constitute a distinct structure within M31. Rubin and Ford (1970) estimate that only 0.015 of the mass of M31 is contained within 400 pc, and most of that is low-mass, late type stars, and yet we find that this region is responsible for one-third of the X-ray emission from M31.

Ten of the 29 HRI Population I sources are associated with bright visible objects. The actual number of such systems must be considerably higher because much of this region is obscured optically. We plan to observe many of these objects optically during the coming months. There also appears to be an association of X-ray sources with dust-lane boundaries; we have not established if this is statistically significant; if we are able to establish this, then it will be reasonable to conclude that these objects are much closer to their places of origin than expected from typical models of their evolution (e.g., van den Heuvel 1978).

V. SUMMARY

We have observed 69 X-ray sources in M31 which appear pointlike at our resolution, and seven additional diffuse or confused regions. We find that the sources can be divided into inner bulge, globular cluster, and Population I objects; these contribute about 0.33, 0.11, and 0.56 of the X-ray flux, respectively. The inner bulge sources in M31 are more luminous than the Population I or globular cluster sources. The Population I sources are associated with bright visible objects, dust, neutral hydrogen, and other Population I tracers. The number and luminosity of the inner bulge X-ray sources in M31 are similar to those observed in the bulge of our own Galaxy, but the spatial extent is much less, those in M31 being contained within 400 pc of the nucleus. The nucleus of M31 is likely to be a 10^{38} ergs s^{-1} X-ray source, which is not unusual, but still a factor of 10^3 brighter than the nucleus of our own Galaxy.

We have benefited from many discussions with Harvey Tananbaum, Ethan Schreier, and Josh Grindlay. J. Bechtold performed most of the optical plate measurements. We are grateful to R. Kirshner and S. D'Odorico and M. Dopita for informing us of their results prior to publication, to W. Sargent and M. Liller for providing optical plate material, and to J. Baldwin for providing the H I map used in Figure 4. Two of us (C. J. and W. F.) thank Dr. Ripley of the Smithsonian Institution for providing travel funds from the secretary's fluid research fund for optical observations.

REFERENCES

Baade, W., and Arp, H. 1964, *Ap. J.*, **139**, 1027.
Berkhuijsen, E. M. 1977, *Astr. Ap.*, **57**, 9.
Clark, G., Doxsey, R., Li, F., Jernigan, F. G., and van Paradijs, J. 1978, *Ap. J. (Letters)*, **221**, L37.
Cooke, B. A., *et al.* 1978, *M.N.R.A.S.*, **182**, 489.
Coombes, F., Encrenaz, P. J., Lucas, R., and Weliachew, L. 1977a, *Astr. Ap.*, **55**, 311.
———. 1977b, *Astr. Ap.*, **61**, L7.
D'Odorico, S., Dopita, M., and Benvenuti, P. 1979, *Astr. Ap. Suppl.*, in press.
Emerson, D. T. 1974, *M.N.R.A.S.*, **169**, 607.
———. 1976, *M.N.R.A.S.*, **176**, 321.
———. 1978, *Astr. Ap.*, **63**, L29.
Epstein, A., *et al.* 1979, in preparation.
Forman, W., Jones, C., Cominsky, L., Julien, P., Murray, S., Peters, G., Tananbaum, H., and Giacconi, R. 1978, *Ap. J. Suppl.*, **38**, 357.
Giacconi, R., *et al.* 1979, *Ap. J.*, **230**, 540.
Grindlay, J. E. 1979, private communication.
Jones, C. 1977, *Ap. J.*, **214**, 856.
Kirshner, R. 1979, in preparation.
Margon, B., Bowyer, S., Cruddace, R., Heiles, C., Lampton, M., and Troland, T. 1974, *Ap. J. (Letters)*, **191**, L117.
Markert, T. H., Canizares, C. R., Clark, G. W., Hearn, D. R., Li, F. K., Sprott, F., and Winkler, F. 1977, *Ap. J.*, **218**, 801.
Morton, D. C., Andereck, C. D., Bernard, D. A. 1977, *Ap. J.*, **212**, 13.
Peterson, C. J. 1978, *Ap. J.*, **221**, 80.
Reber, G. 1940, *Ap. J.*, **91**, 621.
Rodriguez, L., and Chaisson, E. 1979, *Ap. J.*, **228**, 734.
Rubin, V., and Ford, W. K., Jr. 1970, *Ap. J.*, **159**, 379.
———. 1971, *Ap. J.*, **170**, 25.
Rubin, V., Ford, W. K., Jr., and Kumar, C. K. 1973, *Ap. J.*, **181**, 61.
Sargent, W. L. W., Kowal, C. T., Hartwick, F. D. A., van den Bergh, S. 1977, *A.J.*, **82**, 947.
van den Bergh, S. 1975, *Ann. Rev. Astr. Ap.*, **13**, 219.
van den Heuvel, E. P. J. 1978, in *Proc. International School of Physics "Enrico Fermi," Course LXV*, ed. R. Giacconi and R. Ruffini (Amsterdam: North-Holland), p. 828.
van der Kruit, P. C., and Allen, R. J. 1976, *Ann. Rev. Astr. Ap.*, **14**, 417.
van der Kruit, P. C. 1972, *Ap. Letters*, **11**, 173.

A. EPSTEIN, W. FORMAN, R. GIACCONI, C. JONES, W. LILLER, and L. VAN SPEYBROECK: Harvard-Smithsonian Center for Astrophysics, 60 Garden Street, Cambridge, MA 02138

L. SMARR: Department of Astronomy, University of Illinois, Urbana, IL 61801

Reprinted with permission from *Monthly Notices of the Royal Astronomical Society*, 175, *Short Communications*, pp. 29P-34P, R.J. Mitchell, J.L. Culhane, P.J.N. Davison, and J.C. Ives, "Ariel 5 Observations of the X-Ray Spectrum of the Perseus Cluster." © 1976 Blackwell Scientific Publications Ltd.

Mon. Not. R. astr. Soc. (1976) 175, Short Communication, 29P–34P.

ARIEL 5 OBSERVATIONS OF THE X-RAY SPECTRUM OF THE PERSEUS CLUSTER

R. J. Mitchell, J. L. Culhane, P. J. N. Davison and J. C. Ives

Mullard Space Science Laboratory, Department of Physics and Astronomy, University College London, Holmbury St Mary, Dorking, Surrey

(Received 1976 February 12)

SUMMARY

An X-ray spectrum of the Perseus Cluster in the energy range 1·3–16 keV has been obtained with the MSSL collimated proportional counter on *Ariel 5*. An emission feature has been detected at about 7 keV of strength $0·0035 \pm 0·0004$ photon cm^{-2} s^{-1} (equivalent width 360 ± 50 eV). The existence of this feature, which is due to Fe XXV and Fe XXVI transitions, provides strong evidence for the presence of hot plasma in the cluster. In addition the overall spectrum is well described by the bremsstrahlung emitted from an adiabatic hydrostatic atmosphere of hot gas in the gravitational potential well of the cluster.

INTRODUCTION

Following their discovery by the *Uhuru* satellite (Gursky *et al.* 1971; Forman *et al.* 1972), the extended sources of X-ray emission associated with clusters of galaxies have been studied in some detail. The two types of emission mechanism which have been proposed are thermal bremsstrahlung from a hot isothermal gas sphere (e.g. Lea *et al.* 1973) and the inverse Compton radiation from the interaction of the 3 K microwave background photons with a population of relativistic electrons in the cluster (e.g. Brecher & Burbidge 1972). Observations of the X-ray spectra of these clusters, while capable in principle of allowing a choice between the two emission mechanisms, have not yet permitted definite conclusions to be drawn although there has been some evidence in favour of a thermal origin for the X-rays (Solinger & Tucker 1972).

We report here a new measurement of the X-ray spectrum of the Perseus Cluster source. The spectrum which covers the energy range 1·3–16 keV, was obtained by the MSSL proportional counter (experiment C) on *Ariel 5*. There is good evidence for the presence of an emission feature at about 7 keV. In addition it is not possible to provide a good fit to the observed data in this energy range with either a simple thermal or a simple power law spectrum. However, the observed spectrum agrees well with that predicted by the adiabatic gas sphere model of Gull & Northover (1975).

THE OBSERVATIONS

The detector system and its mode of operation were essentially as described by Stark, Davison & Culhane (1976). Spectra are obtained with the aid of an on-board 32-channel pulse height analyser (PHA) which may be operated in either of two gain modes so as to cover the nominal energy band 1·3–25 keV with a region of overlap from 3 to 13 keV. Both gain modes were used in obtaining the spectrum

30P   *R. J. Mitchell* et al.

presented below. PHA counts are integrated on-board the spacecraft for a complete orbit, giving integration times of 30–40 min. The Perseus Cluster was observed between 1975 September $5^d\ 15^h\cdot 9$ and 1975 September $7^d\ 12^h\cdot 0$, for a total of 27 satellite orbits.

The high and low gain data were fitted separately by both thermal and power law spectra. Detector resolution and escape effects were taken account of in the fitting programme. The power law expression was of the usual form while the exponential thermal spectrum included a Gaunt factor $\bar{g}$ for which we used the approximation:

$$\bar{g} = 0.9 \left(\frac{kT}{E}\right)^{(0.3+E/200)}. \qquad (1)$$

The average absorption coefficients of Brown & Gould (1970) were used to represent the interstellar medium. The data did not show a measurable interstellar column density, so its value was fixed at $N_H = 5 \times 10^{20}$ atom cm^{-2} which is consistent with both our data and with the column obtained from 21-cm radio observations for the line of sight to the Perseus Cluster.

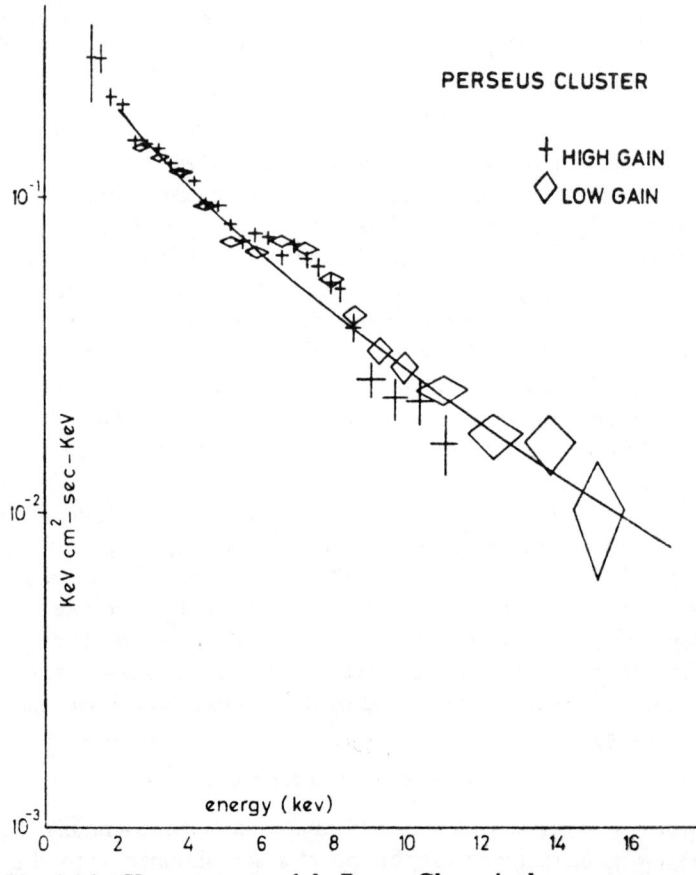

FIG. 1. *The Ariel 5 X-ray spectrum of the Perseus Cluster in the energy range* $1\cdot 3$–$16\ keV$. *The two detector gain modes are selected by ground command. The solid line represents the computed continuous spectrum from a Gull & Northover adiabatic gas sphere with central temperature* $(T(0))$ $32\ keV$ *and* $T_\infty = -4\ keV$. *The emission feature at around* $7\ keV$ *is visible in both gain modes.*

Ariel 5 observations of the Perseus Cluster

The high and low gain spectral parameters differ significantly from each other, indicating that a simple thermal or power law spectrum cannot adequately represent the data over the entire range of observations. Data points representing the energy flux in each channel are shown in Fig. 1, from which the complex nature of the spectrum is apparent. There is good evidence for the presence of an emission feature at around 7 keV. In addition the emission from a hot gas with components at a number of different temperatures is required to explain the data over the 1·3–16 keV range.

The significance of the emission feature may be judged from Fig. 2 where we plot the deviation of the flux in each channel from that given by the best fitting simple continuous spectrum for the 3–12 keV range. The high gain data show an

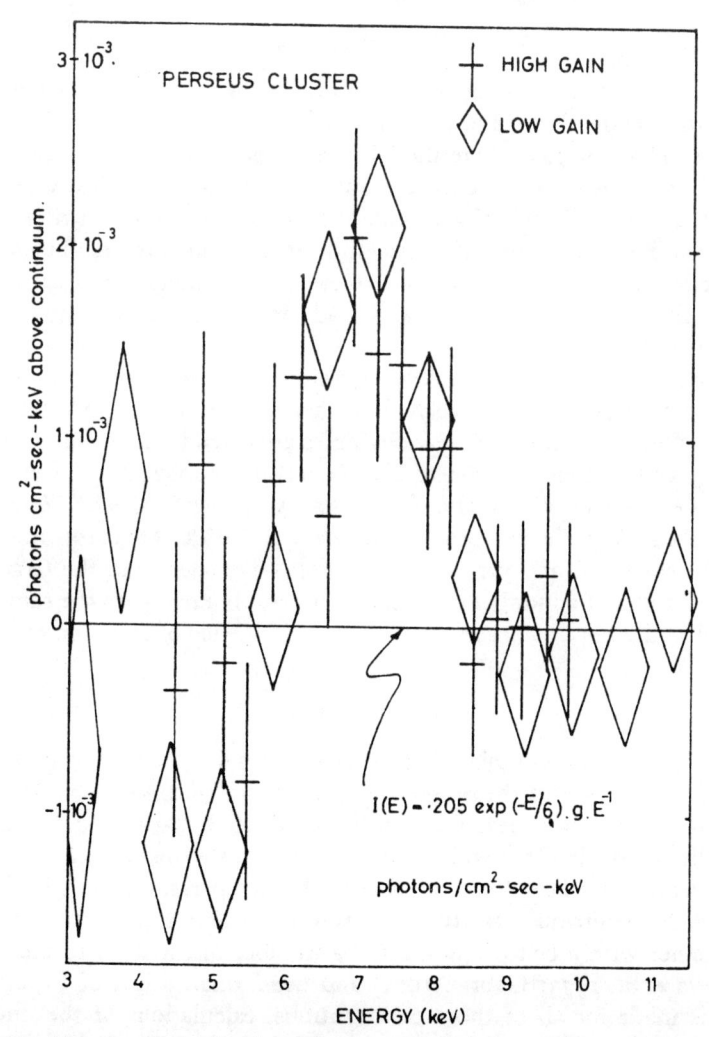

FIG. 2. *The deviation of the flux in each energy channel from that predicted by the best-fitting single temperature continuum is plotted for the range 3–11 keV. Data from both gain modes show a systematic departure from the continuum in the energy range around 6·9 keV. The width of the feature is consistent with that of the gaussian energy resolution function for the detector at 7 keV. Each gain mode shows departures from continuum that are significant at the 5·6 and 6 σ levels.*

excess flux of 3.15×10^{-3} photon cm^{-2} s^{-1}, which is $5.4\,\sigma$ above the continuum. The low gain data give a $6\,\sigma$ signal, with an intensity of $3.56 \cdot 10^{-3}$ photon cm^{-2} s^{-1}. These two independent measurements average to a value of $3.35 \pm 0.4 \times 10^{-3}$ photons cm^{-2} s^{-1} for the flux in the line feature, the line being $7.5\,\sigma$ above the continuum. The corresponding equivalent width is 360 ± 50 eV.

A number of possible continuous spectra were computed following the model of Gull & Northover (1975, GN). This model describes the adiabatic hydrostatic atmospheres of hot gas in the gravitational potential wells of clusters of galaxies. We computed composite thermal continuous spectra in terms of the (GN) parameters $T(0)$, T_∞ and T_0. Here, $T(0)$ is the central temperature of the cluster, T_∞ is a constant of integration whose value and sign specify whether the hot gas is bound to the cluster or extends to infinity and

$$T_0 = T(0) - T_\infty. \qquad (2)$$

A somewhat similar description of the nature of the hot gas in clusters has been published independently by Lea (1975).

The best adiabatic gas sphere model is represented by the solid line in Fig. 1. This computed spectrum, which has parameter values $T_0 = 28$ keV and $T(0) = 24$ keV (i.e. $T_\infty = -T_0/7$) keV was obtained by summing the contributions to the emission from regions of the cluster gas at different temperatures. The summation was performed for 10^6 K temperature intervals. An integration of the observed continuum flux in the 1.5–6 keV band leads to a *measured* emission feature to continuum ratio of 0.02.

Although it is difficult to resolve emission features in proportional counter detectors because of their relatively poor energy resolution, 7 keV features have been detected in three other sources. Examples are given in the work of Serlemitsos *et al.* (1975) and Sanford, Mason & Ives (1975) on the Cyg X-3 spectrum in the work of Serlemitsos *et al.* (1973) and Davison, Culhane & Mitchell (1976) on the spectrum of Cas A while the latter authors have also detected iron line emission in the X-ray spectrum of Tycho's SN. *Ariel 5* X-ray spectra of the Crab Nebula, of Cygnus X-1 and of a number of weaker sources, obtained with the same detector as was used in the Perseus observation show no evidence of emission features.

DISCUSSION

The detection of an emission feature at about 7 keV in the Perseus spectrum provides good evidence for the presence of hot plasma in this source. Although the spectrum in Fig. 1 is best described by the adiabatic hot gas model a statistically acceptable fit is provided at energies above 4 keV by a thermal continuous spectrum at a temperature of 6 keV (66×10^6 K). At this temperature, the 7 keV emission feature includes contributions from the resonance and forbidden transitions of Fe XXV together with a contribution from a number of satellites to the resonance line (Grineva *et al.* 1973; Gabriel 1972) and the Lyman-α line of Fe XXVI. When allowance is made for all of these contributions, calculations of the line to continuum ratio for a 60–70×10^6 K plasma (Mewe 1972; Tucker & Koren 1971) suggest that the ratio should be $\simeq 0.07$ for a cosmic iron abundance ($N(\text{Fe})/N(\text{H})$) of 5×10^{-5}. Thus the observed value of 0.02 would indicate that iron is underabundant by a factor of 4 for the entire cluster source. However, the adiabatic gas sphere model suggests that the hot intracluster plasma is mainly intergalactic material which has fallen into the cluster potential well. If this is so, it would be rather surprising to find significant concentrations of heavy elements in such material. It is possible that active galaxies in the cluster, such as NGC 1275, could enrich the gas in their immediate neighbourhood with iron and other heavy elements. While it is difficult to estimate how much of the intracluster gas has been so enriched, it is possible to consider an extreme case in which the iron line emission originates only in the neighbourhood of NGC 1275. Earlier work (Fabian *et al.* 1974; Wolff *et al.* 1976) has shown that about 10–20 per cent of the total flux from the cluster is associated with NGC 1275. If only the hot material near the galaxy is enriched, then the iron abundance in the neighbourhood of NGC 1275 could be up to 2·5 times greater than the cosmic value assumed above.

It was remarked earlier that simple thermal or power law spectra are unable to fully explain the observations which require the presence of gas having a range of temperatures. It will be of interest to correlate the presence or absence of active galaxies and the distribution of galaxies in the cluster with the X-ray spectral features observed and with the shape of the continuous spectrum. A more detailed discussion of these aspects for a number of the brighter X-ray cluster sources is in preparation. However, high spatial resolution maps of the X-ray cluster sources with sufficient spectral resolution to detect emission lines will ultimately be required if we are to separate the emission associated with individual active galaxies from that of the cluster as a whole.

ACKNOWLEDGMENTS

We are grateful to Professor R. L. F. Boyd, CBE, FRS, for his support and encouragement during this project. We also thank the staffs of MSSL and of the Appleton Laboratory *Ariel 5* control centre for assistance with data reduction and in-flight operation. We acknowledge helpful discussions with Dr S. Gull and Dr K. Northover prior to the publication of their work. RJM acknowledges an SRC research studentship.

REFERENCES

Brecher, K. & Burbidge, G. R., 1972. *Nature*, **237**, 440.
Brown, R. L. & Gould, R. J., 1970. *Phys. Rev. D*, **1**, 2252.
Davison, P. J. N., Culhane, J. L. & Mitchell, R. J., 1976. *Astrophys. J. (Letters)*, in press.
Fabian, A. C., Zarnecki, J. C., Culhane, J. L., Hawkins, F. J., Peacock, A., Pounds, K. A. & Parkinson, J. H., 1974. *Astrophys. J. (Letters)*, **189**, L59.
Forman, W., Kellogg, E., Gursky, H., Tananbaum, H. & Giacconi, R., 1972. *Astrophys. J.*, **178**, 309.
Gabriel, A. H., 1972. *Mon. Not. R. astr. Soc.*, **160**, 99.
Grineva, Yu. I., Karev, V. I., Korneev, V. V., Krutov, V. V., Mandelstam, S. L., Vainstein, L. A., Vasilyer, B. N. & Zhitnick, I. A., 1973. *Sol. Phys.*, **29**, 441.
Gull, S. F. & Northover, K. J. E., 1975. *Mon. Not. R. astr. Soc.*, **173**, 585.
Gursky, H., Kellogg, E., Murray, S., Leong, C., Tananbaum, H. & Giacconi, R., 1971. *Astrophys. J. (Letters)*, **167**, L81.
Lea, S. M., 1975. *Astrophys. Lett.*, **16**, 141.
Lea, S. M., Silk, J., Kellogg, E. & Murray, S., 1973. *Astrophys. J. (Letters)*, **184**, L105.
Mewe, R., 1972. *Sol. Phys.*, **22**, 439.
Sanford, P. W., Mason, K. O. & Ives, J., 1975. *Mon. Not. R. astr. Soc.*, **173**, 9P.
Serlemitsos, P. J., Boldt, E. A., Holt, S. S., Ramaty, R. & Brisken, A. F., 1973. *Astrophys. J. (Letters)*, **184**, L1.
Serlemitsos, P. J., Boldt, E. A., Holt, S. S., Rothschild, R. E. & Saba, J. L. R., 1975. *Astrophys. J.*, **201**, L9.

Solinger, A. B. & Tucker, W. H., 1972. *Astrophys. J.*, **175**, L107.
Stark, J. P., Davison, P. J. N. & Culhane, J. L., 1976. *Mon. Not. R. astr. Soc.*, **174**, 35P.
Tucker, W. H. & Koren, M., 1971. *Astrophys. J.*, **168**, 283.
Wolff, R. S., Mitchell, R. J., Charles, P. A. & Culhane, J. L., 1976. *Astrophys. J.*, in press.

Reprinted with permission from *The Astrophysical Journal*, 234 pp. L21-L25, C. Jones, *et al.*, "The Structure and Evolution of X-Ray Clusters." © 1979 The American Astronomical Society.

THE STRUCTURE AND EVOLUTION OF X-RAY CLUSTERS

C. Jones, E. Mandel, J. Schwarz, W. Forman, S. S. Murray, and F. R. Harnden, Jr.

Harvard-Smithsonian Center for Astrophysics, Cambridge, Massachusetts

Received 1979 June 27; accepted 1979 August 3

ABSTRACT

Observations with the imaging proportional counter on the *Einstein* Observatory have shown a variety of structures in the X-ray surface-brightness distribution of nearby, rich clusters of galaxies. The nature of the X-ray emission ranges from broad, highly clumped radiation to smooth and centrally peaked radiation. The clusters in which the emission is clumped around individual galaxies tend to be rich in spiral galaxies and to have low X-ray temperatures or two-component spectra and low velocity dispersions. The smooth, centrally peaked clusters are spiral-poor and have high temperatures and large velocity dispersions. The smooth clusters can be subdivided into two classes based on their isothermal core radii. Among the clusters we observed, the surface-brightness distribution of those with core radii $\sim$0.25 Mpc is sharply peaked around a centrally located dominant galaxy, while the emission from other smooth clusters shows a broader distribution (core radii $\sim$0.5 Mpc). We interpret these observations in the context of dynamic cluster evolution. In this interpretation, the broad, highly clumped clusters are in their early evolutionary phases, while the smooth, centrally peaked clusters are in later stages.

Subject headings: galaxies: clusters of — X-rays: sources

I. INTRODUCTION

The first observations of extended X-ray emission associated with rich clusters of galaxies were reported by Gursky *et al.* (1971), Forman *et al.* (1972), and Kellogg *et al.* (1972). Nearby clusters represent the largest class of identified extragalactic objects in both the *Ariel 5* Catalog (Cooke *et al.* 1978) and the Fourth *Uhuru* Catalog (Forman *et al.* 1978a).

With the successful launch of the *Einstein* Observatory, we have used the imaging capabilities of the telescope and the imaging proportional counter (IPC) to study the structure of the X-ray emission from 12 of the nearby rich clusters of galaxies. Although we have planned an extensive program of cluster observations, these first X-ray observations have shown new features in the surface-brightness distributions which have important consequences for our understanding of clusters of galaxies.

II. OBSERVATIONS

Observations of selected clusters have been made with the *Einstein* IPC, described by Giacconi *et al.* (1979a). We have analyzed data only in the energy range from 0.25 to 3.0 keV to avoid contamination by a flight calibration source at higher energies and the enhanced soft X-ray background at low energies.

In Figure 1 (Plate L6) we show the X-ray images for 12 clusters. The projected cluster profiles range from broad to centrally peaked. This figure also shows that the cluster emission is not always spherically symmetric.

Figure 2 (Plate L7) shows iso-intensity contour plots of the X-ray emission superposed on the optical PSS fields for each cluster. From these observations we can group the clusters into four categories based on their X-ray morphology. The X-ray emission from A1367, A2147, and A2634 is broad and highly clumped around individual galaxies. A strikingly different class of clusters includes A85, A478, A1413, A1775, A1795, and A2063, whose X-ray distributions are relatively smooth and sharply peaked around a dominant galaxy at the center of the cluster. The emission from A2256 and A2319 is centrally enhanced and smooth, but less peaked than the second category. A2666 represents the fourth class of cluster emission, that of a cD galaxy in a poor cluster or group. Schwartz *et al.* (1979) have discussed other examples of this type of cluster emission. Although A2666 was denoted as a rich cluster by Abell (1958), Butcher and Oemler (1978) have found that it has too few galaxies to be classified as such. Our observations show weak X-ray emission (1.9 $\times$ 10^{42} ergs s^{-1}) from a region around the cD galaxy which is consistent with a point source (core radius <1$'$.5).

We have measured the radial surface-brightness distribution of the X-rays from these clusters. Figure 3 shows the distribution for eight clusters. To characterize the X-ray extent and to allow for a comparison with optical studies, we have measured the core radii using the surface-brightness distribution from a King approximation to an isothermal sphere model for clusters (Lea *et al.* 1973), $S \sim [1 + (r/r_{\text{core}})^2]^{-5/2}$. However, the emission from several of the clusters is clearly elongated and cannot be adequately characterized by an isothermal sphere. Therefore, for each cluster we also have measured its greatest extent and the extent perpendicular to this maximum by finding the FWHM of narrow slices of emission through the cluster center. The ratio

PLATE L6

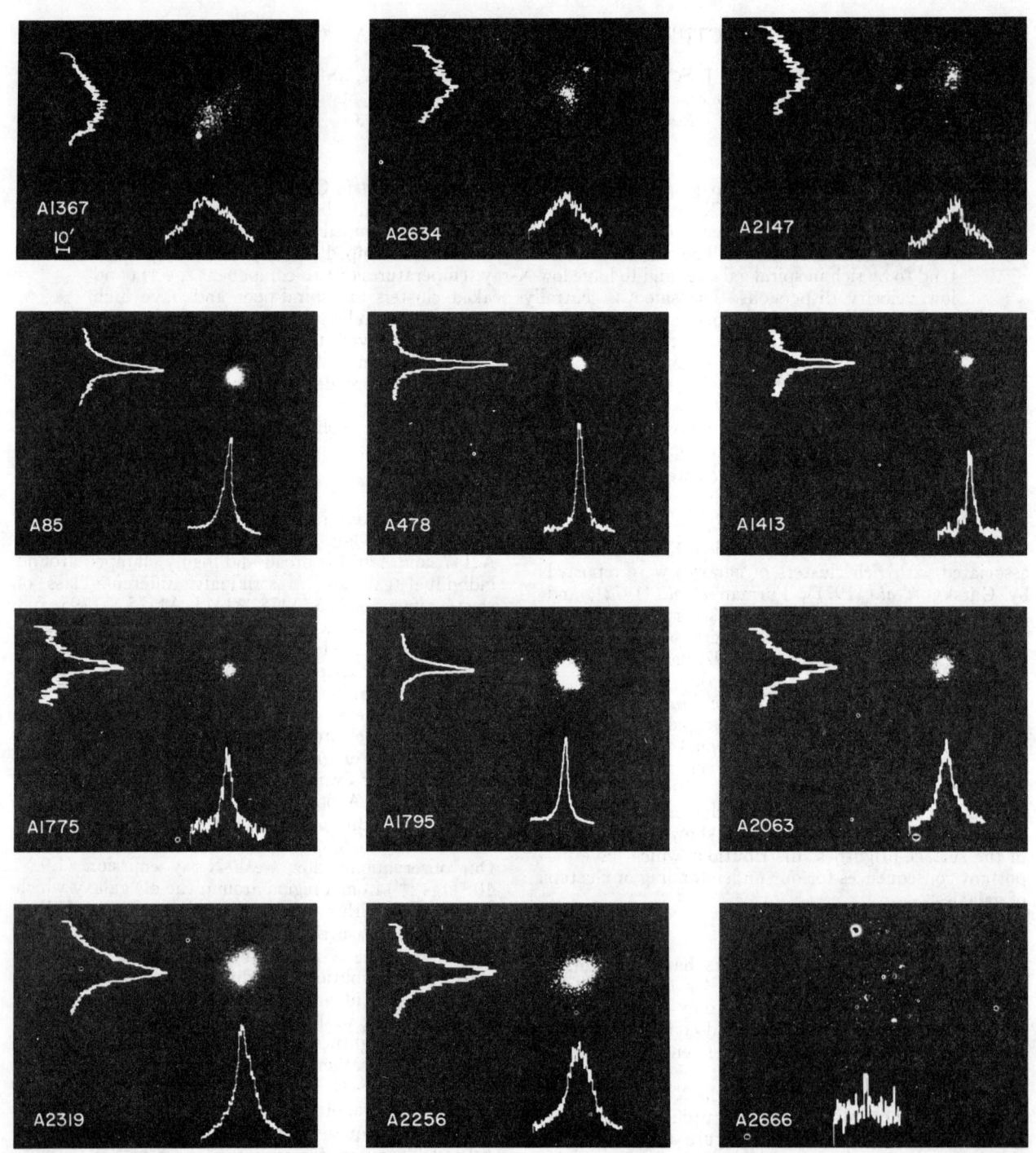

FIG. 1.—IPC images for the 12 clusters (A85, A478, A1367, A1413, A1775, A1795, A2063, A2147, A2256, A2319, A2634, and A2666). We have projected the cluster X-ray emission onto the horizontal and vertical axes to illustrate better the cluster profile. For A2666, the projection was made only for the horizontal axis. The clusters are shown grouped into the four cluster categories discussed in the text.

JONES et al. (see page L21)

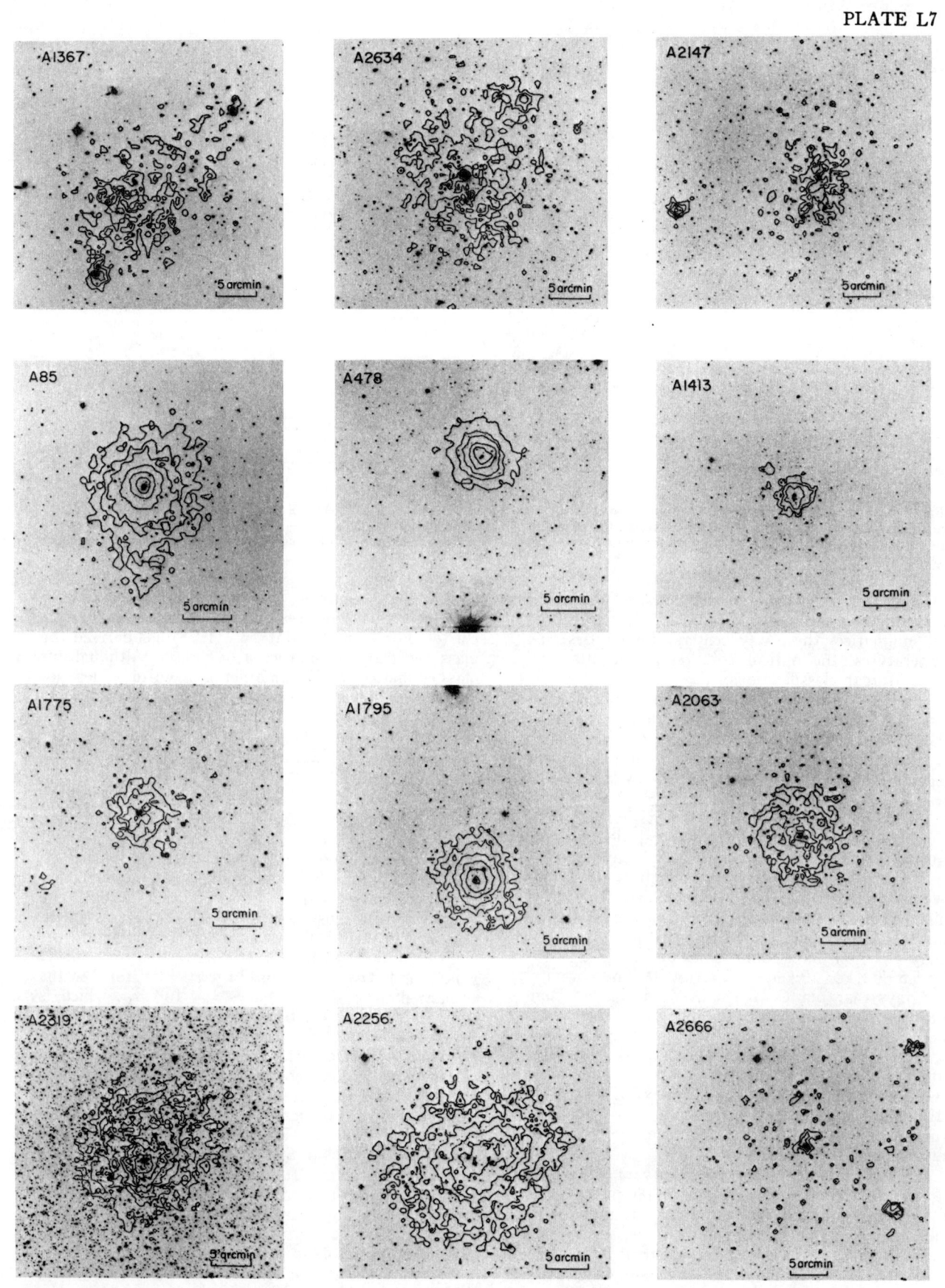

FIG. 2.—Iso-intensity contour plots of the X-ray emission are shown superposed on the optical PSS field for each cluster. The contour levels are arbitrary.

JONES et al. (see page L21)

L22　　　　　　　　　　　　　　　　JONES ET AL.

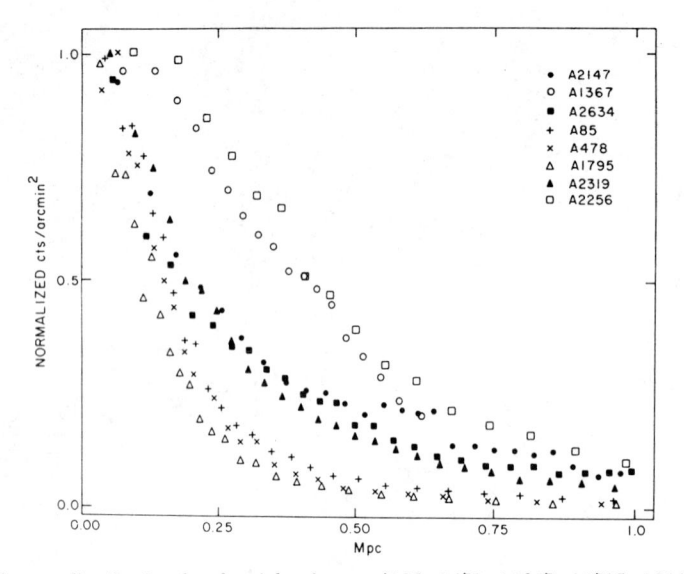

Fig. 3.—Radial surface brightness distribution for the eight clusters (A85, A478, A1367, A1795, A2147, A2256, A2319, and A2634). The angular extent, r (arcmin), has been converted to linear size, R (Mpc); $R(\text{Mpc}) = 1.75\, r\, z/(1+z)^2$ with $q_0 = 1$ and $H_0 = 50$ km s^{-1} Mpc^{-1} (Sandage 1961). In this graph we have normalized each cluster to the same peak central surface brightness to emphasize their different extents.

of the maximum and minimum extents, as well as the core radii, are given in Table 1 for each cluster. Table 1 also summarizes the X-ray centers, luminosities, and temperatures, the optical core radii, redshifts, and Bautz-Morgan classifications.

Individual galaxies are observed to be X-ray emitters in several clusters. The two radio galaxies, 3C 264 = NGC 3862 in A1367 and WE 1601+16W3 in A2147, are the brightest X-ray galaxies observed in their respective clusters. The 0.5–3.0 keV X-ray luminosities are $2.0 \pm 0.2 \times 10^{42}$ ergs s^{-1} and $3.7 \pm 0.4 \times 10^{42}$ ergs s^{-1} for 3C 264 and WE 1601+16W3, respectively. The strong radio emission from these galaxies suggests unusual activity, and therefore, higher X-ray luminosities than for the other cluster members. In A1367 several other galaxies are associated with regions of enhanced X-ray emission. The two galaxies NGC 3842 and NGC 3841, which do not lie near the center of the X-ray emission, have luminosities of $3.4 \pm 1.3 \times 10^{41}$ and $5.5 \pm 1.4 \times 10^{41}$ ergs s^{-1}, about 100 times brighter in X-rays than our Galaxy or Andromeda, but comparable in X-ray luminosity to M86 in the Virgo cluster (Forman et al. 1979). In a recent radio study of A1367, Gavazzi (1978) found weak radio emission associated with NGC 3842, but no detectable radiation from NGC 3841. Although the spiral galaxy ZW 1141.2+2015, which is a few arcmin from NGC 3841, also was found to emit at radio frequencies, we observe no X-ray enhancement.

The similarity in X-ray luminosity of the A1367 galaxies and M86 suggests that the M86 phenomenon of extended emission around normal galaxies is not unique, and may be common in many clusters in early stages of their evolution. Other apparent X-ray fluctuations in A1367 do not correspond to bright galaxies. If there is a relation between the mass of the galaxy and its X-ray emission, these enhancements may indicate the presence of previously undetected mass in the cluster. For example, Mathews (1978) has derived the mass necessary to bind gas to a galaxy. Although his mass estimates vary by an order of magnitude, depending on various parameters including the surface-brightness profile, they can roughly indicate the relative masses for various galaxies. We apply the equation for the mass given by Mathews,

$$M = 2 \sim 3 \times 10^{13} \left(\frac{T}{3 \times 10^7 \text{ K}}\right)\left(\frac{R}{100 \text{ kpc}}\right) M_\odot,$$

to galaxies in different types of clusters. If we use the M86 parameters for core radius and temperature (Forman et al. 1979), the mass needed to bind the clumps in other clusters is $\sim 10^{12} M_\odot$. For clusters with centrally peaked X-ray emission we can estimate the mass around cD galaxies if we assume that it is the cD which binds the cluster gas. In particular, for A85 the estimated mass would be $2 \sim 3 \times 10^{14} M_\odot$, which is comparable to that determined by Mathews for M87.

For five of the brighter clusters we have measured the energy spectra within a 9' radius from the X-ray center. The Einstein energy range allows us to determine the low-energy absorption, but does not permit a reliable temperature measurement if the temperature is above a few keV. The spectra were measured in annuli with widths of 3' to ensure that less than 10% of the flux is scattered from any of the IPC pulse height channels. Since the deep survey (Giacconi et al. 1979b) and the cluster fields are all at high galactic latitudes, we have assumed that the deep survey X-ray background is applicable to the cluster fields.

Figure 4 shows the spectra for the clusters A85, A1795, and A1367 in detector counting rates. The

TABLE 1
X-Ray and Optical Parameters of Clusters

Cluster	X-Ray Center	z	X-Ray Luminosity[a] (ergs s^{-1})	X-Ray Core Radius	Ellipticity	Optical Core Radius	BM Class[LV]	Velocity Dispersion[FD]	X-Ray Temperature
A85	0^{h}39^{m}18^s.6, $-9°34'.1$	0.050[LV]	2.38×10^{44} 5.23×10^{44}	3'.25 (0.26 Mpc)	1.0±0.1	...	I	...	>8.5[JF]keV
A426	4 10 40.3, 10 21.0	0.0184[C] 0.09[BS]	7.17×10^{44} 1.07×10^{45}	15 (0.47)[KM] 2.0 (0.27)	1.0±0.1	7.2[B] (0.22 Mpc) 1.5±0.6[BS] (0.20)	II-III I[BS]	1396±140	6.8±0.1[M] >4.5[JF]
A478									
A1367	11 42 8, 20 1.7	0.0205[LV]	5.24×10^{42} 4.53×10^{43}	≥15.0 (0.52)	...	10[B] (0.34)	II-III	634[YV]	1.3 (+1.2, −0.6)[M]
A1413	11 52 42.0, 23 40.4	0.1426[K]	5.50×10^{44} 8.80×10^{44}	1.8 (0.33)	1.0±0.1	3.0[D] (0.57)	I	...	...
A1656	13 39 27.2, 26 37.5	0.023[LV]	1.26×10^{44} 2.97×10^{44}	16 (0.62)[KM] 3.2 (0.38)	1.0±0.1	6.6[B] (0.25) 2.4[B] (0.26)	II I	900±63	...
A1775		0.080[LV]							
A1795	13 46 33.6, 26 50.4	0.0630[LV]	5.01×10^{44} 8.52×10^{44}	2.2 (0.21)	1.1±0.1	2.6[B] (0.25)		...	...
A2063	15 20 37.2, 8 47.1	0.042[LV]	6.95×10^{43} 2.22×10^{44}	4.5 (0.30)	1.1±0.1	3.5[Ba] (0.24)	II	...	...
A2147	15 59 57.1, 16 5.4	0.0377[LV]	2.73×10^{43} 1.73×10^{44}	5.4 (0.33)	1.5±0.1	...	III	1120±150[M]	2±1 (+>12 keV)[P] component
A2256	17 6 44.7, 78 42.4	0.0603[FD]	1.54×10^{44} 5.32×10^{44}	5.96 (0.56)	1.7±0.1	2.1[B] (0.20)	II-III	1274 (+229, −280)	7.0 (+3.0, −2.0)[M]
A2319	19 19 38.0, 43 50.6	0.0549[LV]	2.29×10^{44} 7.15×10^{44}	5.3 (0.458)	1.2±0.1	2.5[B] (0.22)	II-III	1627 (+195, −244)[b]	12.5 (+7.0, −4.0)[M]
A2634	23 35 55, 26 47.1	0.0307[LV]	3.39×10^{42} 2.20×10^{43}	12.4 (0.63)	1.4±0.1	...	II	...	...
A2666	23 48 29, 26 52.6	0.0273[V]	1.87×10^{42}	≲1.5 (0.065)	...	...	I	...	...

[a] The upper figures are the X-ray luminosities within 3'.
The lower figures are the X-ray luminosities within 11'.

[b] Assuming only one cluster. If A2319B is separate, ΔV becomes = 873 (+131, −148).

REFERENCES.—LV = Leir and van den Bergh 1977. BS = Bahcall and Sargent 1977. JF = Jones and Forman 1978. C = Corwin 1974. FD = Faber and Dressler 1976. P = Pravdo et al. 1979. M = Mushotzky et al. 1978. B = Bahcall 1975. D = Dressler 1976. Ba = Baier 1976. V = Vidal and Peterson 1975. YV = Yahil and Vidal 1977. KM = Kellogg and Murray 1974. K = Kristian et al. 1978.

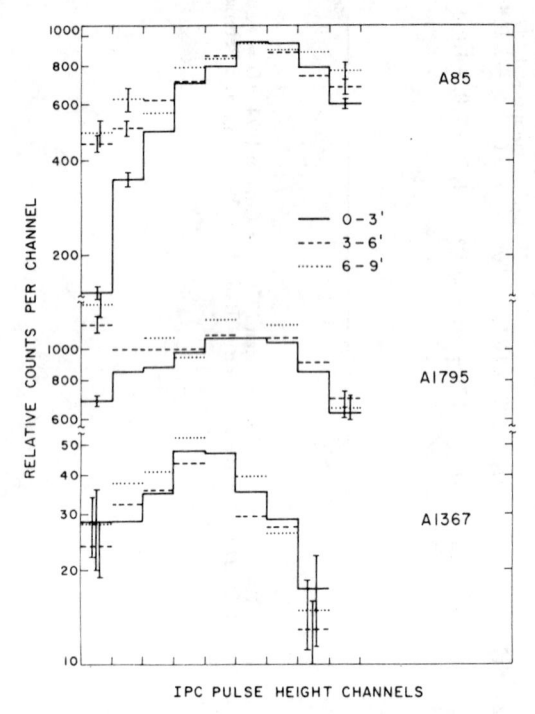

FIG. 4.—Spectra for the three clusters A85, A1795, and A1367 are shown in 3' annuli from the cluster center. The counts for each of the IPC pulse-height channels are shown for the energy range from 0.25 to 3 keV. Background has been subtracted but neither the response of the mirror nor that of the detector has been removed.

apparent increased absorption at the center of A85 may be interpreted as due to a temperature gradient and subsequent changes in line emission by a cool gas (approximately a few keV) contained in the potential of the cD, similar to the situation around M87 (Fabricant et al. 1978). Alternatively, the apparent cutoff may indicate that there is a contribution to the X-ray emission from the core of the cD galaxy. We also measured the spectra of A2256 and A2319 and found no significant changes with radius within the central 9'.

III. DISCUSSION

Our observations of clusters have shown that the structure of their X-ray emission can range from broad and highly clumped to smooth and centrally peaked. These observations can be interpreted using dynamic cluster evolution. Through numerical simulations, Peebles (1970), Aarseth (1969), and White (1976) have shown that a cluster which begins as a large cloud of galaxies will collapse and reach equilibrium with an extended halo around a dense core. Optical cluster studies have suggested that the "regular" clusters (Abell 1975) which have high central galaxy concentrations and a small fraction of spiral galaxies are dynamically more evolved than "irregular" clusters which are not centrally condensed and have a large spiral fractions (see Abell 1975, Bahcall 1977a, and Hausman and Ostriker 1978 for detailed discussions of correlations of cluster properties).

Broad, highly clumped cluster X-ray emission could be understood as a result of an early evolutionary stage, when the gas escaping the galaxies is bound more by the gravitational potential of the individual galaxies than by the relatively weak potential field of the cluster. If we use the velocity dispersion of the galaxies in the cluster as a measure of the gravitational potential, then the low velocity dispersions of the Virgo cluster and A1367 (Yahil and Vidal 1977) indicate a weak cluster field. Clusters in an early stage would be expected to have a lower density of hot intracluster gas, and therefore a higher fraction of spirals than in more evolved clusters in which ram pressure or evaporative processes would strip the galaxies of their interstellar matter, thereby transforming spirals into S0's (Gunn and Gott 1972; Gisler 1976; Cowie and Songaila 1977). Bahcall (1977b) has determined the percentage of spirals in A1367 to be 40%, which is considerably higher than the $\sim 10\%$ spirals she found in Coma and Perseus. It is interesting to note that A2147, whose radial distribution is less extended than that of A1367, has a spiral fraction of 27% (Bahcall 1977b) and may be intermediate between the first extended phase of cluster evolution and the compact stage. Detailed spectral observations of Virgo and A2147 (Mushotzky et al. 1978; Pravdo et al. 1979) show two-component X-ray spectra. *Einstein* imaging observations indicate that the cool gas in these clusters is clumped and probably surrounded by the hotter component (Forman et al. 1979).

As the cluster evolves, a high density core or subcluster is formed, thereby enhancing the chances for building a cD galaxy, either by dynamical friction, which causes bright galaxies to lose energy and spiral in toward the cD (Ostriker and Tremaine 1976), or by the tidal stripping of galaxy halos which fall into the cluster center and transform a centrally located galaxy into a cD (Richstone 1976). The clusters in our rather limited sample which show strongly peaked X-ray radiation all have dominant central galaxies. As shown in Table 1, these clusters have greater central surface brightnesses than do the less evolved clusters. This increased surface brightness is probably due to the higher gas density and temperature at the cluster center as the gas becomes bound by the cluster potential.

When the cluster reaches equilibrium, the X-ray emission should be smooth and should follow the cluster potential as traced by the galaxies. Observationally, the evolved clusters may be divided into two classes, one with the X-ray emission sharply peaked around a dominant galaxy, and the other with emission less peaked. The clusters in the first group have isothermal core radii of ~ 0.25 Mpc, while those in the second, which includes the Coma cluster, have core radii of ~ 0.5 Mpc. Their other X-ray and optical properties are similar. In general, their X-ray gas temperatures are high (~ 10 keV) and their surface brightness distribution is smooth. Optically they have high velocity dispersions and low spiral fractions.

Perrenod (1978) has suggested that as clusters evolve, their X-ray luminosity and temperature should increase as the cluster potential deepens. The *Einstein* observations, combined with the good correlations, between decreasing spiral fraction, increasing central density, and increasing X-ray luminosity found by Bahcall (1977b), support the predicted changes in luminosity. In addition, Jones and Forman (1978) showed that the X-ray temperatures of clusters generally increased with X-ray luminosity.

In summary, we have found from these first observations of the structure of clusters with the *Einstein* Observatory that the nature of the X-ray emission is complex and varies from one cluster to another. The clusters whose emission is clumped around individual galaxies or groups of galaxies tend to be rich in spiral galaxies and to have X-ray temperatures in the few kilovolt range or two-component spectra and low velocity dispersions. For many of the clusters, the emission is irregular and cannot be described by the simple, spherically symmetric models for a hot isothermal or adiabatic cluster gas. For these clusters, the low-density, intracluster gas is strongly influenced by the potential of individual galaxies. The smooth, centrally peaked clusters are spiral-poor, have higher temperatures (~ 10 keV), and larger velocity dispersions.

An additional consequence of these observations is that the previous very extended X-ray emission reported by Forman *et al.* (1978b) for the two clusters, A2634 and A1367, no longer need be interpreted as evidence for a massive halo. Instead, the extent can be understood as broad, highly clumped emission produced by a relatively small total mass of gas.

The observed complexity and variety of X-ray structure, combined with optical properties such as the spiral fraction and central density, supports the theoretical cluster evolutionary models. However, since clusters do not all evolve at the same rate, because of differences in cluster properties such as richness and density (Ostriker 1978), we are able to observe clusters at the same epoch but in a variety of evolutionary phases. The X-ray properties of clusters, including their structure and luminosity, appear to be strongly influenced by their dynamical evolution.

We are grateful to the many people who have been essential to the success of the *Einstein* Observatory. In particular, we thank Riccardo Giacconi, Leon Van Speybroeck, and Harvey Tananbaum for their contributions. In addition, we wish to acknowledge the efforts of the production processing staff. We also have benefited from discussions with J. Ostriker, A. Fabian, L. Cowie, and Y. Avni. We thank M. Twomey for her patience and skill in the preparation of this manuscript. This research was sponsored under NASA contract NAS8-30751.

REFERENCES

Aarseth, S. 1969, *M.N.R.A.S.*, 144, 537.
Abell, G. 1958, *Ap. J. Suppl.*, 3, 211.
———. 1975, in *Galaxies and the Universe*, ed. A. Sandage, M. Sandage, and J. Kristian, (Chicago: University of Chicago Press), p. 601.
Bahcall, N. 1975, *Ap. J.*, 198, 249.
———. 1977a, *Ann. Rev. Astr. Ap.*, 15, 505.
———. 1977b, *Ap. J. (Letters)*, 218, L93.
Bahcall, N., and Sargent, W. 1977, *Ap. J. (Letters)*, 217, L19.
Baier, F. W. 1976, *Astr. Nachr.*, 297, 6.
Butcher, H., and Oemler, A. 1978, *Ap. J.*, 226, 559.
Cooke, B. A., *et al.* 1978, *M.N.R.A.S.*, 182, 455.
Corwin, H. 1974, *A.J.*, 79, 1356.
Cowie, L., and Songaila, A. 1977, *Nature*, 266, 501.
Dressler, A. 1976, Ph.D. thesis, University of California, Santa Cruz.
Faber, S., and Dressler, A. 1976, *Ap. J. (Letters)*, 210, L65.
Fabricant, D., Topka, K., Harnden, F. R., Jr., and Gorenstein, P. 1978, *Ap. J. (Letters)*, 226, L107.
Forman, W., Kellogg, E., Gursky, H., Tananbaum, H., Giacconi, R. 1972, *Ap. J.*, 178, 309.
Forman, W., Jones, C., Cominsky, L., Julien, P., Murray, S., Peters, G., Tananbaum, H., Giacconi, R. 1978a, *Ap. J. Suppl.*, 38, 357.
Forman, W., Jones, C., Murray, S., and Giacconi, R. 1978b, *Ap. J. (Letters)*, 225, L1.
Forman, W., Schwarz, J., Jones, C., Liller, W., and Fabian, A. C. 1979, *Ap. J. (Letters)*, 234, L27.
Gavazzi, G. 1978, *Astr. Ap.*, 69, 355.
Giacconi, R., *et al.* 1979a, *Ap. J.*, 230, 540.
Giacconi, R., *et al.* 1979b, *Ap. J. (Letters)*, 234, L1.
Gisler, G. 1976 *Astr. Ap.*, 51, 137.
Gunn, J. E., and Gott, J. R. 1972, *Ap. J.*, 176, 1.
Gursky, H., Kellogg, E., Murray, S., Leong, C., Tananbaum, H. and Giacconi, R. 1971, *Ap. J. (Letters)*, 167, L81.
Hausman, M., and Ostriker, J. 1978, *Ap. J.*, 224, 320.
Jones, C., and Forman, W. 1978, *Ap. J.*, 224, 1.
Kellogg, E., and Murray, S. 1974, *Ap. J. (Letters)*, 193, L57.
Kellogg, E., Tananbaum, H., Giacconi, R., and Pounds, K. 1972, *Ap. J. (Letters)*, 174, L65.
Kristian, J., Sandage, A., Westphal, J. A. 1978, *Ap. J.*, 221, 383.
Lea, S. M., Silk, J., Kellogg, E., and Murray, S. 1973, *Ap. J. (Letters)*, 184, L105.
Leir, A. A., and van den Bergh, S. 1977, *Ap. J. Suppl.*, 34, 381.
Mathews, W. G. 1978, *Ap. J.*, 219, 413.
Mushotsky, R. F., Serlemitsos, P. J., Smith, B. W., Boldt, E. A., and Holt, S. S. 1978, *Ap. J.*, 225, 21.
Ostriker, J. 1978, in *IAU Symposium No. 79, The Large Scale Structure of the Universe*, ed. M. S. Longair and J. Emasto (Dordrecht: Reidel).
Ostriker, J., and Tremaine, S. D. 1976, *Ap. J. (Letters)*, 202, L113.
Peebles, P. J. E. 1970, *A.J.*, 75, 13.
Perrenod, S. 1978, *Ap. J.*, 226, 566.
Pravdo, S., Boldt, E., Marshall, F., McKee, J., Mushotzky, R., Smith, B., and Reichert, G. 1979, *Ap. J.*, submitted.
Richstone, D. O. 1976, *Ap. J.*, 204, 642.
Sandage, A. 1961, *Ap. J.*, 133, 355.
Schwartz, D., Schwarz, J., and Tucker, W. 1979, preprint.
Vidal, N. V., and Peterson, B. A. 1975, *Ap. J. (Letters)*, 196, L95.
White, S. D. M. 1976, *M.N.R.A.S.*, 177, 717.
Yahil, A., and Vidal, N. 1977, *Ap. J.*, 214, 347.

W. FORMAN, F. R. HARNDEN, JR., C. JONES, E. MANDEL, S. S. MURRAY, and J. SCHWARZ: Harvard-Smithsonian Center for Astrophysics, 60 Garden Street, Cambridge, MA 02138

The Astrophysical Journal, 248:47-54, 1981 August 15
© 1981. The American Astronomical Society. All rights reserved. Printed in U.S.A.

Reprinted with permission from *The Astrophysical Journal*, 248 pp. 47-54, A.C. Fabian, E.M. Hu, L.L. Cowie, and J. Grindlay, "The Distribution and Morphology of X-Ray-Emitting Gas in the Core of the Perseus Cluster." © 1981 The American Astronomical Society.

THE DISTRIBUTION AND MORPHOLOGY OF X-RAY-EMITTING GAS IN THE CORE OF THE PERSEUS CLUSTER

A. C. FABIAN,[1] E. M. HU,[2] L. L. COWIE,[3] AND J. GRINDLAY[4,5]

Received 1980 September 23; accepted 1981 March 4

ABSTRACT

A high-resolution ($\sim 4''$) X-ray image of the core of the Perseus cluster obtained with the *Einstein* Observatory is presented. An unresolved source is found coincident with the nucleus of NGC 1275. In the surrounding extended emission, isointensity contours twist at progressively larger radii from a position angle with SE orientation at $19''$ to $100''$ to a W one at $150''$.

Absorption in the optical features at high velocity with respect to NGC 1275, which are thought to be associated with a foreground galaxy, does not produce any detectable X-ray absorption. The emission tends to become asymmetric in the presence of the lower-velocity filaments, but there is no obvious detailed correlation between X-ray enhancements and individual filaments.

The surface brightness is deprojected to obtain density and temperature profiles of the intracluster gas. The temperature decreases from an asymptotic value of $\sim 10^8$ K to less than 4×10^7 K at $10''$ from NGC 1275, and the gas has a relatively short radiative cooling time. The results are consistent with a quasi-hydrostatic radiative accretion flow onto NGC 1275, and the pressure-driven mass inflow onto the central galaxy NGC 1275 is then $\sim 200-400\ M_\odot$ yr^{-1}. The total mass of intracluster gas within $10'$ of NGC 1275 is $\sim 15\%$ of the mass of the gravitating matter. A possible problem in relating the line-of-sight velocity dispersion to the gravitational mass of the cluster is confirmed.

Subject headings: galaxies: clusters of — galaxies: intergalactic medium — X-rays: sources

I. INTRODUCTION

The Perseus cluster of galaxies is one of the brightest cluster X-ray sources (Forman *et al.* 1972). The X-ray emission extends from a local peak within $3'$ of the central galaxy NGC 1275 (Fabian *et al.* 1974; Wolff *et al.* 1976; Helmken *et al.* 1978; Gorenstein *et al.* 1978) out to at least $1°$. (Nulsen and Fabian 1980; Ulmer *et al.* 1980). The central regions of this cluster ($\lesssim 3'$) show evidence for cooling in the rapidly increasing X-ray surface brightness (Gorenstein *et al.* 1978), the presence of spectral lines characteristic of cooler gas ($T \sim 10^6 - 10^7$ K) (Mushotsky *et al.* 1980, Canizares *et al.* 1980), and the extensive optical filamentation surrounding the central galaxy NGC 1275 (Rubin *et al.* 1977; Kent and Sargent 1979).

We have observed the core region of Perseus (radii $\lesssim 10'$) with the High Resolution Image (HRI) on the

[1]Institute of Astronomy, University of Cambridge.
[2]Department of Terrestrial Magnetism, Carnegie Institution of Washington.
[3]Princeton University Observatory.
[4]Center for Astrophysics.
[5]Alfred P. Sloan Foundation Fellow.

Einstein Observatory and confirm the general smooth structure found by the earlier observations listed above. A 12,399 s image was obtained centered on NGC 1275 (R.A. (1950) = $3^h16^m30^s$, decl. (1950) = $41°20'0''$) during a 2 day period in 1979 February. The effective energy range of the HRI when absorption along the line of sight is taken into consideration ($N_H \approx 10^{21}$ cm^{-2}) is around 700 eV to 3 KeV. We present a preliminary discussion of the X-ray morphology of the central regions and compare the structure with that of the optical filamentation systems around NGC 1275. We have deprojected the HRI surface brightness data under various symmetry assumptions and have derived the variation of density and temperature in this region, assuming a thermal origin for the diffuse emission and a form of the gravitational potential. We also find an unresolved source coincident with the nucleus of NGC 1275, which accounts for 46% of the total HRI counts within $1'$ of the nucleus. The corresponding luminosity of this point source is $\sim 6 \times 10^{43}$ ergs s^{-1} (Branduardi-Raymont *et al.* 1981) for an assumed spectrum like that of the cluster (i.e., $kT \approx 7$ keV, $N_H \approx 1.6 \times 10^{21}$ cm^{-2}, cf. Mushotzky *et al.* 1978).

Einstein observations have also been made with the Imaging Proportional Counter (IPC) which has only 1.'7 resolution, as compared with about 4" for the HRI, but does have some energy resolution. These observations are discussed by Branduardi-Raymont *et al.* (1981).

II. MORPHOLOGY

In Figure 1 (Plate 1) we show an unsmoothed photographic image of the X-ray data in 4" square pixels. In Figure 2 (Plate 2) we show the data passed through two Gaussian smoothings with different half-widths and compare these with a narrow band Hα photograph taken by C. R. Lynds. All images have been reduced to the same scale and orientation by playing back digitized information onto Polaroid film using the Princeton PDS system. The same information is shown in schematic form in Figure 3 where the HRI count rate contours are shown together with the position of the low-velocity filament system which occurs at the velocity of NGC 1275 (Rubin *et al.* 1978) and the high-velocity system which is believed to be a foreground galaxy (Rubin *et al.* 1977).

It is interesting that there is an apparent decrease in X-ray surface brightness in the NW quadrant, which also contains most of the "high-velocity" (HV) knots. However, this appears about 2' farther from NGC 1275 than the HV system, and there is no apparent absorption at the filament position. These knots are receding from us at a velocity 3000 km s^{-1} faster than NGC 1275 and most of the Perseus cluster, yet must lie, in

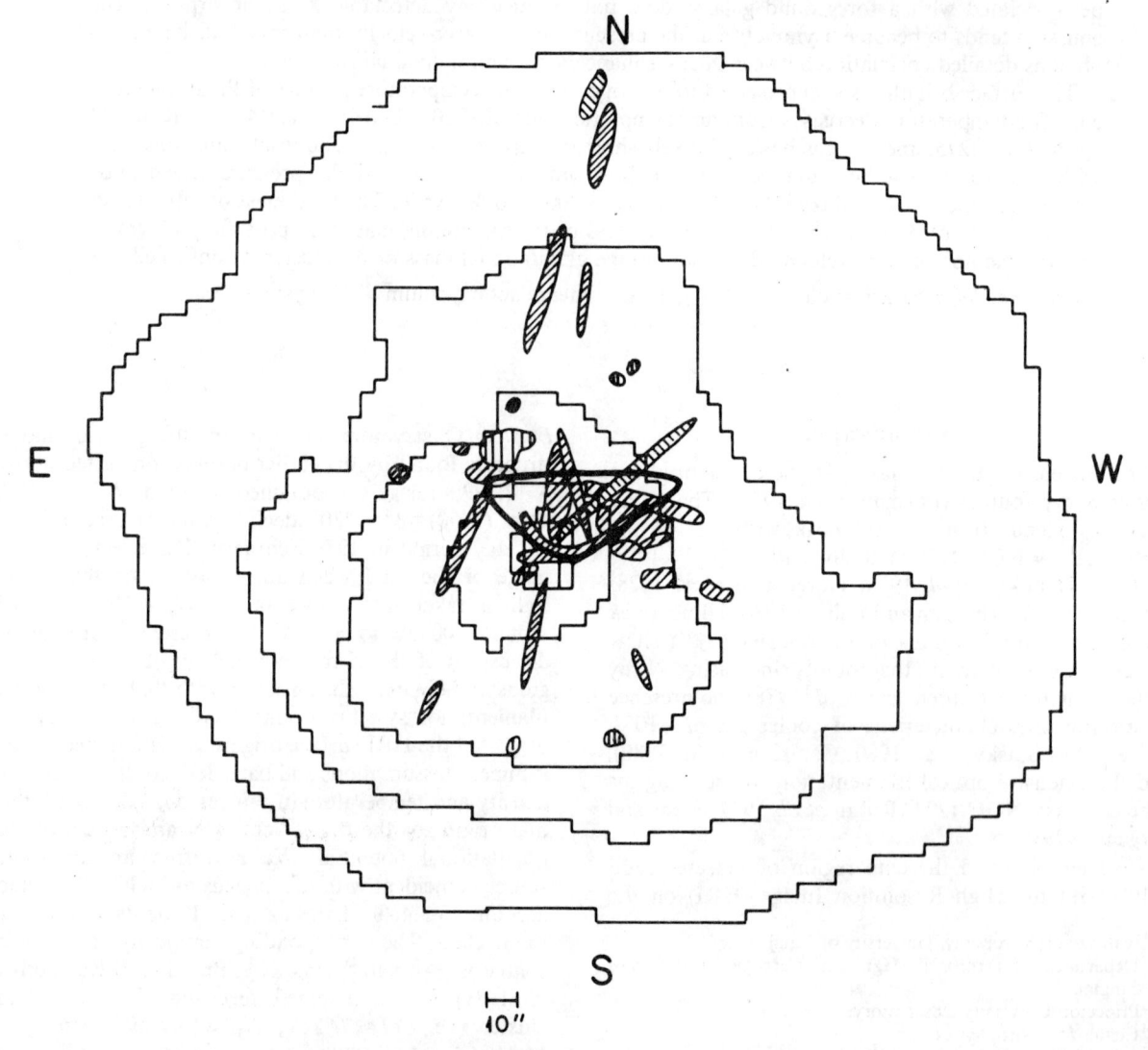

FIG. 3.—A schematic outline of the filamentary structure around NGC 1275 is shown superposed on a plot of the X-ray contours. The low-velocity system described by Rubin *et al.* (1978) is shown schematically by the shaded regions, while the HV system (Rubin *et al.* 1977) is outlined by a thick line. Contour levels in counts per (4"×4" pixel) are (1.8, 3.0, 5.0, 9.0, and 20.0). The NW dip referred to in the text may be seen most clearly in the second and third lowest contour levels.

FIG. 1.—Photographic representations of the *Einstein* HRI data, binned in 4″ square pixels, obtained by playing back the digital data through a PDS system onto Polaroid film. The left-hand plate shows the results with a high gain factor intended to show the general structure, while in the right-hand plate the gain level was decreased to show only the innermost intense regions.

FABIAN *et al.* (*see page* 48)

PLATE 2

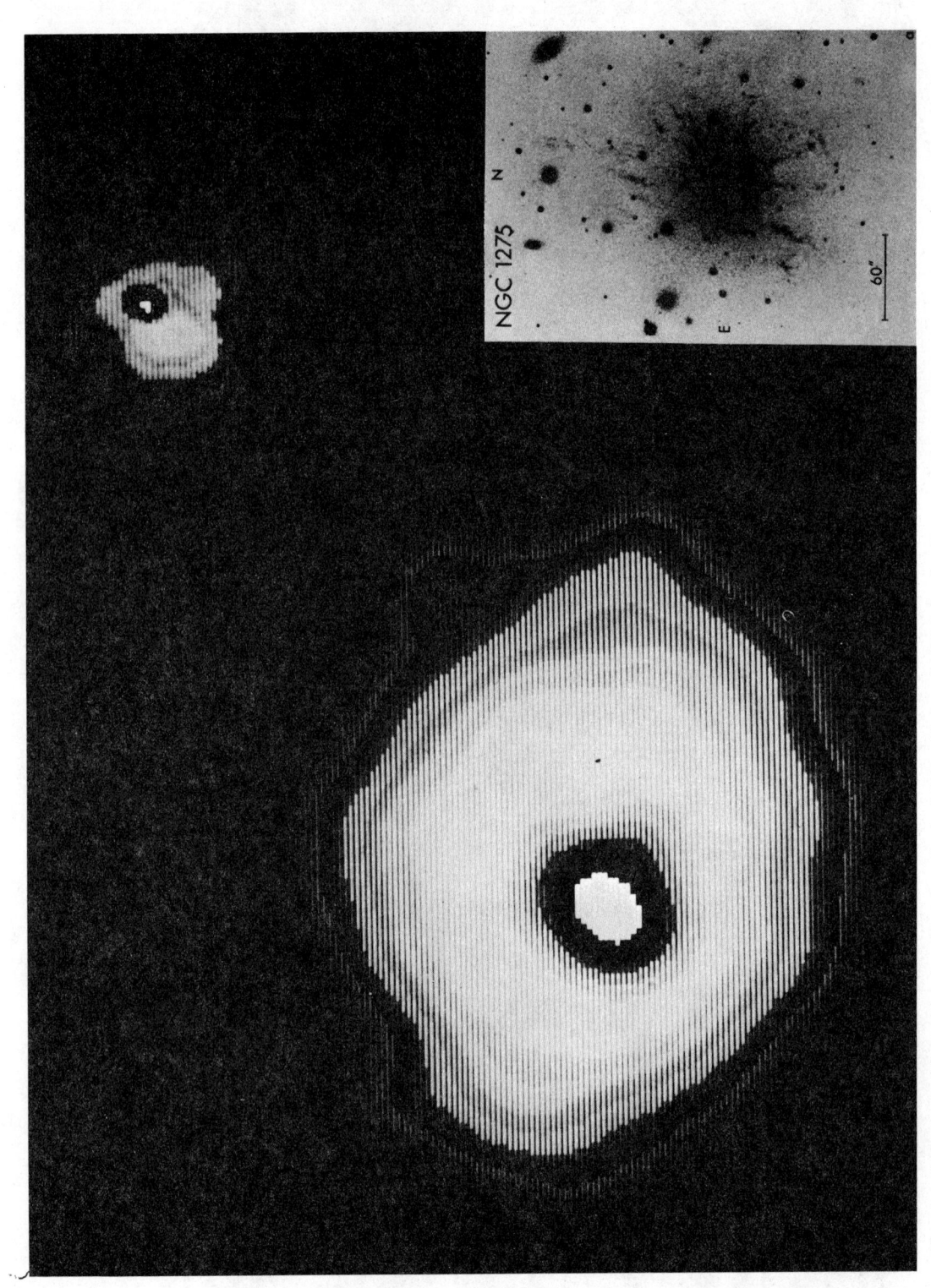

FIG. 2.—Intensity profiles and isophotal contours. In the left-hand image the HRI data has been smoothed through a $\sigma = 32''$ Gaussian and played back on the PDS at high gain to show the outermost regions. The right-hand image has been smoothed at $12''$ and played back at low gain to show the inner regions. In both cases the centermost regions have been overexposed on the photographic representation to show the outer structure more clearly. A square wave offset at alternate groups of four levels shows up the isophotal contours for visual examination or measurement. The inset shows a narrow band Hα photograph of the filament system at the velocity of NGC 1275 (the low-velocity system) taken by C. R. Lynds. The linear scales for all three are identical. The conversion from angular to linear extent is $0.52 h^{-1}$ kpc arcsec^{-1}, where $h = H/50$ km s^{-1} Mpc^{-1}.

FABIAN et al. (see page 48)

part, in front of that galaxy (De Young, Roberts, and Saslaw 1973; Ekers, van der Hulst, and Miley 1976). Optical studies of these knots suggest that they are part of a gas-rich galaxy or cloud falling into the cluster and are probably unrelated to the filamentation around NGC 1275 and are physically separate from it (Rubin et al. 1978; Oort 1976; Adams 1977).

The ram pressure of the intracluster gas acting on such a galaxy could either locally enhance, or diminish, the HRI count-rate emissivity in that region, depending upon the temperature of the compressed, or shocked, gas relative to that of the nearby undisturbed gas. To explain the NW dip in terms of absorption from this object requires us to hypothesize that the high-velocity knots are part of a galaxy, the "gas center" of which is displaced $\sim 1'-2'$ NW from the knots and which is projected in front of NGC 1275 *and* the core of the Perseus cluster. Column densities of hydrogen around 10^{22} cm^{-2} would be required to explain a 50% drop in the count rate if the gas temperature exceeds 3×10^7 K at this radius.

Thermal instabilities within the cooling intracluster gas (Fabian and Nulsen 1977; Mathews and Bregman 1978; Cowie et al. 1980) can lead to the stationary filamentation observed around NGC 1275 (e.g., Lynds 1970; Rubin et al. 1978; Kent and Sargent 1979). We have therefore searched for any correspondence between the distribution of optical filamentation and the X-ray structure. As can be seen from Figure 2, no case can be made for a detailed correlation. However, it should be noted that the large-scale asymmetries of the X-ray emission occur in the region where the LV filaments are present. This is the behavior expected from a thermally unstable cooling flow, where large-scale pressure-driven asymmetry may be expected. This may be a more satisfactory explanation of the NW dip and the spurs in the N, SW, and SE directions.

Further observations of the optical structure are presently in progress, and a more detailed discussion of the morphology will be presented in a subsequent paper (Hu et al. 1981).

III. SPATIAL AND SPECTRAL DEPROJECTION

The discovery of an emission feature due to iron in the X-ray spectrum of the Perseus cluster (Mitchell et al. 1976) established that most of the X-ray emission is thermal bremsstrahlung and line radiation from a hot gas. We have therefore used the count rates from the HRI to produce temperature and density profiles for the gas in the core of the Perseus cluster in a manner similar to that used for the data on the Cas A supernova remnant (Fabian et al. 1980). The further steps taken here involve a spatial deprojection of the source emissivity and inclusion of a pressure gradient. We employ this method, as opposed to model fitting, since it introduces no particular theoretical bias.

The method essentially consists of deriving a temperature T (we assume $T_{\text{electron}} = T_{\text{ion}}$) appropriate to the count rate produced in the *Einstein* detector by a unit volume at the local pressure (here $P = n_e T$):

$$C = \frac{1}{4\pi D^2} \frac{P^2}{T^2} \int \frac{\varepsilon(T, E)}{E} Q(E) \exp -[N_H \sigma(E)] dE$$

$$\equiv \frac{1}{4\pi D^2} \frac{P^2}{T^2} v(T), \qquad (1)$$

where $Q(E)$ is the effective area of the HRI (Giacconi et al. 1979), $\varepsilon(T, E) dE$ is the emissivity of the gas in the band $E \to E + dE$, $N_H \sigma(E)$ is the optical depth, and D is the distance to the cluster. For the temperature range of interest, C is a monotonic function of T and equation (1) may be easily inverted.

We use

$$\varepsilon(T, E) = 2.76 \times 10^{-20} T_e^{1/2} \exp -\frac{E}{kT} + \sum_i A_i \delta(E - E_i) \Lambda_i(T), \qquad (2)$$

where numerical values are in cgs units and $\Lambda_i(T)$ are the emissivities of sulphur, silicon, and iron (Raymond and Smith 1977, but for iron alone we use values due to J. C. Raymond, private communication), weighted by A_i to represent departures from the case of cosmic abundance and ionization equilibrium treated by Raymond and Smith (1977) (A_i was generally fixed at 0.5). The Gaunt factor for the bremsstrahlung component is fixed at 1.2, and the contribution from elements other than hydrogen was assumed to be a factor of 1.4. The interstellar photon absorption cross section is taken as

$$\sigma(E) = 5 \times 10^{-20} (10E)^{-\alpha}, \qquad (3)$$

with $\alpha = 8/3$ if $E < 0.54$ keV and $5/2$ if $E > 0.54$. N_H was taken as 10^{21} cm^{-2}.

In deprojecting the surface brightness to obtain the count rates per unit volume used in the above analysis, two different methods for estimating the emitting volume were used. In the first, count rates were accumulated in 60 concentric annuli of 10″ width centered on NGC 1275. These annuli were subdivided into four sectors. Counts from the outer annulus were used as "background," and a fraction of these counts were removed from all other regions in proportion to their area. This removes the detector and X-ray background contribution, together with a component of the larger-scale cluster emission, all of which are assumed to vary only slowly across the core. Counts from inner annuli are assumed to originate from spherical shells; the contribution from outer spherical shells (within a sector) being progressively subtracted in proportion to their projected area. This procedure was checked by summing the many subvolumes used and comparing this with the total expected volume. We assume spherical symmetry only within each sector, so that regions at the same radius, but different position angle, may have different properties.

The second procedure, recognizing the anisotropy of the distribution fits successive shells determined by a local contour level and the zero count level. The sampled distance at each point in the shell is determined by an assumption of rotational symmetry about an axis either parallel or perpendicular to the longest axis defined by the contour level (prolate or oblate symmetry) and an average count rate emission inside this region calculated. To start the procedure a constant background is subtracted to form the initial zero level and a first shell chosen whose calculated contribution is then debited from the surface brightness and new zero levels and contours calculated. The method then iterates inward. Seven contour levels were used.

The HRI count emission per unit volume calculated by these procedures is shown in Figure 4. In Figure 4a we show the count rates within specific contour levels obtained from the 2D procedure. The average distribution in the four sectors obtained under radial symmetry assumptions are shown in Figure 4b. Comparison of the radial case with the volume count rate versus geometric mean radius for the 2D case demonstrates that the results are very insensitive to the spatial deconvolution procedure and provides confidence in the method. We shall adopt the radial count rate per unit volume versus radius as our basic data set. We note that the count rate per unit volume is roughly proportional to $r^{-\alpha}$ where $\alpha \sim 2.9$ ($2' \leq r \leq 10'$) and 1.3 ($r < 2'$).

To complete the description we require knowledge of the pressure distribution in the gas. Several values of outer pressure ($r = 10'$) were employed: we incorporate a radial pressure gradient in accordance with hydrostatic support:

$$\frac{1}{\rho}\frac{dP}{dr} = \frac{-G4\pi\rho_{gc}a^3}{r^2} f(x), \qquad (4)$$

where $x = r/a$, $f(x) = \ln[x + (1+x^2)^{1/2}] - x/(1+x^2)^{1/2}$, and we have assumed the mass density in the cluster is $\rho_{gc}(1 + r^2/a^2)^{-3/2}$. Equation (4) may be rewritten as

$$\frac{1}{\rho}\frac{dP}{dr} = \frac{9v_\parallel^2}{a} \frac{f(x)}{x^2}, \qquad (5)$$

where $v_\parallel$ is the measured line-of-sight velocity dispersion, since the virial theorem gives $4\pi G\rho_{gc}a^2 = 9v_\parallel^2$ if the velocity distribution is locally isotropic. It is not, however, clear that this relation applies to the core of

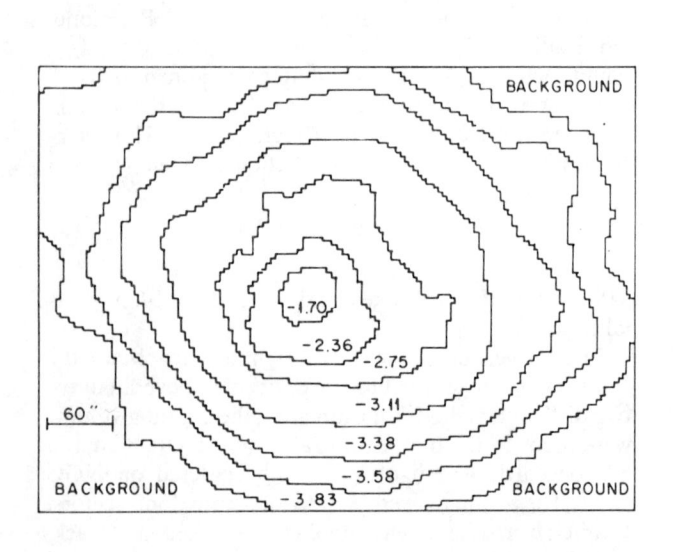

FIG. 4a

FIG. 4b

FIG. 4.—(a) Logarithmic count rates in units of counts per cubic arc second are shown within the contoured annuli (1 cubic arcsec = $0.14h^{-3}$ kpc^3). The results are presented for the case of assumed oblate symmetry. The results obtained for prolate symmetry are basically similar. (b) The relative emissivity in arbitrary units is plotted versus radius from the nucleus of NGC 1275 (radial deconvolution) and versus geometric mean radius of the contour (2D deconvolution). In the latter case we also show a radially fitted background beyond the last contour. The quantity $V(T) = 10^8 v(T)(T)^2$, where $v(T)$ is defined in eq. (1), is plotted as a function of temperature. For the isobaric case the emissivity at a given temperature is proportional to $V(T)$.

the Perseus cluster (Cowie and Binney 1977; Bahcall and Sarazin 1977; Gorenstein *et al.* 1978; Nulsen and Fabian 1980; Branduardi-Raymont *et al.* 1980). The factor of 9 may more reasonably be taken as 3 in order that observations of the large-scale distribution of emission are in agreement with both observed values of $v_\|$ and the overall gas temperature. Such a change could be in part due to a predominance of radial motions of the cluster galaxies or to some of the higher velocity galaxies belonging to a separate small cluster (see Rubin *et al.* 1978). In our calculations, we shall generally use

$$\frac{1}{\rho}\frac{dP}{dr} = \frac{-Kv_\|^2}{a}\frac{f(x)}{x^2}, \qquad (6)$$

with $v_\| = 1420$ km s^{-1}, $a = 250$ kpc, and K is a variable constant.

IV. DISCUSSION

a) Temperature and Density Profiles

Using the method of § III, we find that an outer pressure of 6.5×10^{-11} dynes cm^{-2} yields a temperature of $\sim 8 \times 10^7$ K in the outer $5'-10'$. This temperature is in good agreement with that derived for the whole cluster by Mushotzky *et al.* (1978) and Mitchell *et al.* (1979) from wide-field proportional counter data. Figure 5 shows the effect on the temperature profile of varying K. The value of 3 which agrees with the larger-scale observations gives an approximately constant temperature for the outer half of the core. We adopt this value of K for subsequent detailed discussion. The density profile is shown in Figure 6.

We have used the temperature and density profiles to generate the expected X-ray spectrum for the inner $10'$ spherical region. This is compared with the overall spectrum for the cluster observed with *HEAO 1* (Mushotzky and Smith 1980) in Figure 7. It is also important to stress that the HRI predictions are consistent with observed hardness ratios as measured in the IPC (Fig. 8).[6] The ratios of channels (6–9) to (3–5) (approximately 1.46–3.05 keV to 0.58–1.45 keV) have been predicted in annuli of radii increment $1'$, taking into account the energy resolution of the detector and (energy dependent) point spread function of the IPC. This last process involved a 2D convolution.

---

[6] The HRI count rate from the central $10''$ is too bright to be from gas emission, in hydrostatic balance with its surroundings. It may correspond in part to a point source contribution from the nucleus of NGC 1275 and was treated as a separate component. A contribution due to a 10 keV thermal bremsstrahlung source (essentially a photon power law $\propto E^{-1}$) with variable low-energy cutoff due to extra intervening absorption, and normalized such as to be compatible with the innermost HRI count rate, was added to the energy channels at the center of the IPC model. The subsequent spatial convolution gave the results shown in Fig. 8 with the IPC observations plotted for comparison. The relative hardness of the innermost $1'$ is presumably due to the nucleus.

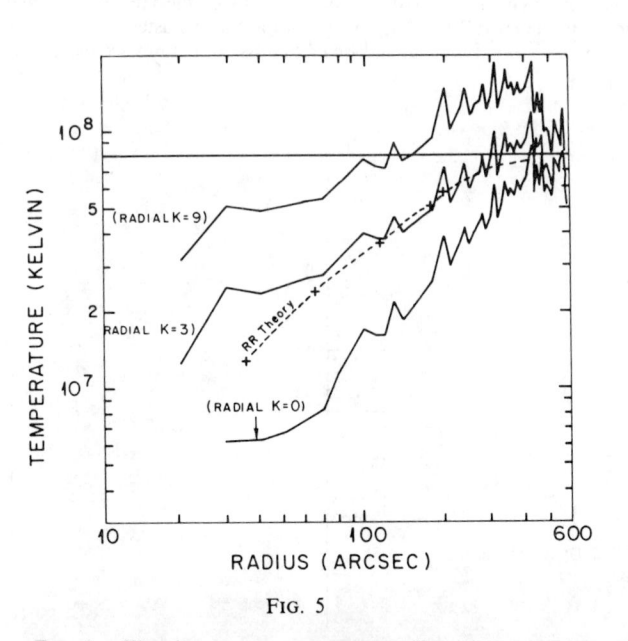

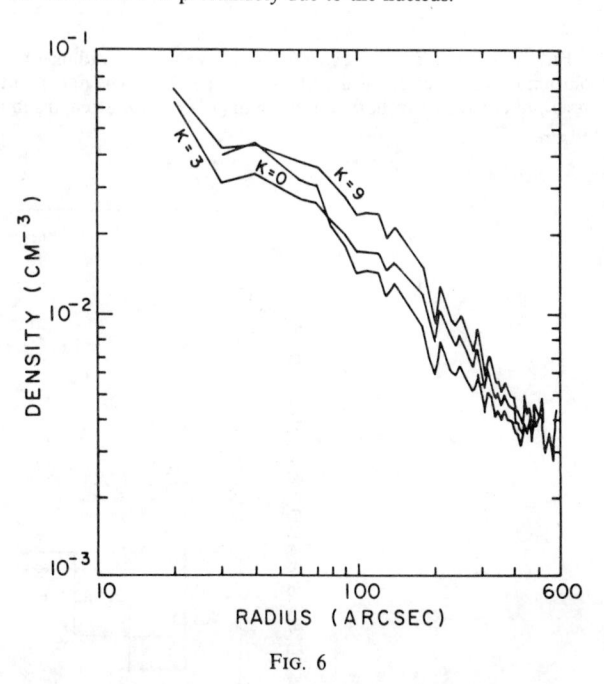

FIG. 5

FIG. 6

FIG. 5.—We plot temperature versus radius under a variety of assumptions concerning the pressure variation. Also shown is the temperature profile of the gas predicted by the radiative regulation theory of the flow (Cowie and Binney 1977) for a core radius of $6'$ and a normalizing temperature of 8×10^7 K. This normalizing temperature corresponds to a value $K=3$. (We have not attempted to extrapolate the cooling flow theory to $T \lesssim 2 \times 10^7$ K, where line cooling becomes important.)

FIG. 6.—Calculated density profiles for the radial deconvolution

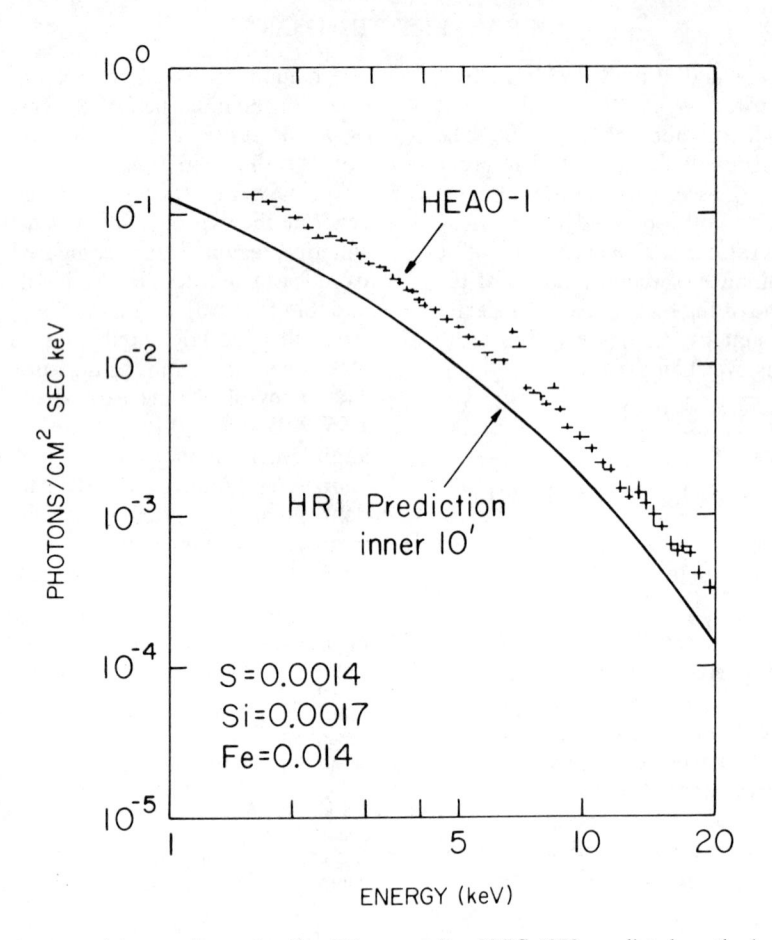

FIG. 7.—Total X-ray spectrum, from a volume of radius 10' surrounding NGC 1275, predicted on the basis of the HRI deprojected solution for intracluster gas density and temperature. The data points are from the *HEAO 1* A-2 spectrum for the whole cluster (~3° field of view) presented by Mushotzky and Smith (1979). Line strengths in photons cm^{-2} s^{-1} for S, Si, and Fe (L) emission features are listed, but not plotted.

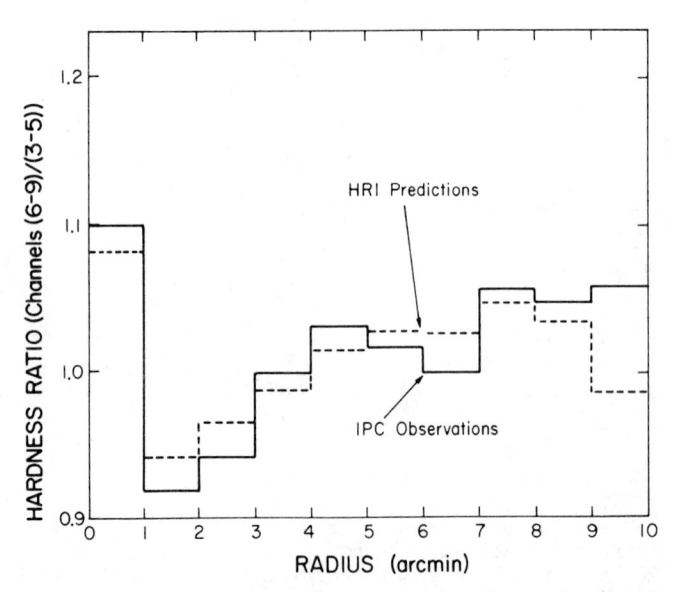

FIG. 8.—IPC hardness ratios predicted by the HRI solution ($K=3$) plotted against radius from NGC 1275 (*dashed line*). The observed values are indicated by the solid lines.

b) Mass Flow Rates

From Figures 5 and 6 we may see that the radiative cooling time

$$t_{\text{cool}} = \frac{nkT}{\Lambda n^2(\gamma-1)\mu} \quad (7)$$

where μ is the mean atomic weight, is less than 10^{10} yr within approximately 5'. While the ratio of the cooling time to the Hubble time has a weak dependence on H_0 ($\sim H_0$), it is clear that through much of the inner region the cooling time is significantly shorter than the Hubble time for all plausible values of H_0.

We can now derive an inward mass flow rate on the assumption that the outer pressure of the cluster causes that gas which is radiatively cooling to flow inward (a pressure-driven accretion flow). Such flows are highly subsonic and the hydrostatic equation assumed in the density deconvolution continues to apply. We use

$$\dot{M} = 4\pi r^2 \rho v \quad (8)$$

and derive the velocity v from the energy equation

$$\rho v \frac{d}{dr}\left[\frac{v^2}{2} + \frac{\gamma}{(\gamma-1)}\frac{kT}{\mu} + \phi\right] = n_e^2 \Lambda(T). \quad (9)$$

We ignore $v^2/2$ and the gravitational potential ϕ with respect to the central term and derive:

$$\dot{M} = \frac{4\pi r^2 \rho}{\gamma} \frac{T}{t_{\text{cool}}} \left(\frac{dT}{dr}\right)^{-1}, \quad (10)$$

which is plotted as a function of r in Figure 9. We see that $\dot{M}$ is approximately constant at $\sim 200-400$ $M_\odot$ yr^{-1}. The slope is due in part to our neglect of the ϕ term in equation (9). This is in excellent agreement with the cooling rate inferred from the emission lines strengths from which Mushotzky et al. (1980) give a value of 290 $M_\odot$ yr^{-1}.

A more model-dependent comparison with the temperature profile calculated for a cooling flow solution with ϕ corresponding to a value $K=3$ is made in Figure 5. The agreement with the observed temperature profile is excellent.

c) Uncertainties

It is difficult to obtain errors on the quantities which we derive from the deprojection. Values at inner radii depend upon values as outer radii. We have assessed the stability of the solution by randomly perturbing the input raw data. Each input count rate was increased (or decreased) by an amount chosen at random from a Poisson distribution of mean and variance equal to that count rate. This was repeated three times, and the results, whilst differing in detail, showed the same trends and gave very similar inner values.

We have also investigated different gravitational potentials. In particular, we have increased the cluster core radius by a factor of 2, and have separately included a galaxy potential of similar form to the cluster (eq. [7]) with $v \sim 250-400$ km s^{-1} and $a \sim 200-400$ pc. There is no substantial change in our results. The increased central pressure induced by the galaxy potential increases the central gas temperature by a factor of up to $\sim 50\%$.

Finally, we have inferred an upper bound to the gravitational potential by assuming that the gas temperature is constant (cf. Fabricant, Lecar, and Gorenstein 1981). The potential is determined via equation (4) where P, rather than T, is determined from the observations. Once again, large excursions occur as we derive $\Delta(\ln P)$. Consequently we only show (Fig. 10) individual points for the inner region and have used a power-law fit to P (or n_e) in the outer 5' to obtain the smooth dashed curves. Such a power law automatically implies that $M(r) \propto r$ for an isothermal hydrostatic atmosphere. It is clear from Figure 10 that isothermal atmospheres require considerably more mass in the inner regions.

The total mass of gas ($\sim 1.0 \times 10^{13}$ $M_\odot$) within radii of 10' is a significant fraction of the mass of the gravitating matter in our models ($\sim 7.7 \times 10^{13}$ $M_\odot$). This mass of gas is relatively insensitive to the model potential and assumed volumes (see Fabian et al. 1980 for a discussion of the limitations of this method).

V. CONCLUSIONS

Using the broad band HRI surface brightness data, we have derived approximate radial dependences for the physical state of the gas accreting onto NGC 1275. We have assumed only a particular form for the X-ray emission process (optically thin bremsstrahlung plus line emission) and a form for the pressure gradient. The results are supported by the approximate agreement with both the observed total line emission fluxes and the continuum spectra. Finally, it is especially interesting that, within the cluster core, this analysis shows that the data are consistent with the radiatively regulated accretion flow model and that the accretion rate $\dot{M}$ onto NGC 1275 is then roughly constant with radius at $\sim 200-400$ $M_\odot$ yr^{-1}.

This work was supported in part by National Aeronautics and Space Administration grants NGL31-001-007 and NAS8-33345. A. C. F. thanks the Radcliffe trust for support. We particularly wish to thank Ed Turner for assistance in using the Princeton PDS, L. Van Speybroeck for assistance with the production of the HRI image, C. R. Lynds for generously providing digitized tracings of his narrow band photography of NGC 1275, and Vera Rubin, Rick Mushotzky, Paul Nulsen, and G. Stewart for very interesting comments and discussion.

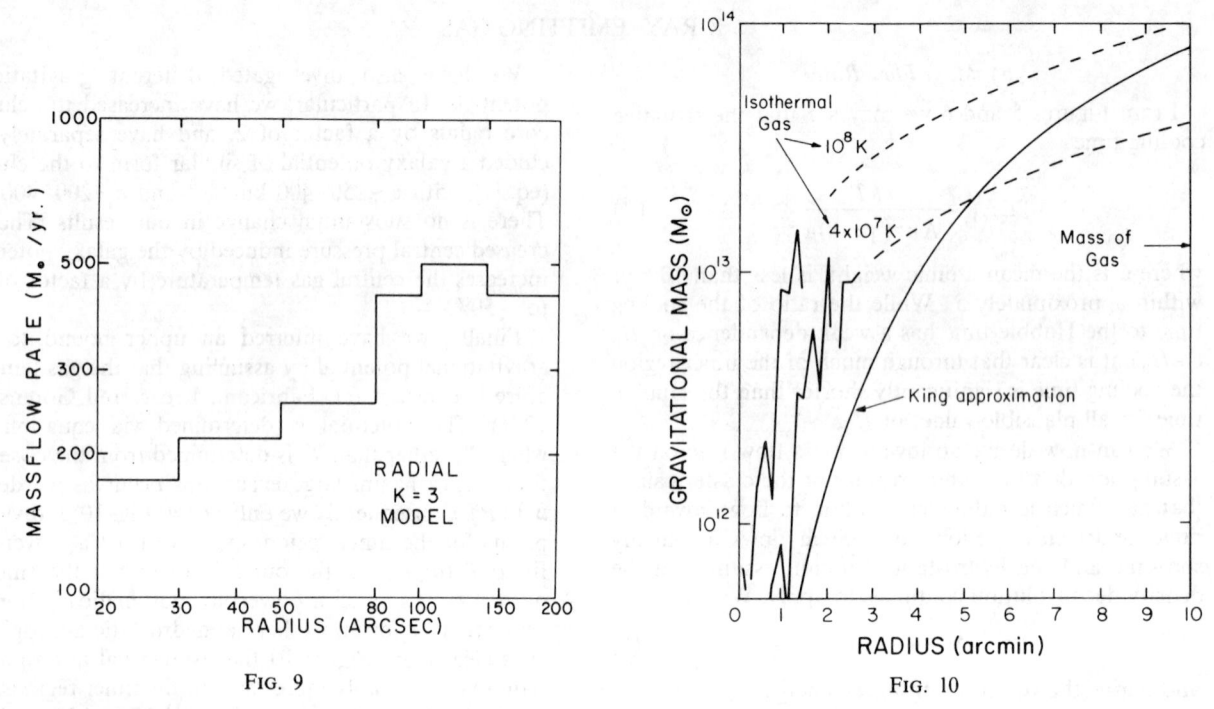

FIG. 9.—The mass flow rate calculated according to the method described in the text is shown as a function of angular radius. The flow rate scales as h^{-2}.

FIG. 10.—The gravitational mass required by the King approximation and $K=3$ is shown by the smooth curve. This is the potential assumed in generating the earlier figures. Forcing the gas to be isothermal and then solving for the potential via the hydrostatic equation gives the jagged/dashed curves shown. We also show by an arrow the total mass of gas obtained to $R=10'$ from the present data. The scale is the same as that of the gravitational mass.

REFERENCES

Adams, T. F. 1977, *Pub. A.S.P.*, **89**, 488.
Bahcall, J. N., and Sarazin, C. 1977, *Ap. J. (Letters)*, **213**, L99.
Branduardi-Raymont, G., Fabricant, D., Gorenstein, P., Grindlay, P., Soltan, A., and Zamorami, G. 1981, *Ap. J.*, in press.
Canizares, C. R., Berg, C., Clark, C., Jernigan, J. G., Kruss, G., Markert, T. H., Schattenburg, M., and Winkler, P. F. 1980, in *Highlights of Astronomy* (Dordrecht: Reidel).
Cowie, L. L., and Binney, J. 1977, *Ap. J.*, **215**, 723.
Cowie, L. L., Fabian, A., and Nulsen, P. E. J. 1980, *M.N.R.A.S.*, **191**, 399.
De Young, D. S., Roberts, M. S., and Saslaw, W. C. 1973, *Ap. J.*, **185**, 809.
Ekers, R. D., van der Hulst, J. M., and Miley, G. K. 1976 *Nature*, **262**, 369.
Fabian, A. C., Zarnecki, J. C., Culhane, J. L., Hawking, F. J., Peacock, A., Pounds, K. A., and Parkinson, J. H. 1974, *Ap. J. (Letters)*, **189**, L59.
Fabian, A. C., and Nulsen, P. E. J. 1977, *M.N.R.A.S.*, **180**, 479.
Fabian, A. C., Willingale, R., Pye, J. P., Murray, S. S., and Fabbiano, G. 1980, *M.N.R.A.S.*, **193**, 175.
Fabricant, D., Lecar, M., and Gorenstein, P. 1981, *Ap. J.*, **241**, 552.
Forman, W., Kellogg, E., Gursky, H., Tananbaum, H., and Giacconi, R. 1972, *Ap. J.*, **178**, 309.
Giacconi, R., et al. 1979, *Ap. J.*, **230**, 540.
Gorenstein, P., Fabricant, D., Topka, K., Harnden, F. R., and Tucker, W. H. 1978, *Ap. J.*, **225**, 718.

Helmken, H., Delvaille, J. P., Epstein, S., Geller, M. J., and Schnopper, H. W., 1978, *Ap. J. (Letters)*, **221**, L43.
Hu, E. M., Cowie, L. L., Fabian, A., and Grindlay, J. 1981, in preparation.
Kent, S. M., and Sargent, W. L. W. 1979, *Ap. J.*, **230**, 667.
Lynds, R. 1970, *Ap. J. (Letters)*, **159**, L151.
Mathews, W. G., and Bregman, J. N. 1978, *Ap. J.*, **224**, 308.
Mitchell, R. J., Culhane, J. L., Davison, P. J. N., and Ives, J. C. 1976, *M.N.R.A.S.*, **175**, 29P.
Mushotzky, R., Serlemitsos, P., Smith, B., Boldt, E., and Holt, S. 1978, *Ap. J.*, **225**, 21.
Mushotzky, R. F., and Smith, B. W. 1980, in *Highlights of Astronomy* (Dordrecht: Reidel).
Mushotzky, R. F., Holt, S., Serlemitsos, P., and Boldt, E. 1980, preprint.
Nulsen, P. E. J., and Fabian, A. C. 1980, *M.N.R.A.S.*, in press.
Oort, J. H. 1976, *Pub. A.S.P.*, **88**, 591.
Raymond, J. C., and Smith, B. W. 1977, *Ap. J. Suppl.*, **35**, 419.
Rubin, V. C., Ford, W. K., Peterson, C. J., and Oort, J. H. 1977, *Ap. J.*, **211**, 693.
Rubin, V. C., Ford, W. K., Peterson, C. J., and Lynds, C. R. 1978, *Ap. J. Suppl.*, **37**, 235.
Ulmer, M. P., et al. 1980, *Ap. J.*, **236**, 58.
Wolff, R. S., Mitchell, R. J., Charles, P. A., and Culhane, J. L. 1976, *Ap. J.*, **208**, 1.

L. L. COWIE: Princeton University Observatory, Peyton Hall, Princeton, NJ 08544

A. C. FABIAN: Institute of Astronomy, University of Cambridge, Madingley Road, Cambridge CB3 0HA, England

J. GRINDLAY: Center for Astrophysics, 60 Garden Street, Cambridge, MA 02138

E. M. HU: Carnegie Institution of Washington, Department of Terrestrial Magnetism, 5241 Broad Branch Road, N.W., Washington, DC 20015

THE ASTROPHYSICAL JOURNAL, 234:L39–L43, 1979 November 15
© 1979. The American Astronomical Society. All rights reserved. Printed in U.S.A.

Reprinted with permission from *The Astrophysical Journal,* 234, pp. L39-L43, E.J. Schreier, et al., "*Einstein* Observations of the X-Ray Structure of Centaurus A: Evidence for the Radio-Lobe Energy Source." © 1979 The American Astronomical Society.

EINSTEIN OBSERVATIONS OF THE X-RAY STRUCTURE OF CENTAURUS A: EVIDENCE FOR THE RADIO-LOBE ENERGY SOURCE

E. J. SCHREIER, E. FEIGELSON, J. DELVAILLE, R. GIACCONI, J. GRINDLAY, AND D. A. SCHWARTZ
Harvard-Smithsonian Center for Astrophysics

AND

A. C. FABIAN
Institute of Theoretical Astronomy, University of Cambridge
Received 1979 June 25; accepted 1979 August 1

ABSTRACT

The X-ray source at the center of the radio galaxy Centaurus A has been resolved into the following components with the imaging detectors on board the *Einstein* X-ray Observatory: (1) a point source coincident with the infrared nucleus; (2) diffuse X-ray emission coinciding with the inner radio lobes; (3) a 4′ extended region of emission about the nucleus; and (4) an X-ray jet between the nucleus and the NE inner radio lobe. The 2×10^{39} ergs s^{-1} detected from the radio lobes probably arises from inverse Compton scattering of the microwave background. The average magnetic field in the SW lobe is determined to be ≥ 4 microgauss. The extended region may be due to emission by a cloud of hot gas, cosmic-ray scattering, or stellar sources. The jet provides strong evidence for the continuous resupply of energy to the lobes from the nucleus. A consistent model is presented for the X-ray jet, an optical jet, and the inner radio lobe.

Subject headings: galaxies: nuclei — radio sources: galaxies — X-rays: sources

I. INTRODUCTION

The structure of Centaurus A (NGC 5128) has been well studied at optical, radio, and, in less detail, X- and γ-ray wavelengths. The observed X-ray low-energy cutoff of greater than 3 keV implied that the X-rays were emitted at the nucleus of the galaxy (Tucker *et al.* 1973); the variability of the hard X-rays (e.g., Winkler and White 1975; Lawrence, Pye, and Elvis 1977) required a compact source. Thus the observed X-ray emission was consistent with a point source at the nucleus of the galaxy. An extended component to the hard X-ray emission has been reported from *SAS 3* data (Delvaille, Epstein, and Schnopper 1977). However, *HEAO 1* scanning modulation collimator data contradict this and are consistent with a point source above 2 keV, with a 99% confidence upper limit of 10^{41} ergs s^{-1} for the emission from an extended source (Doxsey *et al.* 1978). Further contributions of the X-ray observations to the theoretical interpretation of the source have been limited to comparisons of flux and variability at various wavelengths (see Grindlay 1975; Mushotzky *et al.* 1978; Beall *et al.* 1978).

We have now observed the central region of Cen A with the *Einstein* X-ray Observatory, using both the high-resolution imager (HRI) and the imaging proportional counter (IPC) as well as the monitor proportional counter (MPC). A description of the observatory can be found in Giacconi *et al.* (1979). The MPC (similar to *Uhuru* in area, spectral response, and field of view) shows an equivalent counting rate of 3.6×10^{-10} ergs cm^{-2} s^{-1}, 2–10 keV (15 UFU), and a power-law number spectrum with slope $n = 1.7$ and cutoff $E_a = 3.8$, consistent with published spectra (see Mushotzky *et al.* 1978). The imaging detectors show four distinct spatial components to the X-ray emission: a compact source at the nucleus, excess emission coincident with the inner radio lobes, an extended component with about 2′ radius centered on the nucleus, and a jetlike feature located 1′ to the NE of the nucleus. The compact nuclear source corresponds to the previously known variable hard X-ray source, and the possibility of inverse Compton X-ray emission in the inner radio lobes has long been recognized. The NE component, an X-ray jet or new inner lobe, represents a new phenomenon. We suggest that this feature provides evidence for a continuous supply of energy from the nucleus to the inner radio lobe.

II. OBSERVATIONS

The HRI was used to observe Cen A for a total of 13,200 s on 1979 January 15. The 25′ field containing some 36,400 photons was centered 10′ north of the nucleus as a result of an operational error, leading to a several arcsec degradation of the nominal 4″ resolution due to coma. An obvious strong source is centered 0″.7 from the infrared nucleus (Kunkel and Bradt 1971). The statistical uncertainty is 0″.5, and systematic aspect error is estimated to be less than 2″. In addition, the image shows a jetlike feature to the NE of the nucleus. An iso-intensity contour map of the field (Fig. 1) reveals the feature to be clumpy, with its principal component centered 59″ ± 2″ from the nucleus at a position angle 58° east of north. Other fainter components are seen 1′.5 from the nucleus in the same direction.

L39

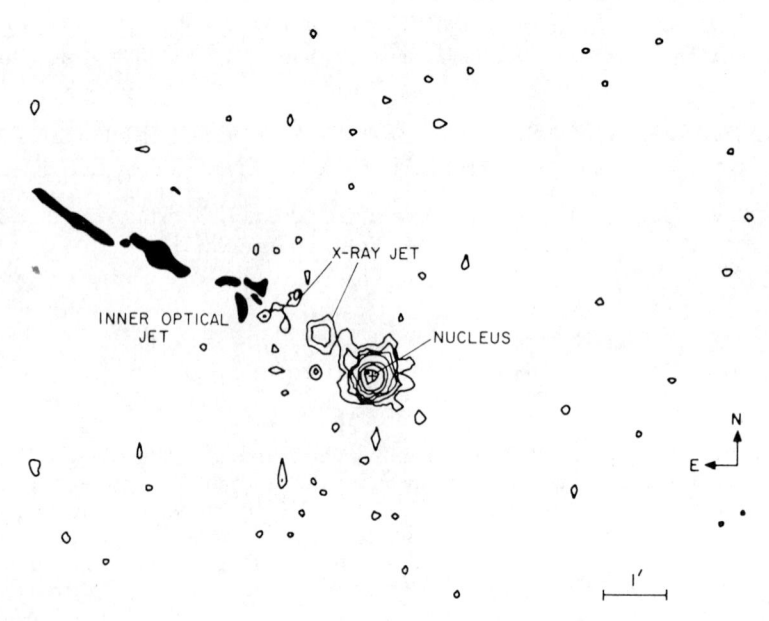

FIG. 1.—Iso-intensity contour map of the HRI image around the nucleus. The X-ray jet to the NE is clearly seen. The dark shapes to the NE show the positions of diffuse features in the optical jet discovered by Dufour and van den Bergh (1978).

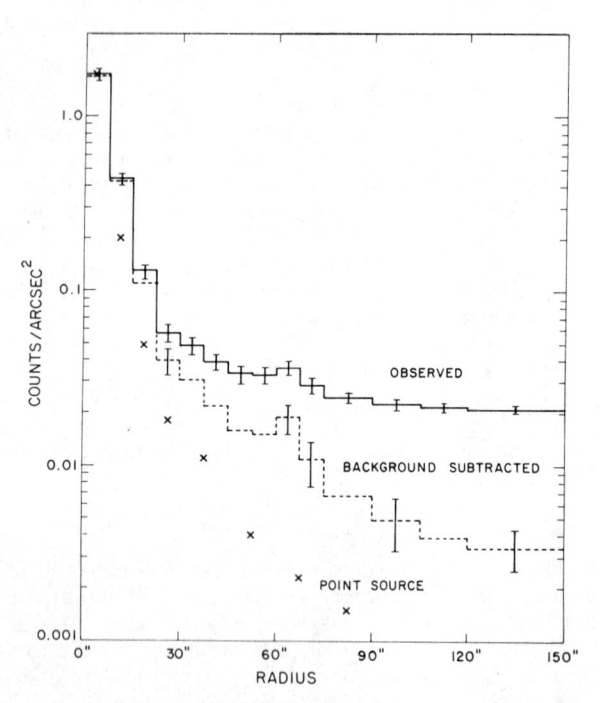

FIG. 2.—Radial distribution of surface brightness (HRI counts arcsec^{-2}), with and without subtraction of background. The X's show the radial distribution of a 3 keV calibration point source seen at the same distance off-axis as Cen A.

The radial surface-brightness distribution of the observed and background-subtracted HRI data is shown in Figure 2. There is emission in excess of the known point-source response, also shown, extended at least 2' from the nucleus. The small bump at 1' is the NE jet. We estimate the luminosity of the extended source by subtracting the predicted point-source contribution from the total flux; the calibration point response function (for an image 10' off-axis) is normalized by folding the observed MPC flux through the telescope and HRI response functions. Assuming in this fashion that the point source dominates the hard, cutoff emission observed by the MPC, we predict a total flux in the HRI from the point source to be 0.04 ± 0.01 counts s^{-1}, within a 150" radius. Subtracting this from the total (background-subtracted) flux, we find an excess of 0.04 ± 0.01 counts s^{-1} within a 150" radius due to the extended component. The quoted errors are due to estimated uncertainty in the point-source spectrum and instrument response. Although this excess is sensitive to the local background level, which varies by about 5% over the field, we find a positive detection of the extended component at the 3.5 σ level even for the highest observed background level. Furthermore, if the extended source contributes to the MPC flux, we have overestimated the point-source contribution to the HRI; our extended source contribution is then a lower limit. The new northeast feature represents an excess flux of 0.0029 ± 0.0005 counts s^{-1}, about 10% of the extended emission or 4% of the total 0.3–3 keV emission.

The IPC data were accumulated on 1979 February 4 with an exposure of 13,100 s. The emission is dominated by the central point source with a counting rate of 1.9 counts s^{-1}. The 1'.5 resolution of the IPC is not adequate to distinguish the point, extended, and X-ray jet components seen in the HRI. However, the image appears elongated to the NE and SW and correlated with the inner radio lobes (Fig. 3 [Plate L10]). Figure 4 shows the azimuthal distribution of surface brightness in an annulus between 3' and 5' from the nucleus. Excess emission is seen in the directions corresponding

PLATE L10

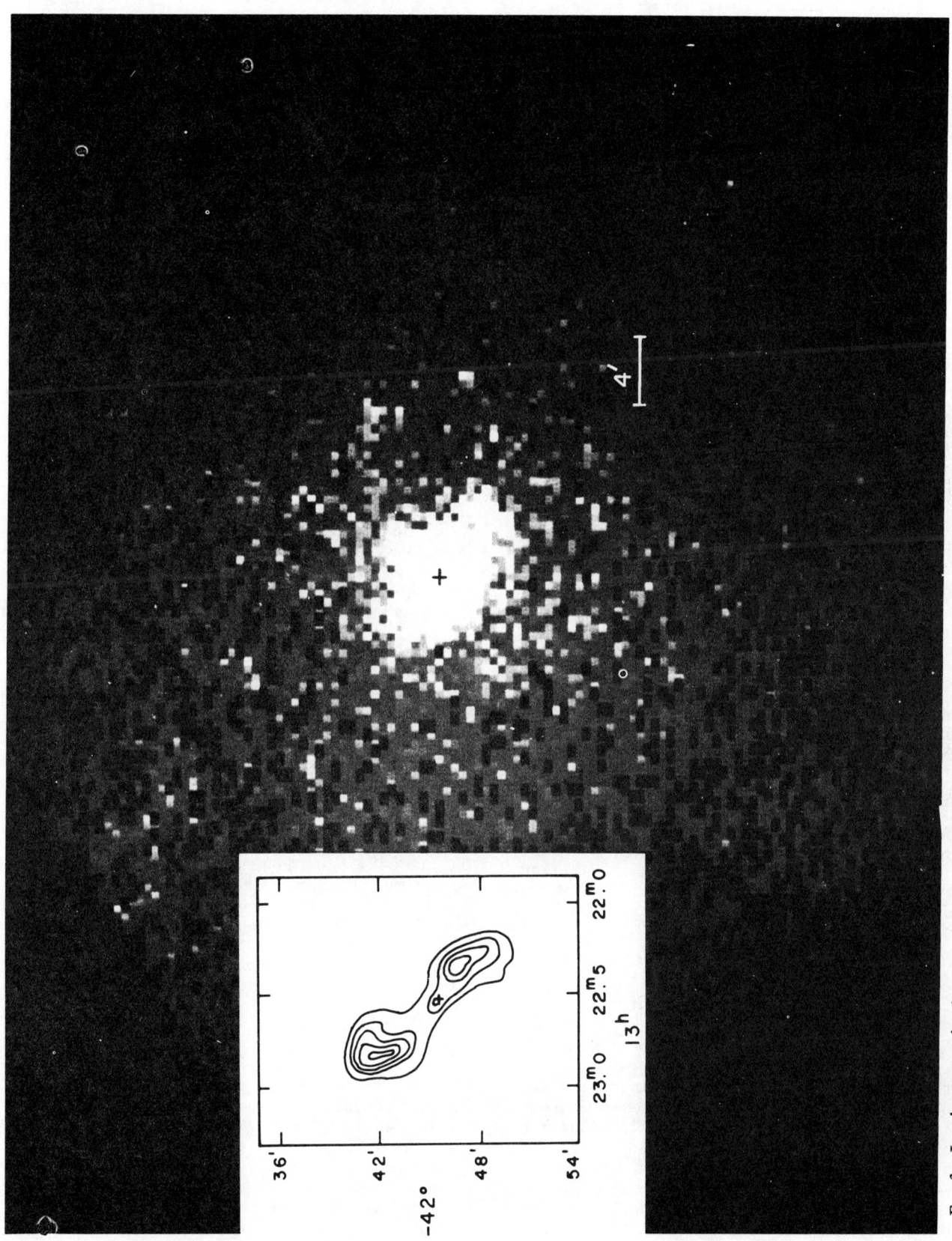

FIG. 3.—Imaging proportional counter (IPC) image of the central region of Cen A, showing the excess counts 3′–5′ SW of the nucleus. Radio contours of the inner radio lobes at 1.4 GHz from Christensen et al. (1976) are shown to the same scale as the X-ray image (inset).

SCHREIER et al. (see page L40)

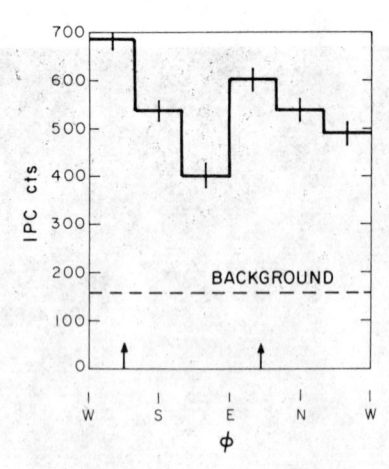

FIG. 4.—Azimuthal distribution of IPC counts in an annulus 3′–5′ from the nucleus. The arrows show the positions of the inner radio lobes.

to the inner lobes, as indicated by the arrows. A total of 0.00565 ± 0.00017 counts s^{-1} arcmin^{-1} (source and background) is seen at those positions; the average for the entire annulus is 0.00464 ± 0.00007 counts s^{-1} arcmin^{-1}. This corresponds to a detection of the inner lobes at the 5.6 σ level. The excess emission associated with the lobes is approximately 0.026 counts s^{-1}, with the SW lobe being about twice as strong as the NE lobe. The extent of the SW X-ray lobe is at least 1′. The X-ray emission is centered 4′ from the nucleus, coincident with the 408 MHz radio centroid (Cameron 1969), but displaced 1′.5 from the 1.4 GHz radio centroid (cf. Fig. 3).

Preliminary analysis of the IPC pulse-height analyzer (PHA) data was performed both for the central source out to 3′ and for the excess emission associated with the lobes. The data from the central source are consistent with the highly cut-off hard X-ray source and the MPC spectrum. The PHA data from the region of the inner lobes, using the remainder of the data in that annulus as background, are best fitted by a power-law number index $n = 1.0$, when a cutoff due to $N_H = 1.2 \times 10^{21}$ cm^{-2} of galactic absorption is imposed (Daltabuit and Meyer 1972). The statistical significance is not high and the data are consistent with the same number index ($n = 1.7$) as the point source.

III. DISCUSSION

a) The Nuclear Component

The central compact component is clearly identifiable with the hard, cut-off variable X-ray source. The HRI counting rate and the IPC spectrum are consistent with the detector responses to a source with the flux and spectrum seen with the MPC. If we use $n = 1.66$ and $E_a = 3.7$ keV (see Mushotzky et al. 1978), the observed flux corresponds to a luminosity of 1.0×10^{43} ergs s^{-1} in the 0.1–100 keV band. The intrinsic luminosity of the point source would be 1.7×10^{43} ergs s^{-1} in this band, with approximately 6×10^{42} ergs s^{-1} being absorbed near the nucleus. The intrinsic luminosity in a nominal HRI band of 0.3–3 keV is 2.8×10^{42} ergs s^{-1}; we supply this for comparison with the features discussed below, where we do not know the spectra and do not have independent higher energy measurements of the flux.

These luminosities, and all those following, apply if NGC 5128 is at a distance of 5 Mpc. This has been the most commonly assumed distance, dating from the work of Burbidge and Burbidge (1959), although certain assumptions in that work (such as $H_0 = 75$ km s^{-1} Mpc^{-1} and a galaxy luminosity function limit of $M_v = -22.5$) may no longer hold. However, recent work on dwarf irregular galaxies of the Centaurus group suggest a distance of 4–6 Mpc (Webster et al. 1979). In addition, considerations of H II region sizes, in conjunction with a review of other recent work, led Dufour et al. (1979) to suggest a distance of 6 ± 2 Mpc. We therefore use the classical value of 5 Mpc, and allow an uncertainty of a factor of 2 in luminosity.

b) The Inner Radio Lobes

The X-ray luminosity of the stronger SW lobe is 2×10^{39} ergs s^{-1} (0.3–3 keV), assuming a power law with $n = 1.7$ and galactic absorption. If we assume the observed X-ray emission is produced by inverse Compton scattering of the radio electrons off the microwave background, we can calculate the mean magnetic field of the lobe via the standard formalism (e.g., Tucker 1975). Using a radio spectrum $S_\nu = 1.9 \times 10^4 \nu_{MHz}^{-0.73}$, which holds for $85 \leq \nu_{MHz} \leq 1415$ (Christensen et al. 1976), and a microwave temperature $T = 2.7$ K, we derive a magnetic field strength $B = 4$ microgauss. This is 10–20 times lower than the estimated equipartition field strength. The field may be as low as 3 microgauss if the radio spectrum flattens at frequencies $\lesssim 20$ MHz (see Shain 1958). The field may be considerably stronger than our estimate if other mechanisms (e.g., inverse Compton scattering off starlight photons, or thermal bremsstrahlung from a hot plasma) contribute to the X-ray emission.

c) The Extended Component

Assuming a power-law spectrum with $n = 1.7$ and no intrinsic absorption, we calculate a luminosity of 2×10^{40} ergs s^{-1} (0.3–3 keV) for the extended component detected in the HRI image. This is 1% of the intrinsic point-source luminosity (most of which is not seen in this band due to strong absorption). Several interpretations of the extended component can be considered. (i) Electron scattering of the intrinsic point-source X-ray flux by a surrounding cloud (see Fabian 1977) appears unlikely. It would require a geometry where most of the cloud is illuminated directly by the point source; the high absorption we observe would then perhaps be due to an accretion disk aligned with the dust lane and obscuring our line of sight. However, the size of the observed emission region (~6 kpc) exceeds that of the dust lane and requires that the cloud be at X-ray emitting temperatures if pressure-supported. A cloud of the necessary density (~1 particle cm^{-3}) to provide the observed flux by scattering would emit more X-rays by bremsstrahlung than are observed. (ii) Thermal emission from $\sim 6 \times 10^8 \, M_\odot$ of hot gas in the potential well of the Cen A galaxy is a possibility by the preceding argument. (iii) The luminosity could be due to the integrated emission of discrete X-ray sources. The size of the region divided by the number of resolution elements gives a lower limit of about 300 for the number of discrete sources; the average source luminosity would be less than 10^{38} ergs s^{-1}. (iv) Another possible mechanism involves diffusion of the intrinsic cosmic-ray electrons out of the central source. These could then undergo inverse Compton scattering with starlight or the nuclear infrared emission. This might be consistent with the bridge of radio emission between the inner radio lobes (Berlin et al. 1975).

Further observations to better define the extent and spectrum of the extended component are intended to help interpret this emission.

d) The Inner Jet

The new feature observed 1' to the NE of the nucleus of Centaurus A is most remarkable in the context of the radio and optical structure of the source. It is aligned with, but just within, the optical jet found by Dufour and van den Bergh (1978), shown superposed on the HRI contour map in Figure 1. This in turn is reasonably well aligned with the inner radio lobe and extends out to the Hα emitting filaments found by Blanco et al. (1975). The main component appears about 0.4 kpc (15") in size; it has a greater X-ray surface brightness that the radio lobes, which are too diffuse to be apparent in the HRI image. The alignment argues against infalling or rotationally supported material. We believe that the feature provides evidence for the existence of a jet of matter from the nucleus, consistent with the need to power the inner radio lobes (see Blandford and Rees 1978).

Although one can consider various detailed models for the jet, the presence of X-ray emission near to the nucleus and optical emission farther out is suggestive of a thermal model. Using this as a working hypothesis, we assume that the power in such a jet must be at least equal to that required to replenish the inner lobe, $P \geq 5 \times 10^{40}$ ergs s^{-1}. Equating the power with the kinetic energy flux of a matter stream, we calculate a mass transfer rate

$$\dot{M} = 10^{26} v_8^{-2} P_{41.7} \epsilon_{0.1}^{-1} \text{ g s}^{-1} ,$$

where $v_8 = v/10^8$ cm s^{-1}, $P_{41.7} = P/5 \times 10^{41}$ ergs s^{-1}, and an efficiency $\epsilon_{0.1} = \epsilon/0.1$ has been assumed for the conversion of kinetic energy in the jet into relativistic particles at the radio lobe. Continuity arguments can be used to calculate the particle density in a conical jet at a distance r (kpc) from the source:

$$n = \frac{\dot{M}}{m_H v \pi (r\theta/2)^2} = 1.3 P_{41.7} v_8^{-3} r^{-2} \theta^{-2}_{0.25} \epsilon_{0.1}^{-1} \text{ cm}^{-3} ,$$

where $\theta_{0.25}$ is the opening angle in units of 0.25 rad, the estimated angle subtended by the X-ray jet at the nucleus (15" at a distance of 60"). Assuming that the X-rays are due to thermal bremsstrahlung, the luminosity of the jet between radii of 1 and 2 kpc (the approximate jet extent) is

$$L_x \approx 10^{-27} \int_{r=1}^{r=2} n^2 \pi r^2 \theta^2 T^{1/2} dr$$

$$= 4 \times 10^{39} T_7^{1/2} P^2_{41.7} v_8^{-6} \theta^{-2}_{0.25} \epsilon_{0.1}^{-2} \text{ ergs s}^{-1} .$$

If we assume a temperature of 10^7 K, and galactic absorption, we convert the observed flux of 0.0029 counts s^{-1} (0.3–3 keV) to a luminosity of $\sim 3 \times 10^{39}$ ergs s^{-1} (0.3–3 keV), in good agreement with the above expression. Line emission may be important if $T_7 \sim 1$.

The strongest dependence of the luminosity expression is on the flow velocity. Velocities of about 10^8 cm s^{-1} have been inferred from the structure and dynamics of the jet in 3C 31 (Blandford and Icke 1978) and from depolarization measurements of 3C 449 (Perley, Willis, and Scott 1979). We can also estimate the flow velocity from the opening angle

$$v = c\theta^{-1} ,$$

$$v_8 = 1.2 T_7^{1/2} \theta^{-1}_{0.25} ,$$

where c is the sound speed in the gas. These direct and indirect estimates suggest that $v_8 \approx 1$, lending support to this origin for the X-ray emission in the jet.

The cooling time for matter in the jet, due to bremsstrahlung radiation, is

$$t_c = \frac{2 \times 10^{11} T^{1/2}}{n}$$

$$= 4 \times 10^7 T_7^{1/2} P^{-1}_{41.7} v_8^3 r^2 \theta^2_{0.25} \epsilon_{0.1} \text{ yr} .$$

Additional cooling due to line emission and/or inhomogeneities in the stream may decrease the cooling time by an order of magnitude, leading to cooling-time estimates as low as 10^6 years. This should be compared with the flow time of matter from the nucleus to the jet $t_f = r/v = 10^6 \, r_{\rm kpc} v_8^{-1}$ yr. Thus the possibility of thermal instability exists, suggesting that the inner optical jet is due to emission from matter in the stream which has cooled and is moving ballistically outward ($\sim 10^6 \, M_\odot / 10^6$ years). This interpretation is consistent with optical observations (Dufour and van den Bergh 1978; Graham 1979; Osmer 1978). The stream would have to avoid cooling too soon (radii less than 1 kpc), which might imply greater velocities in the stream closer to the nucleus. It should be noted that most of the energy in the above model for the jet resides in a dense subrelativistic plasma. A mechanism for reacceleration of electrons is required to provide for synchrotron emission in the lobe. The above cooling time scales indicate that little X-ray emission is produced within the lobes by bremsstrahlung. A higher and possibly variable stream velocity may lead to shocks and X-ray emission as suggested for the M87 jet (Rees 1978; Blandford and Königl 1979), but our observations are consistent with much of the energy being fed into the northern inner radio lobe via a dense subrelativistic plasma. The lack of an observed jet to the south may be due to a factor ~ 2 higher velocity, or to an intrinsic variability in the jet production mechanism.

IV. CONCLUSIONS

The images of the central region of Centaurus A have resolved the X-ray structure into four components: the hard, cut-off point source; extended emission coincident with the inner radio lobes; a several kpc region of emission surrounding the nucleus; and a jetlike structure located between the nucleus and an inner radio lobe. The clear alignment of the NE X-ray, optical, and radio features argues in favor of continuous or repetitive ejection of material from the nucleus. In addition, the self-consistent calculation in which hot gas emits X-rays and then cools to emit visible light provides a physically plausible model. A better understanding of the features of this nearby giant radio galaxy will be important for the interpretation of emission from more distant radio galaxies and active galactic nuclei.

We thank H. Tananbaum for useful discussions, and acknowledge the assistance of the data processing staff, and of J. Peritz in the preparation of the manuscript. This work was supported under NAS8-30751.

REFERENCES

Beall, J. H., et al. 1978, *Ap. J.*, **219**, 836.
Berlin, A. B., et al. 1975, *Soviet Astr. Letters*, **1**, 234.
Blanco, V. M., Graham, J. A., Lasker, B. M., and Osner, P. S. 1975, *Ap. J. (Letters)*, **198**, L63.
Blandford, R. D., and Icke, V. 1978, *M.N.R.A.S.*, **185**, 527.
Blandford, R. D., and Königl, A. 1979, *Ap. Letters*, **20**, 15.
Blandford, R. D., and Rees, M. J. 1978, *Phys. Scripta*, **17**, 265.
Burbidge, E. M., and Burbidge, G. R. 1959, *Ap. J.*, **129**, 271.
Cameron, M. J. 1969, *Proc. Astr. Soc. Australia*, **1**, 229.
Christensen, W. N., Frater, R. N., Watkinson, A., O'Sullivan, J. D., and Lockhart, I. A. 1976, *M.N.R.A.S.*, **181**, 183.
Daltabuit, E., and Meyer, S. 1972, *Astr. Ap.*, **20**, 415.
Delvaille, J. P., Epstein, A., and Schnopper, H. W. 1978, *Ap. J. (Letters)*, **219**, L81.
Doxsey, R., Bradt, H., Gursky, H., Schwartz, D. A., and Schwarz, J. 1978, *Bull. AAS*, **10**, 390.
Dufour, R. J., and van den Bergh, S. 1978, *Ap. J. (Letters)*, **226**, L73.
Dufour, R. J., van den Bergh, S., Harvel, C. A., Martins, D. H., Schiffer, F. H., III, Talbot, R. J., Jr., Talent, D. L., and Wells, D. C. 1979, *A.J.*, **84**, 284.

Fabian, A. C. 1977, *Nature*, **269**, 672.
Giacconi, R., et al. 1979, *Ap. J.*, **230**, 540.
Graham, J. A. 1979, preprint.
Grindlay, J. E. 1975, *Ap. J.*, **199**, 49.
Kunkel, W. E., and Bradt, H. V. 1971, *Ap. J. (Letters)*, **170**, L7.
Lawrence, A., Pye, J. P., and Elvis, M. 1977, *M.N.R.A.S.*, **181**, 93P.
Mushotzky, R. F., Serlemitsos, P. J., Becker, R. H., Boldt, E. A., and Holt, S. S. 1978, *Ap. J.*, **220**, 790.
Osmer, P. S. 1978, *Ap. J. (Letters)*, **226**, L79.
Perley, R. A., Willis, A. G., and Scott, J. S. 1979, preprint.
Rees, M. J. 1978, *M.N.R.A.S.*, **184**, 61P.
Shain, C. A. 1958, *Australian J. Phys.*, **11**, 517.
Tucker, W. H. 1975, *Radiation Processes in Astrophysics* (Cambridge: MIT Press).
Tucker, W., Kellogg, E., Gursky, H., Giacconi, R., and Tananbaum, H. 1973, *Ap. J.*, **180**, 715.
Webster, B. L., Goss, W. M., Hawarden, T. G., Longmore, A. J., and Mebold, U. 1979, *M.N.R.A.S.*, **186**, 31.
Winkler, P. R., Jr., and White, A. E. 1975, *Ap. J. (Letters)*, **199**, L139.

Note added in proof.—It should be noted that the positions of the optical features listed in Table 2 of Dufour and van den Bergh (1978) differ from those in their Figure 3 by a factor of ~ 2. We have used those in the table. There would be no significant effect on our conclusions if we used the figure directly.

J. DELVAILLE, E. FEIGELSON, R. GIACCONI, J. GRINDLAY, E. J. SCHREIER, and D. A. SCHWARTZ: Harvard-Smithsonian Center for Astrophysics, 60 Garden Street, Cambridge, MA 02138

A. C. FABIAN: Institute of Theoretical Astronomy, University of Cambridge, Madingley Road, Cambridge CB3 0HA, England

THE ASTROPHYSICAL JOURNAL, 262:L17–L21, 1982 November 15
© 1982. The American Astronomical Society. All rights reserved. Printed in U.S.A.

Reprinted with permission from *The Astrophysical Journal*, 262 pp. L17-L21, Y. Avni and H. Tananbaum, "On the Cosmological Evolution of the X-Ray Emission from Quasars." © 1982 The American Astronomical Society.

ON THE COSMOLOGICAL EVOLUTION OF THE X-RAY EMISSION FROM QUASARS

Y. AVNI[1] AND H. TANANBAUM
Harvard-Smithsonian Center for Astrophysics
Received 1982 June 30; accepted 1982 August 4

ABSTRACT

We derive the average dependence of the ratio of X-ray luminosity to optical luminosity as a function of redshift and optical luminosity for quasars and find that the explicit dependence of this ratio is predominantly on optical luminosity. For a wide class of models for the cosmological evolution of quasars, our results imply that the evolution of the X-ray luminosity function is weaker than the evolution of the optical luminosity function. We indicate implications for physical models of quasars and for global properties of the quasar population.

Subject headings: cosmic background radiation — cosmology — luminosity function — quasars — X-rays: general

I. INTRODUCTION

X-ray emission from quasars is common and strong (Tananbaum *et al.* 1979). The dependence of quasar X-ray luminosity on redshift, optical luminosity, and radio luminosity is a fundamental datum for constructing physical models of quasars, as well as for understanding global properties of the quasar population. These global properties, which include the evolution of quasars on a cosmological time scale, determine the contribution of quasars to the X-ray background (Zamorani *et al.* 1981), the effects of quasars on the intergalactic medium (e.g., Field and Perrenod 1977), and the relation of high-luminosity quasars to lower luminosity active galactic nuclei (AGNs) (e.g., Avni 1978).

The first study of a sample of previously known quasars with the *Einstein Observatory* (Tananbaum *et al.* 1979) has suggested that, on the average, quasars with higher optical luminosity, L_o, have higher X-ray luminosity, L_x. A further study of a larger sample (Zamorani *et al.* 1981) has established, in addition, that the ratio L_x/L_o depends on radio luminosity, L_R, and also that higher values of z or L_o, or both, imply lower values of L_x/L_o. In this *Letter*, we derive the average dependence of L_x/L_o on z and L_o for optically selected quasars (§ II). We use in our analysis a variant of our general method (Avni *et al.* 1980) for deriving luminosity functions from samples containing both flux measurements and flux upper limits. For a wide class of models for the cosmological evolution of quasars, our results imply that the evolution of the quasar X-ray luminosity function is weaker than the evolution of the optical luminosity function (§ III). We discuss briefly some implications of such a difference in evolution (§ IV).

[1] Also from Weizmann Institute of Science, Rehovot, Israel.

II. AVERAGE DEPENDENCE OF L_x/L_o ON z AND L_o

We consider a composite sample of optically selected quasars (including AGNs),[2] that were later observed with the *Einstein Observatory*. The sample contains: (*a*) 48 optically selected quasars from Zamorani *et al.* (1981); (*b*) 15 optically selected Seyfert 1 (≤ 1.5) AGNs, with measured nuclear optical magnitudes, from Kriss, Canizares, and Ricker (1980); and (*c*) 10 quasars, which form an optically complete sample at $B = 19.2$, from Marshall *et al.* (1983*b*). The distribution of these quasars in redshift and optical luminosity is similar to distributions found in optically selected complete samples. The sample is not complete in the optical; however, since we extract the explicit dependence of L_x/L_o on z and L_o, completeness in the optical is not required. Out of the total of 73 quasars, 41 were detected in X-rays and have known X-ray fluxes, while 32 have only flux upper limits.

We follow Tananbaum *et al.* (1979) and denote the monochromatic luminosity at 2 keV by L_x and the monochromatic luminosity at 2500 Å by L_o (both in ergs s^{-1} Hz^{-1}), and define $\alpha_{o,x} = -\log(L_x/L_o)/2.605$. We use $H_0 = 50$ km s^{-1} Mpc^{-1} and $q_0 = 0$. To find the average dependence of $\alpha_{o,x}$ on cosmological epoch and on optical luminosity, we assume the simplest, linear expansion of $\alpha_{o,x}$ as a function of fractional look-back time $\tau(z)$ [$= z/(1+z)$ for $q_0 = 0$] and of $\log L_o$:

$$\alpha_{o,x} = A_z[\tau(z) - 0.5] + A_o(\log L_o - 30.5) + A + R, \quad (1)$$

where 0.5 and 30.5 are (approximately) the central values of $\tau(z)$ and $\log L_o$, respectively, in the sample; A_z,

[2] Since the quasars were selected regardless of their radio properties, any dependence on L_R is automatically statistically integrated upon.

A_o, and A are the coefficients of the expansion; and R represents the residual scatter of $\alpha_{o,x}$ around the average dependence.

If all quasars in the sample had been detected, a standard least squares regression analysis could have yielded the coefficients A_z and A_o. Such an analysis assumes implicitly that the residuals R follow a Gaussian distribution. However, not all quasars were detected, and, for 32 of them, only lower limits on $\alpha_{o,x}$ are available. We therefore follow the general, maximum-likelihood method that we have developed previously (Avni et al. 1980) for deriving luminosity functions from samples containing detections and bounds.[3] We use a generalized regression analysis appropriate for this type of sample, assuming that the residuals R follow a Gaussian distribution with mean zero and variance σ^2.

The results of this analysis for the dependence of $\alpha_{o,x}$ on $\tau(z)$ and $\log L_o$ are given in Figure 1, where we plot contours of equal $S = -2 \ln$ (likelihood) in the plane of the parameters A_z and A_o (see Avni 1976). We find that the best estimates are $A_z = -0.01$ and $A_o = 0.118$, which indicates that the explicit dependence of $\alpha_{o,x}$ is predominantly on $\log L_o$.[4] However, within the 2 σ uncertainty, a *joint* dependence of $\alpha_{o,x}$ on $\tau(z)$ and $\log L_o$ cannot be excluded. In any case, the parameters A_z and A_o are very strongly correlated (due to the intrinsic correlation between $\tau(z)$ and $\log L_o$ in the sample). This means that the total dependence of $\alpha_{o,x}$ on $\tau(z)$ or $\log L_o$, or on both, is well determined from the data, but the way this dependence is split between $\tau(z)$ and $\log L_o$ is not so well determined. This strong correlation between A_z and A_o is important for our considerations below (§ III). Moreover, we establish that $\alpha_{o,x}$ cannot be independent of both $\tau(z)$ and $\log L_o$ ($A_z = 0, A_o = 0$) with a significance of $1 - (2 \times 10^{-7})$, and that L_x cannot be independent of both $\tau(z)$ and $\log L_o$ ($A_z = 0, A_o = 1/2.605$) with a significance of $1 - (2 \times 10^{-13})$.

Having found the best estimate for the *average* dependence of $\alpha_{o,x}$ on $\tau(z)$ and $\log L_o$, we now derive the distribution of the residuals around the average dependence. We define the rescaled residuals by $r = R/\sigma$, where R is given by equation (1), and the parameters are assigned their best estimates: $A_z = -0.01$, $A_o = 0.118$,

---

[3] We refer to this method as the DB method, for detections and bounds.

[4] An analysis of the radio selected 3CR quasar sample (Tananbaum et al. 1983), finds that the dependence of $\alpha_{o,x}$ on $\log L_o$ is significant while the dependence on z is not significant. A similar analysis was recently carried out by Zamorani (1982), who assumed *a priori* that $A_z = 0$ and excluded three optically selected, radio-loud quasars from Zamorani et al. (1981), while we include all optically selected quasars. Reichert et al. (1982) have also considered the dependence of L_x/L_o on L_o and z, but their work is biased by not taking into account X-ray upper limits and by including X-ray selected quasars with quasars previously selected by other means.

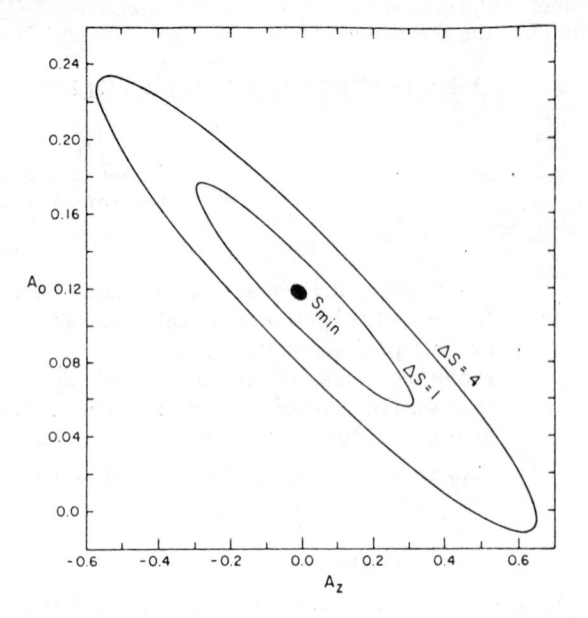

FIG. 1.—Contours of equal $S = -2 \ln$ (likelihood) for the average dependence of $\alpha_{o,x}$ on $\tau(z)$ and $\log L_o$. The projections of these contours on each of the A_z and A_o axes yield the 68% and 95% confidence limits for each of the parameters separately.

$A = 1.50$, and $\sigma = 0.1975$. For the detected quasars, the values of r are known. For the undetected quasars, lower limits for r are known. We now use the *nonparametric*, unbinned version of the DB method (see n. 3 and Avni et al. 1980) and calculate the cumulative distribution function of r, shown in Figure 2. In deriving the average dependence of $\alpha_{o,x}$ on $\tau(z)$ and $\log L_o$ and in deriving the distribution of the residuals, we have used two steps which are not rigorously consistent. In the first step we assumed that the distribution of the residuals is Gaussian and derived the average dependence. In the second step we used that average dependence and derived nonparametrically the distribution of the residuals. To justify this procedure *a posteriori*, we compare the derived cumulative distribution of r with a cumulative Gaussian distribution of mean zero and variance 1, which is also plotted in Figure 2. While the derived distribution seems to be somewhat skew (having a sharper cutoff at low $\alpha_{o,x}$ and a longer tail at high $\alpha_{o,x}$ relative to the Gaussian), the two distributions agree very well in their means and widths, which are the important elements affecting the regression analysis. Thus, for the present sample, the simplest version of the generalized regression analysis is adequate.

The distribution of r, in conjunction with the parameters of the average dependence, provides an approximation for the X-ray *conditional* luminosity function. The bivariate optical–X-ray luminosity function can be written in general in the form

$$dN = [\psi_0(z, L_o) \, dV(z) \, dL_o][\phi_x(L_x | z, L_o) \, dL_x], \quad (2)$$

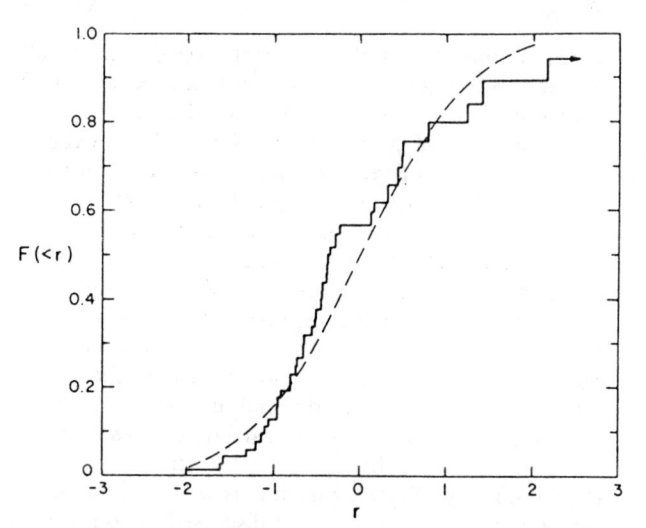

FIG. 2.—The cumulative distribution of the rescaled residuals of $\alpha_{o,x}$ around the average dependence of $\alpha_{o,x}$ on $\tau(z)$ and $\log L_o$ (*solid line*). This distribution, in conjunction with the average dependence, approximates the X-ray conditional luminosity function. Also shown is a cumulative Gaussian distribution with mean zero and variance 1 (*broken line*).

where dN is the expected number of quasars in a differential element of volume and luminosity, $\psi_o(z, L_o)$ is the optical luminosity function, and $\phi_x(L_x | z, L_o)$ is the X-ray conditional luminosity function that describes the distribution of L_x at any given values of z, L_o. This latter conditional luminosity function is the one derived from the present sample. At any given z and L_o, the distribution of L_x is fully defined by the distribution of $\alpha_{o,x}$, which in turn is fully defined by the distribution of r. The observed distribution of $\alpha_{o,x}$ around the average dependence is somewhat wider than the intrinsic distribution because of measurement uncertainties (and, possibly, time variations). We estimate that measurement uncertainties for L_x and L_o contribute ~ 0.01 to the width of the observed distribution of $\sigma \approx 0.20$. Hence Figure 2 provides an approximation to $\phi_x(L_x | z, L_o)$.

III. COSMOLOGICAL EVOLUTION

Any dependence of L_x on z and L_o determines a relation between the cosmological evolution of the optical luminosity function and the evolution of the X-ray luminosity function. We now explore the relation implied by the results of § II.

We assume that the average dependence of $\alpha_{o,x}$ on $\tau(z)$ and $\log L_o$ is given by equation (1) and that the residuals R have no further dependence on $\tau(z)$ and $\log L_o$. It then follows that the variable ξ defined by

$$\xi = L_x / [L_o^\lambda \exp[\eta \tau(z)]], \quad (3)$$

where

$$\lambda = 1 - 2.605 A_o, \quad \eta = -2.605 \ln(10) A_z, \quad (4)$$

is independent of z and L_o, and that the X-ray conditional luminosity function, $\phi_x(L_x | z, L_o) dL_x$, can be written as a function of ξ alone. On the average, then, $L_x \propto L_o^\lambda \exp[\eta \tau(z)]$. (For the best estimates, $L_x \propto L_o^{0.7}$.)

We now assume that the optical luminosity function, $\psi_o(z, L_o)$, satisfies pure luminosity evolution (e.g., Mathez 1976, 1978; Braccesi *et al.* 1980; Marshall *et al.* 1983*a*). We denote by $g_o(z)$ the optical evolution function. It then follows that the X-ray luminosity function, $\psi_x(z, L_x)$, also satisfies pure luminosity evolution, and that the X-ray evolution function, $g_x(z)$, is related to the optical evolution function through

$$g_x(z) = [g_0(z)]^\lambda \exp[\eta \tau(z)]. \quad (5)$$

We have thus derived a relation between the X-ray and optical evolution functions, with parameters λ and η which are dictated by the average dependence of $\alpha_{o,x}$ on $\tau(z)$ and $\log L_o$. (This relation holds when $\tau(z)$ in eqs. [1] and [5] is *any* function of z, not just the fractional look-back time.)

If we represent $g_o(z)$ as an exponential of look-back time, $g_o(z) = \exp[\gamma_o \tau(z)]$, as is commonly done, then it follows that $g_x(z)$ has the same functional form, $g_x(z) = \exp[\gamma_x \tau(z)]$, and that the evolution parameters are related through

$$\gamma_x = \lambda \gamma_o + \eta = \gamma_o - 2.605(\gamma_o A_o + \ln(10) A_z). \quad (6)$$

By combining this relation with the S contours given in Figure 1, one can calculate γ_x (and its uncertainty) for any assumed value of γ_o. For $\gamma_o = 7$ (Marshall *et al.* 1983*a*), we find $\gamma_x = 4.9$ (1 σ: 4.0–5.7; 2 σ: 3.0–6.5). Thus, we predict that the cosmological evolution of the X-ray luminosity function is weaker than the evolution of the optical luminosity function (at better than 95% confidence for the evolution model considered if $\gamma_o \geq 6$). The best estimate of 2.1 for the difference $(\gamma_o - \gamma_x)$ between the evolution parameters is substantial. The predicted difference is not very sensitive to the value of γ_o, being 2.4 for $\gamma_o = 8$.

We note that the strong correlation between A_z and A_o, as exhibited in Figure 1, plays an important role in deriving these results. In general, γ_x depends on both A_z and A_o. However, the results of our regression analysis imply that, while the separate uncertainties in A_z and in A_o are quite substantial, A_z and A_o are jointly confined within a rather narrow, elongated region. The allowed, correlated variations in A_z and A_o induce only small variations in γ_x. The parameter γ_x does not depend on A or on the shape or width of the distribution of the residuals R. Estimates of other quantities, which depend on A_z and A_o differently, or on A or the distribution of residuals, may be subject to very large uncertainties.

Also note that a difference between γ_o and γ_x is derived even when $A_z = 0$, that is, even if there is no *explicit* dependence of L_x on $\tau(z)$. This can be understood by noting that, for optical, pure luminosity evolution, L_o increases as a function of z. This increase of L_o implies an increase in L_x which is, however, slower than the increase in L_o.

While our quantitative results apply specifically to pure luminosity evolution, qualitatively they hold for a wide range of evolutionary models. If the cosmological evolution of the density of high L_o quasars is stronger than that of low L_o quasars,[5] then the average dependence of $\alpha_{o,x}$ on $\tau(z)$ and log L_o derived in § II implies that the evolution of the X-ray luminosity for the quasar population (as measured, e.g., by the volume emissivity) is weaker than that of the optical luminosity.[6]

IV. DISCUSSION

We have derived the average dependence of the ratio of X-ray luminosity to optical luminosity on redshift and optical luminosity for optically selected quasars. The numerical details of this dependence may well change when larger samples become available and more complicated functional dependences can be analyzed. In particular, it will be important to study optically selected, optically complete samples, that are also observed in X-rays, since such samples will not be affected by possible "hidden" (presently unrecognized) dependences on other quasar properties. It will also be valuable to find the joint dependence of L_x on redshift, optical luminosity, *and* radio luminosity, incorporating data for radio selected quasars. The results of the present study, however, do determine rather well the total dependence of L_x on z and L_o and indicate that the explicit dependence of L_x is predominantly on L_o. The results have implications for physical models of quasars. For example, if the X-rays are primary radiation and the optical radiation results from processed X-rays, then, for high-luminosity quasars, such a processing is relatively more efficient on the average. Additionally, these results may provide insight into the relations between thermal and nonthermal emission from quasars (Tananbaum *et al.* 1983) or provide constraints for models invoking emission from an accretion disk around a massive black hole.

The average dependence, $L_x(z, L_o)$, derived here implies, for a wide class of evolution models, that the X-ray evolution is weaker than the optical evolution. The current measure of the X-ray evolution from the *Einstein Observatory* Medium Survey (Maccacaro *et al.* 1983) is consistent with this result. Our present results and their subsequent refinement can be used to provide constraints for evolution models and for more detailed representations of $\varphi_o(z, L_o)$ and of $\varphi_x(z, L_x)$. A cosmological evolution of the X-ray luminosity function which is weaker than that of the optical luminosity function has several implications for the quasar population:

1. Any calculation of the contribution of quasars to the X-ray background from a local X-ray luminosity function should use the appropriate evolution law. A calculation of this contribution that starts from the optical luminosity and evolution functions should use the appropriate X-ray conditional luminosity function.[7]

2. We have previously argued (Avni 1978) (on the basis of the X-ray background) that the evolution of low-luminosity "Seyfert galaxies" is weaker than that of high-luminosity "quasars" and discussed the qualitative continuity of properties between low-luminosity AGNs and high-luminosity AGNs. In retrospect, what we have actually shown then is that the allowed *X-ray* evolution of low-luminosity "Seyfert galaxies" is weaker than the *optical* evolution found for high-luminosity "quasars." A weaker X-ray evolution of quasars, as suggested by our present results, makes the continuity of properties between "Seyfert galaxies" and "quasars" even tighter. In fact, it is even possible that the evolution of AGNs with low luminosity at the present epoch is not very different from that of high-luminosity quasars. Any estimate of the contribution of such "Seyfert galaxies" to the X-ray background should fully take into account such a possible evolution.

3. The cosmological evolution of the *bolometric* luminosity function of quasars may also be different from the evolution of the optical luminosity function. Attempts to derive the cosmological evolution of the quasar population from physical models for the evolution of individual quasars should take this possibility into account.

4. Effects of quasars on the intergalactic medium at early epochs have to be reevaluated, since the bolometric volume emissivity may evolve differently from the optical luminosity function.

Y. A. wishes to thank R. Giacconi, H. Tananbaum, G. Field, and A. Cameron for making his visit at the Center for Astrophysics possible and very enjoyable. We thank P. Giommi, H. Marshall, and G. Zamorani for useful discussions. This research was supported by NASA contract NAS 8-30751.

---

[5] This is satisfied by pure luminosity evolution and is generally indicated by considerations of the quasar optical number count and of the X-ray background.

[6] Of course, for pure density evolution of the optical luminosity function (now ruled out observationally; see, e.g., Marshall *et al.* 1983*a*), $A_z = 0$ would imply identical optical and X-ray rates of evolution.

[7] A first, approximate step in this direction was taken by Zamorani *et al.* (1981) in their estimate of the population average of $\alpha_{o,x}^{\text{eff}} = 1.45$. For the nominal optical evolution model and luminosity function described by Marshall *et al.* (1983*b*), combined with our best estimates for A_z, A_o, A, and σ, one obtains $\alpha_{o,x}^{\text{eff}}$ for $z \leq 3.5$ (H. Marshall, private communication).

REFERENCES

Avni, Y. 1976, *Ap. J.*, **210**, 642.
———. 1978, *Astr. Ap.*, **63**, L13.
Avni, Y., Soltan, A., Tananbaum, H., and Zamorani, G. 1980, *Ap. J.*, **238**, 800.
Braccesi, A., Zitelli, V., Bonoli, F., and Formiggini, L. 1980, *Astr. Ap.*, **85**, 80.
Field, G. B., and Perrenod, S. C. 1977, *Ap. J.*, **215**, 717.
Kriss, G. A., Canizares, C. R., and Ricker, G. R. 1980, *Ap. J.*, **242**, 492.
Maccacaro, T., Avni, Y., Gioia, I. M., Giommi, P., Griffiths, R. E., Leibert, J., Stocke, J., and Danziger, J. 1983, *Ap. J.* (*Letters*), submitted.
Marshall, H. L., Avni, Y., Tananbaum, H., and Zamorani, G. 1983a, *Ap. J.*, submitted.
Marshall, H. L., Tananbaum, H., Huchra, J. P., Braccesi, A., Zitelli, V., and Zamorani, G. 1983b, *Ap. J.*, submitted.
Mathez, G. 1976, *Astr. Ap.*, **53**, 15.
———. 1978, *Astr. Ap.*, **68**, 17.
Reichert, G. A., Mason, K. O., Thorstensen, J. R., and Bowyer, S. 1982, *Ap. J.*, **260**, 437.
Tananbaum, H., *et al.* 1979, *Ap. J.* (*Letters*), **234**, L9.
Tananbaum, H., Wardle, J. F. C., Zamorani, G., and Avni, Y. 1983, *Ap. J.*, submitted.
Zamorani, G. 1982, *Ap. J.* (*Letters*), **260**, L31.
Zamorani, G., *et al.* 1981, *Ap. J.*, **245**, 357.

Y. AVNI and H. TANANBAUM: Harvard-Smithsonian Center for Astrophysics, 60 Garden Street, Cambridge, MA 02138